*PROGRESS IN*

# Molecular Biology and Translational Science

*Volume 102*

*PROGRESS IN*

# Molecular Biology and Translational Science

## Cellular RNA Interference Mechanisms

*edited by*

Dirk Grimm

*Department of Infectious Diseases, Virology,*
*Cluster of Excellence CellNetworks*
*University of Heidelberg*
*Heidelberg, Germany*

*Volume 102*

ELSEVIER

AMSTERDAM • BOSTON • HEIDELBERG • LONDON
NEW YORK • OXFORD • PARIS • SAN DIEGO
SAN FRANCISCO • SINGAPORE • SYDNEY • TOKYO
Academic Press is an imprint of Elsevier

Academic Press is an imprint of Elsevier
32 Jamestown Road, London, NW1 7BY, UK
Radarweg 29, PO Box 211, 1000 AE Amsterdam, The Netherlands
225 Wyman Street, Waltham, MA 02451, USA
525 B Street, Suite 1900, San Diego, CA 92101-4495, USA

This book is printed on acid-free paper. ∞



**Library of Congress Cataloging-in-Publication Data**
A catalog record for this book is available from the Library of Congress

**British Library Cataloguing in Publication Data**
A catalogue record for this book is available from the British Library

ISBN: 978-0-12-415795-8
ISSN: 1877-1173

For information on all Academic Press publications
visit our website at elsevierdirect.com

Printed and bound by CPI Group (UK) Ltd, Croydon, CR0 4YY

Transferred to digital print 2012

# Contents

# Contributors

*Numbers in parentheses indicate the pages on which the authors' contributions begin.*

**Ben Berkhout,** Laboratory of Experimental Virology, Department of Medical Microbiology, Center for Infection and Immunity Amsterdam, Academic Medical Center, University of Amsterdam, Amsterdam, The Netherlands (141)

**Julia J.M. Eekels,** Laboratory of Experimental Virology, Department of Medical Microbiology, Center for Infection and Immunity Amsterdam, Academic Medical Center, University of Amsterdam, Amsterdam, The Netherlands (141)

**Roberto Fiore,** Interdisziplinäres Zentrum für Neurowissenschaften, SFB488 Junior Group, Universität Heidelberg, and Institut für Neuroanatomie, Universitätsklinikum Heidelberg, Im Neuenheimer Feld 345, Heidelberg, Germany (47)

**Victoria A. Green,** Antiviral Gene Therapy Research Unit, Department of Molecular Medicine and Haematology, University of the Witwatersrand, Johannesburg, South Africa (11)

**Dirk Grimm,** University of Heidelberg, Cluster of Excellence CellNetworks, Department of Infectious Diseases, Virology, Heidelberg, Germany (165)

**Catherine L. Jopling,** School of Pharmacy, Centre for Biomolecular Sciences, University of Nottingham, Nottingham NG7 2RD, United Kingdom (101)

**Sharof Khudayberdiev,** Interdisziplinäres Zentrum für Neurowissenschaften, SFB488 Junior Group, Universität Heidelberg, and Institut für Neuroanatomie, Universitätsklinikum Heidelberg, Im Neuenheimer Feld 345, Heidelberg, Germany (47)

**Stuart Knowling,** Department of Molecular and Experimental Medicine, The Scripps Research Institute, La Jolla, California, USA (1)

**Andrew P. Lewis,** School of Pharmacy, Centre for Biomolecular Sciences, University of Nottingham, Nottingham NG7 2RD, United Kingdom (101)

**Stefan Mockenhaupt,** University of Heidelberg, Cluster of Excellence CellNetworks, Department of Infectious Diseases, Virology, Heidelberg, Germany (165)

**Kevin V. Morris,** Department of Molecular and Experimental Medicine, The Scripps Research Institute, La Jolla, California, USA (1)

**Ashley P.E. Roberts,** School of Pharmacy, Centre for Biomolecular Sciences, University of Nottingham, Nottingham NG7 2RD, United Kingdom (101)

**Reuben Saba,** Interdisziplinäres Zentrum für Neurowissenschaften, SFB488 Junior Group, Universität Heidelberg, and Institut für Neuroanatomie, Universitätsklinikum Heidelberg, Im Neuenheimer Feld 345, Heidelberg, Germany (47)

**Nina Schürmann,** University of Heidelberg, Cluster of Excellence CellNetworks, Department of Infectious Diseases, Virology, Heidelberg, Germany (165)

**Gerhard Schratt,** Interdisziplinäres Zentrum für Neurowissenschaften, SFB488 Junior Group, Universität Heidelberg, and Institut für Neuroanatomie, Universitätsklinikum Heidelberg, Im Neuenheimer Feld 345, Heidelberg, Germany (47)

**Marc S. Weinberg,** Antiviral Gene Therapy Research Unit, Department of Molecular Medicine and Haematology, University of the Witwatersrand, Johannesburg, South Africa (11)

# Preface

Despite centuries filled with biological discoveries that shaped our fundamental view and understanding of nature and that refined our approaches to distinguish and treat human diseases, few findings made such a pivotal dual impact as the one that RNA interference (RNAi)—an evolutionarily conserved cellular means of specific and potent gene silencing—is active in mammals. Indeed, what we have witnessed over the past decade since RNAi's initial description in a nematode is nothing short of a revolution in our awareness of the amazing intricacy underlying gene regulation in higher organisms, and yet we are probably only beginning to appreciate the inherent beauty and complex diversity of the numerous relevant mechanisms. Concurrently, we are experiencing constant shifts in our perception of the molecular processes governing human diseases that offer a myriad of exciting and promising novel avenues to expand our repertoire of diagnostic and prognostic tools and to further enlarge our clinical arsenal of potent intervention strategies.

In this issue of *Progress in Molecular Biology and Translational Sciences*, a collection of six chapters provides an insight into the amazing sophistication and versatility underlying "cellular RNAi mechanisms" in healthy or diseased humans, hoping to leave the reader with an impression of the sometimes breathtaking speed at which the entire field is moving and incessantly pushing the boundaries of biology and medicine like no other before.

Starting off the issue is a chapter by Stuart Knowling and Kevin V. Morris, which reviews one of the latest and most unanticipated discoveries in our area, namely the fact that small RNAs not only regulate gene expression on the posttranscriptional level (the canonical RNAi pathway) but also exert their control at the stage of transcription. Even more, we are just now starting to realize that small RNAs not only suppress but can also activate gene expression, by means that still escape our current understanding but will certainly provide intriguing research topics for years to come. As is typical for this review series as well as for the field of RNAi, Knowling and Morris finally round up their chapter by highlighting the promises to translate these new discoveries into novel future biomedical concepts ideally allowing us to control human gene expression at yet another important level.

Following up on this introductory overview is a second complementing chapter by Victoria Green and Marc Weinberg, which delves even deeper into the astonishingly large variety of molecular mechanisms that potentially govern these noncanonical epigenetic RNAi-related regulatory processes in human cells. While today we can hardly fully appreciate the surprisingly vast array of noncoding RNA-based networks acting at this fundamental level of gene expression, Green and Weinberg already present their insightful vision of the significant possible implications of these new findings for our basic understanding of the function of human cells as well as for the implementation of entirely novel biomedical strategies in future drug and gene therapy development.

Leaving the cellular nucleus and entering gene regulation in the cytoplasm, Roberto Fiore, Sharof Khudayberdiev, Reuben Saba, and Gerhard Schratt next review another flourishing and exciting area of current RNAi research which is posttranscriptional gene silencing by miRNAs in the central nervous system. Also here, our understanding of how RNAi shapes and controls central cellular processes such as neuronal differentiation, synaptogenesis and plasticity is in its infancy at best, as are our insights into how RNAi is involved in higher cognitive functions and the etiology of neurological diseases. All these are utmost fascinating and captivating questions that the authors will address and whose answers will continue in the following years to revolutionize and overthrow our current picture of cellular RNAi mechanisms in the human nervous system in health and disease.

Complementing these three reviews of endogenous RNAi processes in human cells, two consecutive chapters then enter the thrilling field of pathogenic viruses and assess their relationship with cellular RNAi mechanisms from the molecular biology and translational sciences standpoints. In the first, Ashley P.E. Roberts, Andrew P. Lewis, and Catherine L. Jopling will initially provide a broad and comprehensive overview over virus structure and function, in general, before moving into the incredible intricacies of virus–host interactions involving miRNAs and human RNAi machinery. One special example that will then be discussed in particular detail is the interplay of human hepatitis C virus (HCV) with a liver-specific miRNA the unconventionality of which has consistently amazed the RNAi and virus communities over the past few years and will hopefully excite the readers of this chapter as well.

Taking over from this overview over how viruses exploit and modulate the human RNAi machinery, Julia Eekels and Ben Berkhout next highlight how one can turn this cellular mechanism against the intruder and develop clinical RNAi-based strategies that not only aim to battle and eradicate the AIDS virus HIV-1 but concurrently also rob it of any chance to ever escape from therapy by natural mutation. Therefore, the authors first thoroughly review the plethora of strategies and tools that have recently been discussed, developed, and validated to achieve these goals, before they describe combinatorial RNAi targeting of multiple essential viral and cellular genes as the latest and most promising therapeutic approach to tackle and forever eliminate the HIV pathogen from infected patients.

Finishing off this issue is a chapter from our own group that once again documents the remarkably tight affiliation of basic and applied RNAi research by reviewing our emerging picture of cellular RNAi mechanisms potentially dysregulated in human cancers and other devastating diseases, and by concomitantly discussing the novel therapeutic concepts and translational avenues that fortuitously continue to surface and materialize with each new report. Extending the preceding chapters, special focus will also be put on viruses not only because they represent beautiful authentic examples for molecular means to perturb, hijack, and reprogram cellular processes but because viruses are also being engineered as RNAi-expressing tools to treat human diseases that can occasionally turn into our worst enemy and unintentionally overthrow the human RNAi machinery. The related recent observations and vital lessons learned will be discussed, before Stefan Mockenhaupt, Nina Schürmann, and Dirk Grimm round up the chapter and issue with an outlook into what may be the next important steps to even further advance our understanding and utilization of cellular RNAi mechanisms and to ensure that the whole field will positively continue to prosper and thrive for many more years to come.

DIRK GRIMM

# Epigenetic Regulation of Gene Expression in Human Cells by Noncoding RNAs

Stuart Knowling and Kevin V. Morris

*Department of Molecular and Experimental Medicine, The Scripps Research Institute, La Jolla, California, USA*



Emerging evidence has begun to suggest that a vast array of noncoding RNAs is operative in human cells, with some containing the ability to directly modulate gene transcription. While observations of noncoding-RNA-based epigenetic regulation of gene expression were in the past relegated to imprinted or X-linked genes, it is now becoming apparent that several different genes in differentiated cells may be under some form of RNA-based regulatory control. Studies have begun to discern certain aspects of an underlying mechanism of action whereby noncoding RNAs modulate gene transcription. Much of the evidence suggests that noncoding RNAs are functional in controlling gene transcription by the targeted recruitment of epigenetic silencing complexes to homology-containing loci in the genome. The results of these studies, as well as the implications that a vast array of noncoding-RNA-based regulatory networks may be operative in human cells, are discussed. Knowledge of this emerging RNA-based epigenetic regulatory network has implications in cellular evolution as well as in an entirely new area of pharmacopeia.

## I. Introduction to ncRNA

The central dogma of molecular biology states that DNA sequence information tends to flow in one direction from a protein-coding region of the genome, through transcription to mRNA, and then, finally, is translated to protein. Every step along the way is assisted by proteins; therefore it is easy to see why it is described as a protein world. Recently though, new techniques such as genomic tiling arrays and cDNA sequencing, as used in RNA Seq, have

Progress in Molecular Biology
and Translational Science, Vol. 102
DOI: 10.1016/B978-0-12-415795-8.00003-9

provided a surprising view of how little of the human genome is actually used for protein coding (~2% of the eukaryotic genome encodes protein-coding genes—mRNA). The advent of these new DNA sequencing technologies has allowed RNA transcription in eukaryotic genomes to be assessed with higher accuracy and better resolution than ever before. The ENCODE (encyclopedia of DNA elements) project used deep sequencing of human cells to identify all functional elements in the human genome sequence by the analysis of a small fraction of RNAs (< 200 nucleotides—nt).[1] Approximately 90% of the genome is transcribed,[2] generating a large number of RNA transcripts that are non-protein-coding RNAs (ncRNAs). The results of ENCODE show that RNA can be processed via an unknown mechanism to yield complex populations of both short and long RNAs, which overlap both the 5′ and 3′ ends of protein-coding transcripts.[1]

The procession of longer RNA to shorter RNA to yield unique secondary and tertiary structures that may participate in cellular processes has long been known in the presence of[3] and absence of protein.[4,5] This knowledge is currently being exemplified in cellular and pharmaceutical processes by RNA aptamers.[6] The production of these long or short ncRNAs effectively splits DNA transcription into two camps: coding and noncoding transcripts,[7] where traditional noncoding transcripts are viewed to be ribosomal RNA (rRNA) and transfer RNA (tRNA), both of which play essential functions in protein translation. Although there is disagreement as to whether these ncRNAs are transcriptional "noise," there is growing evidence that ncRNAs (miRNAs, small RNAs) play an important role in cellular function[7] as is the case for the aforementioned tRNA and rRNA. Therefore, ncRNA raises the enticing possibility that the large number of ncRNAs being discovered play key roles in cellular control. Recently, evidence has begun to emerge suggesting that ncRNAs play a role in controlling gene transcription through the targeted recruitment of epigenetic silencing complexes to particular loci (reviewed in Refs. 8,9) and through degradation of transcriptionally active mRNAs via STAU-1-mediated RNA decay.[10] Staufen 1 is a protein that binds to double-stranded RNA and mediates (STAU1)-mediated messenger RNA decay (SMD). SMD degrades active mRNAs that contain 3′ untranslated regions (3′UTRs), which contain a STAU1 binding site (of approximately 688 nucleotides and consisting of a 19-base pair stem with a 100-nt apex).[11] Although the STAU1 binding site is not present in all the 3′UTRs of all mRNAs, it was recently shown that a cytoplasmic, polyadenylated lncRNA (ncRNA 1/2-sbsRNAs) that contains Alu elements can form imperfect base pairing with Alu elements in the 3′UTR of mRNA, forming an STAU-1 binding site and thus targeting it for SMD. Interestingly, the ncRNA 1/2-sbsRNAs was found in every tissue examined by Gong *et al.*[10] but only in the cytoplasmic fractions of HeLa cells. 1/2-sbsRNA1 was found not to be a substrate for the enzymes di.

cer 1 (DICER1) or argonaute 2 (AGO2), and thus represents an unexpected strategy that cells can use to recruit proteins to mRNAs and facilitate their decay. Only 23% of ncRNA contain Alu elements, so it is possible that further pathways exist for regulation of dsRNA by dsRNA-binding proteins.

## II. Epigenetics and ncRNAs

Epigenetics is the study of the underlying changes in phenotype that are caused by alterations to the expression of the genome by chemical modification of the DNA molecule but without changing the overall DNA sequence. Although several forms of epigenetic regulation exist, the two main forms of interest for this review are the (1) addition of chemical groups to specific bases, as with DNA methylation and (2) the local alterations of histones, such as targeted methylation at particular lysines that affects the accessibility of the surrounding genomic DNA to the transcriptional machinery. Other modifications of interest to this review are the splicing of RNA to form novel structures and double-stranded RNA, as with RNA interference (RNAi) pathways.[12]

The current view of genetics is dominated by Darwinism, but epigenetic control of the genome offers a "touch of soft Lamerckism"[7] in that epigenetic changes caused by environmental processes are capable of being passed on to daughter cells. It has been shown that epigenetic changes to the *IGF2* gene that were caused by prenatal exposure to the Dutch Hunger Winter in the 1944–1945 famine are persistent across familial generations six decades later,[13] an effect that has also been noted in mice.[14,15] The prospect of the ability to impact further generations of genome via epigenetic methods run counter to Darwinian existence but suggests that environmental changes indeed affect the expression of DNA and can be passed onto offspring. The comprehensive role that ncRNA plays in this epigenetic control remains to be seen, but if the recent observations are any indication from a limited number of studies,[16–23] the notion is that much of the ncRNAs in human cells might be active regulators involved in controlling gene expression via the targeted recruitment of epigenetic complexes to various loci in the genome.

The obvious question is whether there is a link between the large amount of transcribed ncRNA and the regulation of genome modification via epigenetics. The fact that almost four times the amount of ncRNA is transcribed, compared to protein-coding RNA,[24] suggests there must be a function for these ncRNAs: that is, this is an immense cost to the cell with regard to energy expenditure. Expressed ncRNAs show clear evolutionary conservation[25] and many emanate from gene promoter regions, which tend to be more conserved than protein-coding genes,[26] thus suggesting a level of retention in the machinery of the cell and a possible role in gene level control, possibly via the recruitment of

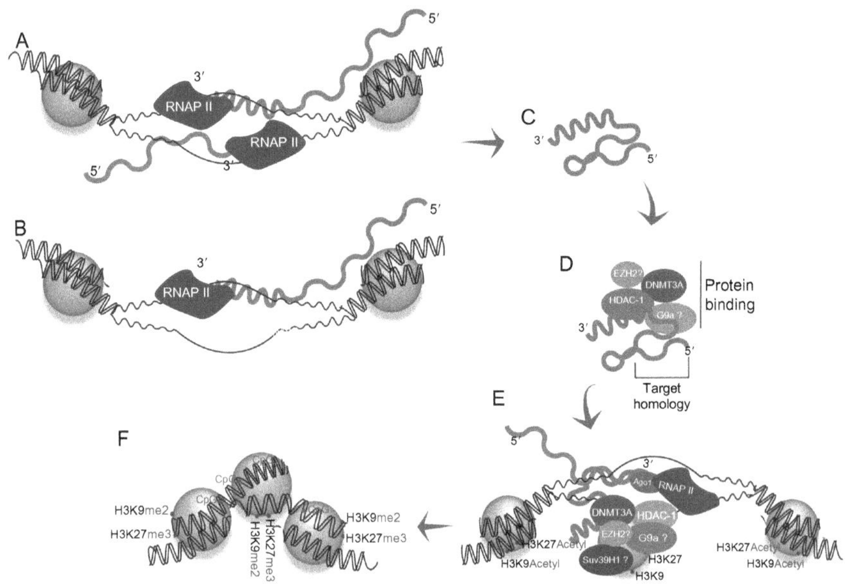


FIG. 1. Endogenous ncRNA-directed epigenetic regulation of gene transcription. (A) Long antisense ncRNAs can be expressed in *cis* or (B) *trans*, where they (C) fold into complex structures that can (D) contain both target homology to other loci or the bidirectionally transcribed loci in the genome as well as possess an ability to bind epigenetic remodeling proteins. The ncRNA-directed epigenetic complexes could then (E) target the silencing of the corresponding homology-containing loci, which can result in (F) heterochromatinization of the targeted site. (See Color Insert.)

promoter. When these antisense ncRNAs were suppressed, concordant regulation of Nanog was observed.[40] Discordantly regulated genes are regulated in the inverse manner to concordantly regulated genes and are reviewed in Refs. 19, 41.

Several insights have been obtained from these works. Taken together, the Oct-4 discordant and Nanog concordant model systems might be telling us something important about endogenous RNA circuits and gene regulation. No doubt, when one considers now the plethora of ongoing RNA interactions in the cell, such as these observations whereby psuedogene-expressed antisense ncRNAs are actively regulating their respective mRNA-expressing counterpart in either a positive or negative manner, it indicates a far more complex landscape of RNA-based regulation than has been previously appreciated.

It is now clear that there is significant and growing evidence that small and long antisense ncRNAs are able to direct the epigenetic control of gene transcription. Such a model was recently proposed for the PRC2-associated lincRNA TUG1,[29] a gene that is upregulated in response to DNA damage through the tumor suppressor gene p53. Knockdown of TUG1 by siRNA led to an upregulation of 120 genes, suggesting that PRC2 and TUG1 (and other lincRNA) repress transcription via directing silencing to specific loci in the cell.

It has also been noted that, although many lincRNAs act in a repressive manner by chromatin modification, it is possible that many genes could be upregulated in a similar manner by chromatin modification[42]; this seems a plausible occurrence based on the striking number of observations to date demonstrating RNA-based regulation of gene expression (reviewed in Ref. 43). Such a notion runs counter to the dogma arguing that transcription is predominantly controlled by transcription factors and other DNA-binding proteins. However, it may also be that the localization of transcription factors and larger protein-activating complexes at particular loci in the genome also require the action of lincRNAs. This has been suggested recently for the p53-activated lincRNA, lincRNA-p21.[44] The promoter region of lincRNA-p21 contains a high number of p53-binding motifs and is significantly induced upon p53 activation. Upon activation, lincRNA-p21 plays a significant role in apoptosis of the cell, acting as a downstream repressor in p53 transcriptional response. lincRNA-p21 was shown to interact with the DNA-binding protein hn-RNP-K, indicating that lncRNA may alter the binding specificity of DNA-binding proteins in order to alter their target specificity, subsequently altering gene expression.

Regardless, recent observations have begun to craft a modified version of previously held dogma. This new model intercalates the action of ncRNAs in the control of gene expression and is based on a growing body of evidence. Such observations suggest only one plausible notion: that there are multiple layers of RNA-based regulation, be they direct RNA/RNA-, RNA/prote-, or RNA/DNA-based. The basal strata of RNA-directed regulation of gene expression are most likely at the act of transcription, as the RNA requires transcription in its essence of function. This basal layer of RNA-based regulation may have some striking similarities to and functional properties involved in the mechanism that dictates euchromatin versus heterochromatin transitions in the nucleus. Or, alternatively, they have the same mechanism of action. Certainly, work with imprinted genes and X-inactivation, which involve significant compaction of the ncRNA-targeted loci, would suggest that many similar components are involved in the basal modes of action.[37,45]

The regulation of transcription, via ncRNA-targeted epigenetic controlling mechanisms, might in fact be the essence of endogenous gene regulation and the fundamental core regulator of protein production. One could envision that the stoichiometric ratios of particular ncRNAs are active in regulating gene transcription, but also that these ncRNAs are themselves regulated by other ncRNAs and RNA-binding complexes. Such a concept adds so many layers of complexity that one has to envision the cell as a larger "ecosystem" of interactions. While such speculation cannot in principle be experimentally validated, as the complexity within the cell would no doubt be infinite, there are some aspects of this regulation that are noteworthy: namely, the notion that modulating the expression paradigm of particular ncRNA regulatory networks might allow an ability to modulate both single gene/protein expression and possibly yet-to-be-discovered larger complex networks regulated by a single or a few ncRNAs or the particular single gene/protein. Such knowledge could in theory allow increased gene-specific targeting capabilities to be developed as well as studies on larger gene regulatory networks to be carried out. No doubt, we will, in the not too distant future, be faced with the notion that larger systems have the capability of functioning in an autonomous and unique manner far different than each of the constituent parts. Such a notion will provide information on the essence of gene regulation and possibly some clues as to how natural selective pressures are operative in living systems on genomic evolution.

## Acknowledgments

The Morris lab is funded by Grants NIH R01 HL083473, NIH R01 AI084406, NIH R01 CA151574, and NIH R01 CA153124.

## References

1. Affymetrix ENCODE Transcriptome Project & Cold Spring Harbor Laboratory ENCODE Transcriptome Project. Post-transcriptional processing generates a diversity of 5′-modified long and short RNAs. *Nature* 2009;**457**:1028–32.
2. Wilhelm BT, Marguerat S, Watt S, Schubert F, Wood V, Goodhead I, et al. Dynamic repertoire of a eukaryotic transcriptome surveyed at single-nucleotide resolution. *Nature* 2008;**453**:1239–43.
3. Stark BC, Kole R, Bowman EJ, Altman S. Ribonuclease P: an enzyme with an essential RNA component. *Proc Natl Acad Sci USA* 1978;**75**:3717–21.
4. Kruger K, Grabowski PJ, Zaug AJ, Sands J, Gottschling DE, Cech TR. Self-splicing RNA, autoexcision and autocyclization of the ribosomal RNA intervening sequence of Tetrahymena. *Cell* 1982;**31**:147–57.
5. Robertson DL, Joyce GF. Selection in vitro of an RNA enzyme that specifically cleaves single-stranded DNA. *Nature* 1990;**344**:467–8.
6. Bunka DH, Platonova O, Stockley PG. Development of aptamer therapeutics. *Curr Opin Pharmacol* 2010;**6**:557–62.
7. Costa FF. Non-coding RNAs: meet thy masters. *Bioessays* 2010;**32**:599–608.
8. Morris KV. Long antisense non-coding RNAs function to direct epigenetic complexes that regulate transcription in human cells. *Epigenetics* 2009;**4**:296–301.
9. Morris KV. Non-coding RNAs, epigenetic memory and the passage of information to progeny. *RNA Biol* 2009;**6**:242–7.

10. Gong C, Maquat LE. lncRNAs transactivate STAU1-mediated mRNA decay by duplexing with 3′ UTRs via Alu elements. *Nature* 2011;**470**:284–8.
11. Kim YK, Furic L, Parisien M, Major F, DesGroseillers L, Maquat LE. Staufen1 regulates diverse classes of mammalian transcripts. *EMBO J* 2007;**26**:2670–81.
12. Morris KV, Rossi JJ. Antiviral applications of RNAi. *Curr Opin Mol Ther* 2006;**8**:115–21.
13. Heijmans BT, Tobi EW, Stein AD, Putter H, Blauw GJ, Susser ES, et al. Persistent epigenetic differences associated with prenatal exposure to famine in humans. *Proc Natl Acad Sci USA* 2008;**105**:17046–9.
14. Rassoulzadegan M, Grandjean V, Gounon P, Vincent S, Gillot I, Cuzin F. RNA-mediated non-mendelian inheritance of an epigenetic change in the mouse. *Nature* 2006;**441**:469–74.
15. Wagner KD, Wagner N, Ghanbarian H, Grandjean V, Gounon P, Cuzin F, et al. RNA induction and inheritance of epigenetic cardiac hypertrophy in the mouse. *Dev Cell* 2008;**14**:962–9.
16. Camblong J, Iglesias N, Fickentscher C, Dieppois G, Stutz F. Antisense RNA stabilization induces transcriptional gene silencing via histone deacetylation in *S. cerevisiae*. *Cell* 2007;**131**:706–17.
17. Cho DH, Thienes CP, Mahoney SE, Analau E, Filippova GN, Tapscott SJ. Antisense transcription and heterochromatin at the DM1 CTG repeats are constrained by CTCF. *Mol Cell* 2005;**20**:483–9.
18. Ebralidze AK, Guibal FC, Steidl U, Zhang P, Lee S, Bartholdy B, et al. PU.1 expression is modulated by the balance of functional sense and antisense RNAs regulated by a shared *cis*-regulatory element. *Genes Dev* 2008;**22**:2085–92.
19. Hawkins PG, Morris KV. Transcriptional regulation of Oct4 by a long non-coding RNA antisense to Oct4-pseudogene 5. *Transcription* 2010;**1**:165–75.
20. Li K, Blum Y, Verma A, Liu Z, Pramanik K, Leigh NR, et al. A noncoding antisense RNA in tie-1 locus regulates tie-1 function in vivo. *Blood* 2010;**115**:133–9.
21. Morris KV, Santoso S, Turner AM, Pastori C, Hawkins PG. Bidirectional transcription directs both transcriptional gene activation and suppression in human cells. *PLoS Genet* 2008;**4**: e1000258.
22. Tufarelli C, Stanley JA, Garrick D, Sharpe JA, Ayyub H, Wood WG, et al. Transcription of antisense RNA leading to gene silencing and methylation as a novel cause of human genetic disease. *Nat Genet* 2003;**34**:157–65.
23. Yu W, Gius D, Onyango P, Muldoon-Jacobs K, Karp J, Feinberg AP, et al. Epigenetic silencing of tumour suppressor gene p15 by its antisense RNA. *Nature* 2008;**451**:202–6.
24. Ponting CP. The functional repertoires of metazoan genomes. *Nat Rev Genet* 2008;**9**:689–98.
25. Guttman M, Amit I, Garber M, French C, Lin MF, Feldser D, et al. Chromatin signature reveals over a thousand highly conserved large non-coding RNAs in mammals. *Nature* 2009;**458**:223–7.
26. Carninci P, Kasukawa T, Katayama S, Gough J, Frith MC, Maeda N, et al. The transcriptional landscape of the mammalian genome. *Science* 2005;**309**:1559–63.
27. Cheng J, Kapranov P, Drenkow J, Dike S, Brubaker S, Patel S, et al. Transcriptional maps of 10 human chromosomes at 5-nucleotide resolution. *Science* 2005;**308**:1149–54.
28. Kondo T, Hashimoto Y, Kato K, Inagaki S, Hayashi S, Kageyama Y. Small peptide regulators of actin-based cell morphogenesis encoded by a polycistronic mRNA. *Nat Cell Biol* 2007;**9**:660–5.
29. Khalil AM, Guttman M, Garber HMM, Raj A, Rivea Morales D, et al. Many human large intergenic noncoding RNAs associate with chromatin-modifying complexes and affect gene expression. *Proc Natl Acad Sci USA* 2009;**106**:11667–72.
30. Heard E. Delving into the diversity of facultative heterochromatin: the epigenetics of the inactive X chromosome. *Curr Opin Genet Dev* 2005;**15**:482–9.
31. Nagano T, Mitchell JA, Sanz LA, Pauler FM, Ferguson-Smith AC, Feil R, et al. The air noncoding RNA epigenetically silences transcription by targeting G9a to chromatin. *Science* 2008;**322**:1717–20.

transcriptional regulation. The therapeutic potential of TGS and TGA is examined, emphasizing the distinct advantages over PTGS approaches with examples of application to cancer and diseases associated with viruses, aberrant splicing, and dysregulated heterochromatin. Finally, the influence of promoter architecture on gene susceptibility to transcriptional regulation is discussed in the light of how this impacts the scope of small RNA-induced transcriptional regulation within the genome.

## I. Introduction

Few scientific discoveries have proved as meaningful to biology and modern medicine as that by Andrew Fire, Craig Mello, and colleagues in 1998, when they demonstrated that double-stranded RNA (dsRNA) can effect powerful inhibition of gene expression.[1] Commonly referred to as RNA interference (RNAi), this central component of eukaryote (and possibly prokaryote) biology has a role in many cellular functions. In mammals, RNAi is triggered by dsRNA precursors, which are enzymatically processed into short RNAs of $\sim$20–30 nucleotides (nt) in length. These short RNAs, which associate with members of the Argonaute (AGO) family of proteins, regulate gene expression at the transcriptional and posttranscriptional level. The most emphatic demonstration of the power of RNAi came from harnessing this pathway to silence specific genes, with subsequent application to reverse genetics and therapeutics. Thus, RNAi has spawned a myriad of important biotechnological and therapeutic advances.

The last decade of mammalian RNAi research has been dominated by approaches that harness posttranscriptional gene silencing (PTGS), mainly by mimicking structural elements associated with the precursors of the canonical microRNA (miRNA) or short interfering RNA (siRNA) biogenesis pathways. It is clear that, in addition to PTGS, small RNAs derived from RNAi and related pathways play an important role in affecting gene expression by modulating chromatin structure, resulting in transcriptional regulation, either by transcriptional gene silencing (TGS) or activation (TGA). In this review, we briefly outline the biogenesis of small RNAs and explore the mechanisms behind the various phenomena attributed to AGO-bound small RNA-mediated transcriptional regulation. The potential of TGS and TGA as therapeutic strategies, which have distinct advantages over PTGS approaches, is examined. Finally, the scope of small RNA-induced transcriptional regulation is discussed.

10. Gong C, Maquat LE. lncRNAs transactivate STAU1-mediated mRNA decay by duplexing with 3′ UTRs via Alu elements. *Nature* 2011;**470**:284–8.
11. Kim YK, Furic L, Parisien M, Major F, DesGroseillers L, Maquat LE. Staufen1 regulates diverse classes of mammalian transcripts. *EMBO J* 2007;**26**:2670–81.
12. Morris KV, Rossi JJ. Antiviral applications of RNAi. *Curr Opin Mol Ther* 2006;**8**:115–21.
13. Heijmans BT, Tobi EW, Stein AD, Putter H, Blauw GJ, Susser ES, et al. Persistent epigenetic differences associated with prenatal exposure to famine in humans. *Proc Natl Acad Sci USA* 2008;**105**:17046–9.
14. Rassoulzadegan M, Grandjean V, Gounon P, Vincent S, Gillot I, Cuzin F. RNA-mediated non-mendelian inheritance of an epigenetic change in the mouse. *Nature* 2006;**441**:469–74.
15. Wagner KD, Wagner N, Ghanbarian H, Grandjean V, Gounon P, Cuzin F, et al. RNA induction and inheritance of epigenetic cardiac hypertrophy in the mouse. *Dev Cell* 2008;**14**:962–9.
16. Camblong J, Iglesias N, Fickentscher C, Dieppois G, Stutz F. Antisense RNA stabilization induces transcriptional gene silencing via histone deacetylation in *S. cerevisiae*. *Cell* 2007;**131**:706–17.
17. Cho DH, Thienes CP, Mahoney SE, Analau E, Filippova GN, Tapscott SJ. Antisense transcription and heterochromatin at the DM1 CTG repeats are constrained by CTCF. *Mol Cell* 2005;**20**:483–9.
18. Ebralidze AK, Guibal FC, Steidl U, Zhang P, Lee S, Bartholdy B, et al. PU.1 expression is modulated by the balance of functional sense and antisense RNAs regulated by a shared *cis*-regulatory element. *Genes Dev* 2008;**22**:2085–92.
19. Hawkins PG, Morris KV. Transcriptional regulation of Oct4 by a long non-coding RNA antisense to Oct4-pseudogene 5. *Transcription* 2010;**1**:165–75.
20. Li K, Blum Y, Verma A, Liu Z, Pramanik K, Leigh NR, et al. A noncoding antisense RNA in tie-1 locus regulates tie-1 function in vivo. *Blood* 2010;**115**:133–9.
21. Morris KV, Santoso S, Turner AM, Pastori C, Hawkins PG. Bidirectional transcription directs both transcriptional gene activation and suppression in human cells. *PLoS Genet* 2008;**4**: e1000258.
22. Tufarelli C, Stanley JA, Garrick D, Sharpe JA, Ayyub H, Wood WG, et al. Transcription of antisense RNA leading to gene silencing and methylation as a novel cause of human genetic disease. *Nat Genet* 2003;**34**:157–65.
23. Yu W, Gius D, Onyango P, Muldoon-Jacobs K, Karp J, Feinberg AP, et al. Epigenetic silencing of tumour suppressor gene p15 by its antisense RNA. *Nature* 2008;**451**:202–6.
24. Ponting CP. The functional repertoires of metazoan genomes. *Nat Rev Genet* 2008;**9**:689–98.
25. Guttman M, Amit I, Garber M, French C, Lin MF, Feldser D, et al. Chromatin signature reveals over a thousand highly conserved large non-coding RNAs in mammals. *Nature* 2009;**458**:223–7.
26. Carninci P, Kasukawa T, Katayama S, Gough J, Frith MC, Maeda N, et al. The transcriptional landscape of the mammalian genome. *Science* 2005;**309**:1559–63.
27. Cheng J, Kapranov P, Drenkow J, Dike S, Brubaker S, Patel S, et al. Transcriptional maps of 10 human chromosomes at 5-nucleotide resolution. *Science* 2005;**308**:1149–54.
28. Kondo T, Hashimoto Y, Kato K, Inagaki S, Hayashi S, Kageyama Y. Small peptide regulators of actin-based cell morphogenesis encoded by a polycistronic mRNA. *Nat Cell Biol* 2007;**9**:660–5.
29. Khalil AM, Guttman M, Garber HMM, Raj A, Rivea Morales D, et al. Many human large intergenic noncoding RNAs associate with chromatin-modifying complexes and affect gene expression. *Proc Natl Acad Sci USA* 2009;**106**:11667–72.
30. Heard E. Delving into the diversity of facultative heterochromatin: the epigenetics of the inactive X chromosome. *Curr Opin Genet Dev* 2005;**15**:482–9.
31. Nagano T, Mitchell JA, Sanz LA, Pauler FM, Ferguson-Smith AC, Feil R, et al. The air noncoding RNA epigenetically silences transcription by targeting G9a to chromatin. *Science* 2008;**322**:1717–20.

32. Rinn JL, Kertesz M, Wang JK, Squazzo SL, Xu X, Brugmann SA, et al. Functional demarcation of active and silent chromatin domains in human HOX loci by noncoding RNAs. *Cell* 2007;**129**:1311–23.
33. Gupta RA, Shah N, Wang KC, Kim J, Horlings HM, Wong DJ, et al. Long non-coding RNA HOTAIR reprograms chromatin state to promote cancer metastasis. *Nature* 2010;**464**:1071–6.
34. Clemson CM, Hutchinson JN, Sara SA, Ensminger AW, Fox AH, Chess A, et al. An architectural role for a nuclear noncoding RNA: NEAT1 RNA is essential for the structure of paraspeckles. *Mol Cell* 2009;**33**:717–26.
35. Fox AH, Lam YW, Leung AK, Lyon CE, Andersen J, Mann M, et al. Paraspeckles: a novel nuclear domain. *Curr Biol* 2002;**12**:13–25.
36. Tsai MC, Manor O, Wan Y, Mosammaparast N, Wang JK, Lan F, et al. Long noncoding RNA as modular scaffold of histone modification complexes. *Science* 2010;**329**:689–93.
37. Latos PA, Barlow DP. Regulation of imprinted expression by macro non-coding RNAs. *RNA Biol* 2009;**6**:100–6.
38. Sheik Mohamed J, Gaughwin PM, Lim B, Robson P, Lipovich L. Conserved long noncoding RNAs transcriptionally regulated by Oct4 and Nanog modulate pluripotency in mouse embryonic stem cells. *RNA* 2010;**16**:324–37.
39. Wahlestedt C. Natural antisense and noncoding RNA transcripts as potential drug targets. *Drug Discov Today* 2006;**11**:503–8.
40. Morris KV. The emerging role of RNA in the regulation of gene transcription in human cells. *Semin Cell Dev Biol* 2011; [Epub ahead of print].
41. Morris KV, Vogt PK. Long antisense non-coding RNAs and their role in transcription and oncogenesis. *Cell Cycle* 2010;**9**:2544–7.
42. Dinger ME, Amaral PP, Mercer TR, Pang KC, Bruce SJ, Gardiner BB, et al. Long noncoding RNAs in mouse embryonic stem cell pluripotency and differentiation. *Genome Res* 2008;**18**:1433–45.
43. Morris KV. RNA-directed transcriptional gene silencing and activation in human cells. *Oligonucleotides* 2009;**19**:299–306.
44. Huarte M, Guttman M, Feldser D, Garber M, Koziol MJ, Kenzelmann-Broz D, et al. A large intergenic noncoding RNA induced by p53 mediates global gene repression in the p53 response. *Cell* 2010;**142**:409–19.
45. Lee JT. As model for RNA's niche in epigenomic regulation. *Cold Spring Harb Perspect Biol* 2010;**2**:a003749.

# Small RNA-Induced Transcriptional Gene Regulation in Mammals: Mechanisms, Therapeutic Applications, and Scope Within the Genome

Victoria A. Green and
Marc S. Weinberg

*Antiviral Gene Therapy Research Unit, Department of Molecular Medicine and Haematology, University of the Witwatersrand, Johannesburg, South Africa*



Argonaute-bound small RNAs, derived from RNA interference and related pathways, are well-known effectors of posttranscriptional gene silencing (PTGS). Yet, these complexes also play an important role in affecting gene expression at the transcriptional level, either by transcriptional gene silencing (TGS) or activation (TGA). Our current understanding of how small RNAs are able to both activate and suppress transcription is unclear. In this review, we briefly outline the biogenesis of small RNAs and explore the mechanisms behind the various phenomena attributed to AGO-bound small RNA-mediated

Progress in Molecular Biology
and Translational Science, Vol. 102
DOI: 10.1016/B978-0-12-415795-8.00005-2




1877-1173/11 $35.00

transcriptional regulation. The therapeutic potential of TGS and TGA is examined, emphasizing the distinct advantages over PTGS approaches with examples of application to cancer and diseases associated with viruses, aberrant splicing, and dysregulated heterochromatin. Finally, the influence of promoter architecture on gene susceptibility to transcriptional regulation is discussed in the light of how this impacts the scope of small RNA-induced transcriptional regulation within the genome.

## I. Introduction

Few scientific discoveries have proved as meaningful to biology and modern medicine as that by Andrew Fire, Craig Mello, and colleagues in 1998, when they demonstrated that double-stranded RNA (dsRNA) can effect powerful inhibition of gene expression.[1] Commonly referred to as RNA interference (RNAi), this central component of eukaryote (and possibly prokaryote) biology has a role in many cellular functions. In mammals, RNAi is triggered by dsRNA precursors, which are enzymatically processed into short RNAs of ~20–30 nucleotides (nt) in length. These short RNAs, which associate with members of the Argonaute (AGO) family of proteins, regulate gene expression at the transcriptional and posttranscriptional level. The most emphatic demonstration of the power of RNAi came from harnessing this pathway to silence specific genes, with subsequent application to reverse genetics and therapeutics. Thus, RNAi has spawned a myriad of important biotechnological and therapeutic advances.

The last decade of mammalian RNAi research has been dominated by approaches that harness posttranscriptional gene silencing (PTGS), mainly by mimicking structural elements associated with the precursors of the canonical microRNA (miRNA) or short interfering RNA (siRNA) biogenesis pathways. It is clear that, in addition to PTGS, small RNAs derived from RNAi and related pathways play an important role in affecting gene expression by modulating chromatin structure, resulting in transcriptional regulation, either by transcriptional gene silencing (TGS) or activation (TGA). In this review, we briefly outline the biogenesis of small RNAs and explore the mechanisms behind the various phenomena attributed to AGO-bound small RNA-mediated transcriptional regulation. The potential of TGS and TGA as therapeutic strategies, which have distinct advantages over PTGS approaches, is examined. Finally, the scope of small RNA-induced transcriptional regulation is discussed.

## II. Mammalian Small RNA Biogenesis Pathways

Short, single-stranded RNAs represent the most abundant and well-characterized species of small RNAs. Appropriating elements of the endogenous canonical miRNA biogenesis pathway (Fig. 1A) has accelerated the development of RNAi tools and therapies. However, several alternative or complementary pathways have also been shown to yield small RNAs that enter into complexes associated with one or more AGO proteins (Fig. 1B).

### A. Canonical miRNA Biogenesis

miRNAs are integral components of the RNAi pathway and consist of a class of short (~22 nt), noncoding RNAs (ncRNAs) derived from longer dsRNA precursors known as primary miRNAs (pri-miRNAs).[2–4] Most pri-miRNAs are embedded within RNA polymerase II (Pol II) transcripts and consist of one or more (polycistronic) short hairpin motifs. These motifs contain a ~33-bp duplex stem region with embedded miRNA/anti-miRNA sequences, a flexible terminal loop, and flanking single-stranded sequences.[5–7] As with most Pol II transcripts, pri-miRNAs are 5′ capped and 3′ polyadenylated. Furthermore, miRNA biogenesis is coregulated with transcription and occurs prior to pre-mRNA splicing.[8,9] miRNA signatures can be found within exonic and intronic regions of coding mRNAs, and also within ncRNAs, including antisense and intergenic transcripts.[10]

Pri-miRNAs are enzymatically processed in two sequential steps: first in the nucleus, and then in the cytoplasm. In the nucleus, the hairpin-like structure of the pri-miRNA is recognized and cleaved by the multiprotein "microprocessor" complex into a shorter 70–80-nt hairpin known as a precursor-miRNA (pre-miRNA).[4,11] At its core, the microprocessor complex comprises the RNase III enzyme Drosha and its partner DGCR8, a dsRNA-binding domain protein. However, several additional auxiliary cofactors have been discovered that enhance Drosha cleavage and processivity in a subset of miRNAs. These include the DEAD box RNA helicases p68 (DDX5) and p72 (DDX17),[12,13] as well as heterogeneous nuclear ribonucleoproteins (hnRNPs).[14–16] While the pri-miRNA structure appears to be a determinant in microprocessor function, several sequence-dependent properties have been identified that also modulate processing. Some factors have been implicated as attenuators of miRNA biogenesis at the Drosha cleavage step, many of which are involved in self-regulatory negative feedback loops.[17] Regardless, once cropped by the microprocessor complex, a nuclear karyopherin, exportin-5, recognizes both the 2-nt 3′ hydroxyl overhangs, generated by RNase III cleavage, as well as 16–18 bp of the pre-miRNA stem. This recognition facilitates Ran-GTP-dependent export to the cytoplasm,[18–21] where the second processing step occurs.

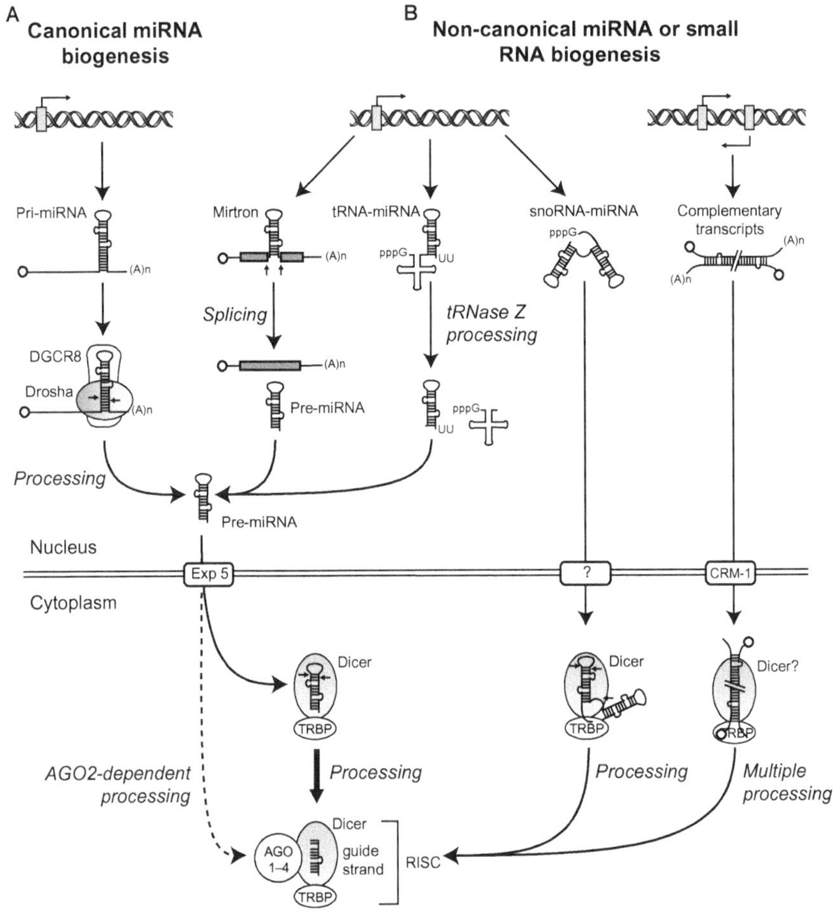


FIG. 1. *Mammalian small RNA biogenesis pathways.* (A) Canonical miRNA biogenesis. miRNAs are encoded in pri-miRNAs, which comprise inverted repeat motifs, usually embedded in Pol II-derived transcripts. Pri-miRNAs are first processed in the nucleus by Drosha/DGCR8 to produce shorter (70–80 nt) hairpin duplexes termed precursor-miRNAs (pre-miRNAs). Pre-miRNAs are exported from the nucleus to the cytoplasm by exportin-5. Here they are cleaved by Dicer/TRBP to produce a ~22-bp staggered miRNA/anti-miRNA duplex with 2-nt 3′ overhangs. Dicer/TRBP loads the miRNA or "guide strand" into an RNA-induced silencing complex (RISC) containing one of Argonautes 1–4 (AGO1–4). (B) Noncanonical miRNA or small RNA biogenesis. For a small minority of miRNAs, short intronic sequences, termed "mirtrons," can be directly processed by the spliceosome into pre-miRNA-like hairpins without requiring Drosha cleavage. Spliced lariats are debranched and likely produce functional pre-miRNAs for export. miRNAs can be embedded in hairpin motifs expressed within tRNA or snoRNA transcripts. The former are likely 3′ processed in the nucleus by tRNase Z prior to export. snoRNA hairpins are exported by an unknown pathway and are directly recognized by Dicer for processing. Lastly, endogenous siRNAs are derived from long hairpin sequences and complementary Pol II transcripts, which, following export by CRM-1, are processed by Dicer into siRNAs.

The pre-miRNA is cleaved by the RNase III enzyme Dicer to produce a ~22-bp, staggered miRNA/anti-miRNA duplex.[22,23] Dicer and its binding partner, TAR RNA-binding protein (TRBP),[24] load one of the strands, referred to as the mature miRNA or "guide strand," into the RNA-induced silencing complex (RISC), which contains members of the AGO family of proteins. AGO proteins comprise PIWI, MID, and Piwi–Argonaute–Zwille (PAZ) domains.[25] The MID and PAZ domains bind and anchor the 5′ and 3′ ends of the guide strand, respectively,[26,27] while the RNase H-like PIWI domain cleaves the complementary target RNA. Interestingly, of the four mammalian AGOs (AGO1–4), only AGO2 has endonuclease or "slicer" activity.[28] In its simplest form, RISC comprises AGO2 and Dicer–TRBP.[29,30] The other strand of the RNA duplex, the "passenger strand," may be cleaved within RISC by AGO2[31–33] or simply released from the complex. Selection of the guide strand is based on asymmetric thermodynamic stability of the RNA duplex ends[34,35]; the strand with the weaker base-pairing at the 5′ end is preferentially incorporated into RISC as a guide. However, guide and passenger strand levels vary significantly between tissues, suggesting that sequence-related thermodynamics is not the only determinant in RISC loading (reviewed in Ref.17).

In classical PTGS, the mature miRNA associates with AGO2-containing RISC and acts as a guide by targeting its "seed regions," which comprise nucleotides two to seven from the 5′ end, to the 3′ untranslated regions (3′UTRs) of mRNAs.[36,37] Typically, miRNAs do not bind their targets with full sequence complementation, and this induces suppression of translation. While there are many mechanisms known to cause translational suppression, lowered mRNA levels appear to account for 84% of miRNA effects.[38] miRNA guide strands with near-perfect sequence complementarity to their cognate targets, which is a feature of siRNAs, effect gene silencing by AGO2-mediated posttranscriptional cleavage of targeted mRNAs.[37]

## B. Noncanonical miRNA and siRNA Biogenesis

Several subclasses of small RNAs, some of which meet the definition of an miRNA, are derived from alternative mammalian biogenesis pathways, many of which make use of specific dsRNA substrates (Fig. 1B). Although the role of some of these noncanonical biogenesis pathways remains unclear, harnessing aspects of these pathways may be useful for future RNAi-based therapeutic applications.

There are several Drosha-independent pathways that produce 60–70-nt pre-miRNA hairpin precursors. Mirtrons are a class of short introns that, following transcription, are spliced, debranched, and folded into pre-miRNA-like mimics, thus bypassing Drosha cleavage.[39,40] Several human small-nucleolar RNAs (snoRNAs) resemble two or more pre-miRNA-like hairpins tethered together within a longer transcript.[41] These are processed into pre-miRNAs by a Drosha-independent, Dicer-dependent mechanism and their downstream miRNAs associate with AGO1 and 2.[41] Moreover, several short RNAs and Dicer-dependent pre-miRNA mimics are derived from ribonuclease cleavage of leader and trailer sequences expressed from tRNA-like structures.[42–45]

Dicer-independent mechanisms also exist for the generation of small RNAs. In studies of mouse oocytes and embryonic stem cells, an abundant class of endogenous siRNAs, or endo-siRNAs, was discovered. These RNAs are thought to be derived from transcripts with long, inverted repeats or from convergent and divergent transcripts of pseudogenes or transposons.[46–48] Apart from blocking retrotransposition, little is known about the function of endo-siRNAs. However, the ubiquity of overlapping transcripts suggests that the number of endo-siRNAs (and their cognate targets) may be substantial within the human genome. Regardless of the origin of small RNAs, they may associate with AGOs and affect gene expression, both at the transcriptional and posttranscriptional levels.

## III. Mechanisms of Small RNA-Induced Transcriptional Regulation

Small RNA molecules bound to AGO proteins have been shown to modulate chromatin and affect gene expression by TGS. First described in plants and yeast,[49–51] TGS in mammalian cells was initially met with controversy, partly due to the paucity of endogenous effectors identified. RNAi pathway effectors can be loaded into a nuclear form of RISC, termed the RNA-induced transcriptional silencing complex (RITS).[52] Like RISC, AGO proteins are the core components of the complex, and targeting is mediated by complementary base-pairing of the RNA guide strand. Except for budding yeast, small RNA-mediated heterochromatin regulation is conserved throughout eukaryotes.

The first report of mammalian TGS utilized exogenous siRNAs to silence expression of an integrated GFP reporter driven by the eukaryotic translation elongation factor 1 alpha (EF1$\alpha$) promoter.[53] Morris and coworkers confirmed silencing at the transcriptional level by performing nuclear run-on analysis, which is the gold standard for differentiating silencing effects mediated by the TGS from those of the PTGS RNAi pathways. The same strategy was employed to obtain TGS at the endogenous E-cadherin (CDH1) promoter, a year later.[54] Both reports implicated a role for epigenetic changes in RNA-induced gene silencing, although neither explained the mechanism by which these changes occurred.

### A. Epigenetic Changes Associated with TGS

The formation of chromatin through the binding of histones to DNA both restricts access to the genetic code and creates a mechanism by which access is regulated. The nucleosome is the basic unit of chromatin, around which DNA is coiled.[55] Each nucleosome consists of a core of two copies of histones H2A, H2B, H3, and H4. Approximately 147 bp of DNA is wrapped around each nucleosome.[56] Between core units lies the linker histone, H1, which may be involved in higher-order compaction of DNA. The histones are subject to posttranslational modifications, including acetylation and methylation, particularly on N-terminal tails that protrude from the core, mediated by histone acetylases/deacetylases (HATs/HDACs) and histone methyltransferases/demethylases (HMTs/HDMTs), respectively.[57] Histone variants, such as H2A. Z and H3.3, provide additional variety to the nucleosome.[52,58] These modifications and variants can recruit specific proteins and also alter the binding affinity between histones and DNA. Binding affinities affect chromatin compaction and, thus, access of the transcription machinery to the DNA. Histone modifications mediate, therefore, epigenetic regulation of transcription and may be silencing and activating, depending on the residue and type of modification.

Another common type of epigenetic regulation involves DNA methylation, particularly at CpG dinucleotides. Most CpG-rich regions (CpG islands) overlap promoters, where DNA methylation is associated with gene silencing.[59] Whilst DNA methylation is largely maintained by DNA methyltransferase (DNMT) 1,[60] a number of proteins mediate *de novo* DNA methylation, including DNMT3A and DNMT3B.[61]

Identification of the protein complexes involved in the epigenetic changes that accompany mammalian TGS is ongoing. In the seminal Morris article,[53] the ability of a promoter-targeted siRNA to induce TGS was reversed by treatment with trichostatin A (TSA) or 5-azacytidine (5-azaC), implicating roles for histone deacetylation and DNA methylation, respectively. Ting *et al.* went further by demonstrating that TGS was accompanied by an enrichment for histone 3 lysine 9 (H3K9) dimethylation,[54] a chromatin modification associated with inactive promoters. Many studies have shown a role of core histone posttranslational modifications in TGS.[54,62–77] Enzymes responsible have been elucidated over the years through their enrichment at targeted promoters, as determined by chromatin immunoprecipitation (ChIP). These include the polycomb group protein EZH2,[64–66] which trimethylates histone 3 lysine 27 (H3K27), HDAC1[64,69], and possibly the H3K9 methyltransferase G9a.[64] Such experiments have also shown that the roles of proteins differ in the establishment and maintenance of TGS.[64]

Although Ting and coworkers observed the induction of silent-state histone modifications at the targeted E-cadherin promoter, DNA methylation was not found at the targeted region.[54] Thus, it was surprising that the first epigenetic modifier described associating with a TGS-inducing RNA was flag-tagged DNMT3A.[72] The controversy associated with the requirement for DNA methylation in TGS has been fuelled by several conflicting reports.[78–81] Work by the Morris group provides an appealing explanation for the disparate data. To allow long-term analysis of RNA-mediated TGS, cells were generated with a tetracycline-inducible short hairpin RNA (shRNA) targeting the ubiquitin C (UBC) promoter.[64] Following a 7-day shRNA induction, the resulting TGS was sustained for over a month, indicating that the effector was required to establish, but not maintain, silencing. In addition, DNA methylation only increased following several days of shRNA expression, and after histone methylation, at a time that correlated with the establishment of long-term silencing. Similar results were obtained in rat hepatic stellate cells: lentivirus transduction was used to generate cells that stably express shRNAs targeting the transforming growth factor beta receptor II (TGFβRII) promoter, and DNA methylation was observed around the target region.[82] Given that most TGS studies employ transient expression of siRNAs, it is not surprising that DNA methylation is not always observed, since it appears that sustained effector expression is necessary. Nevertheless, the importance of DNA methylation in small RNA-induced transcriptional regulation is still unclear: a DNMT3A association with endogenous short RNA guides has not been demonstrated using immunoprecipitation techniques.[83]

Other reasons are likely to exist for the conflicting results, including variation in cell lines, targeted genes, and experimental methodology. In addition, periodic, strand-specific cyclical DNA methylation at active promoters has been described, which might introduce artifactual observations.[84] Therefore, care should be taken to confirm sustained DNA methylation as evidence for TGS.

The relative importance of DNA methylation in TGS is likely to vary between genes. Interestingly, Murayama *et al.* showed that shRNA-induced methylation of a single cytosine residue in the interleukin-2 (IL2) promoter is sufficient to abrogate the binding of the transcription factor (TF) OCT1 and cause histone modifications associated with heterochromatin, leading to decreased gene transcription.[85] Conversely, it is possible to reactivate aberrantly silenced tumor suppressor genes (TSGs) without loss of DNA hypermethylation.[86] Although the effect was only achieved when the HDAC SIRT1 was inhibited, it still brings into question the role of DNA methylation as a "marker" of sustained TGS. Furthermore, others report that some small RNAs induce TGS without accompanying epigenetic changes,[79,80,87,88] and this is discussed in greater detail below.

## B. Promoter-Associated Transcripts

The observation that siRNA-induced silencing at the EF1α promoter was sensitive to the Pol II inhibitor α-amanitin[72] led to the proposal of two models to explain the mechanism of TGS: (a) the siRNA binds to DNA, facilitated by the opening of the DNA duplex by the transcription machinery; and (b) the siRNA binds to nascent promoter-associated RNA. Work by Han *et al.* favored the latter by revealing a requirement for a promoter-associated transcript.[63] The antisense strand of the siRNA was shown to associate with a low-copy EF1α-coding transcript with an extended 5′UTR. In addition, the association was inhibited by RNase A but resistant to RNase H treatment, implicating an RNA–RNA interaction in TGS.

Other studies have also demonstrated a requirement for promoter-associated transcripts for small RNA-mediated gene silencing. Some of these studies have shown that, similar to EF1α, silencing requires sense strand transcripts,[64,75,77] while others have found that antisense transcription through the targeted region is necessary.[62,88] Interestingly, Gonzalez *et al.* showed that siRNA-induced TGS of the regulatory domain (RD) of the INK4b-ARF-INK4a locus required antisense transcription overlapping the RD element.[62] Conversely, miR-17-5p- and miR-20a-mediated TGS in genes with promoters containing cognate seed matches was correlated with levels of sense promoter transcripts.[62] Several reports have shown the presence of both sense and antisense promoter transcripts[62,80] and it has been suggested that bidirectional transcription is a feature of all active promoters.[89,90] Unfortunately, detection of promoter-associated transcripts is not always accompanied by confirmation of their necessity in TGS.[65,74,80] Therefore, the relative contributions of sense and antisense promoter transcripts to RNA-induced TGS require further clarity, particularly as their relative levels may affect outcome.[68,88]

Despite these complications, the findings support a model for TGS in which small RNA and AGO-containing RITS complexes are guided to their targets by complementary base-pairing with low-copy, promoter-associated Pol II transcripts. This serves as a scaffold for the recruitment of epigenetic-modifying protein complexes that favor heterochromatin formation and, therefore, decreased transcription of the targeted gene (Fig. 2A). Experiments in *Schizosaccharomyces pombe* support the ability of nascent RNA transcripts to direct heterochromatin formation: remodeling occurred where RNAi-associated protein complexes were tethered artificially to nascent transcripts.[91] The model has gained general acceptance, facilitated by characterization of endogenous miRNA TGS effectors.[62,65,75] However, a different mechanism for TGS has also become accepted: one that does not involve any epigenetic modifications at all, first described by David Corey's group.[79]

## C. Transcriptional Interference

The Corey group has focused on targeting the region surrounding the transcription start site (TSS), as opposed to upstream regions which were targeted in preceding descriptions of mammalian TGS.[53,54] They reasoned that an open complex is generated at the TSS by the binding of Pol II, rendering it more susceptible to hybridization with oligonucleotides.[92] The small duplex RNA effectors were referred to as antigene RNAs (agRNAs), to

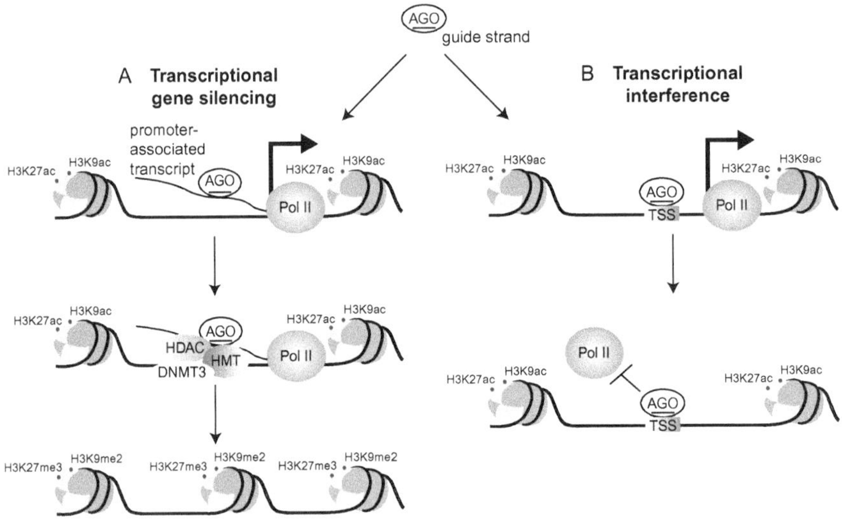


FIG. 2. *Proposed mechanisms of transcriptional gene silencing.* (A) Transcriptional gene silencing. In classical TGS, the small RNA is targeted to a promoter-associated transcript. This recruits epigenetic modifying enzymes, such as histone deacetylases (HDACs), histone methyltransferases (HMTs), and DNA methyltransferase 3 (DNMT3), leading to enrichment of silent-state histone modifications, including histone 3 lysine 9 dimethylation (H3K9me2) and lysine 27 trimethylation (H3K27me3), and heterochromatin formation at the target site. (B) Transcriptional interference. The AGO-bound small RNA target sequences spanning the TSS such that transcription is prevented. The exact mechanism is unclear, but may involve a promoter-associated transcript and/or DNA:RNA triplex formation. The silencing may not be accompanied by epigenetic changes at the targeted promoter such that histone modifications associated with active transcription, including histone 3 lysine 9 and 27 acetylation (H3K9ac and H3K27ac), are maintained.

differentiate them from those targeting mRNAs.[87] Corey's group designed agRNAs targeting the progesterone receptor (PGR) TSS.[79] Surprisingly, the observed silencing in T47D breast cancer cells was potent and not sensitive to either 5-azaC[79] or TSA.[87] This originally led to skepticism that a TGS mechanism was involved, particularly given the lack of nuclear run-on data in the original report.[79] However, the follow-up study did include experiments confirming inhibition at the transcriptional level, albeit only for the PGR gene and not for the huntington (HD) and androgen receptor (AR) genes to which they extended the silencing strategy.[87] This suggests that the inhibition obtained on targeting the PGR TSS is mediated by an alternative mechanism that does not require epigenetic changes, termed transcriptional interference (TI; Fig. 2B).

It appears, therefore, that small duplex RNAs are able to silence transcription of genes by partially distinct mechanisms. Yet, it remains unclear how different the mechanisms of TI and "classical TGS" are from each other.

TI may involve a very different mechanism, such as RNA–DNA–DNA triplex formation blocking the pre-initiation complex (PIC) assembly.[93] Martianov and coworkers have demonstrated that an endogenous ncRNA, transcribed from the DHFR minor promoter, is able to form stable triplex structures with the DHFR major promoter *in vitro*.[94] This may contribute to the ncRNA-induced TI, characterized by PIC dissociation *in vivo*. However, biotin-RNA isolations by Napoli *et al.* revealed an interaction between the antisense strand of the siRNA and a sense noncoding transcript, reminiscent of classical TGS.[80] Additionally, the most recent report from the Corey group does demonstrate increased H3K27 trimethylation and decreased Pol II occupancy at the PGR promoter with a TSS-targeting agRNA,[77] which is in contrast to their previous work. Further studies are required to elucidate whether a universal mechanism exists for TSS-targeting TGS effectors. Moreover, little is known about why the AGO–guide strand–promoter complex does not consistently result in epigenetic changes at the TSS, but does when targeted upstream (Box 1).

Variations in epigenetic outcome may also depend on which AGO is involved. Whilst most reports of classical TGS demonstrate a requirement for AGO1, Corey's group has shown that TI involves both AGO1 and AGO2.[87] Napoli *et al.* showed that TI at the MYC gene did not involve AGO1, only AGO2, although the latter's "slicer" activity was not required.[80] However, AGO2 is involved in heterochromatin maintenance in a Dicer-dependent manner at a heterochromatin domain proximal to the chicken beta-globin locus.[95] Also, when TGS is induced by an shRNA targeting the vascular endothelial growth factor (VEGF-A) gene, AGO2 was recruited to the targeted region, which was accompanied by silent-state histone modifications and increased nucleosome density at both the target site and TSS.[76] Therefore, AGO2 is not exclusively associated with TGS by nonepigenetic mechanisms.

Gonzalez *et al.* proposed that the requirement for different AGO proteins may depend not on the target region but on the degree of complementarity within the effector duplex.[62] A variety of siRNAs were designed with silencing activity: perfectly matched and those with passenger strand bulges due to mismatches at positions 9 and 10. Interestingly, AGO2-null cells lost the ability to mediate silencing by perfectly matched, but not bulged, siRNAs. In addition, ChIP revealed enrichment for AGO1 at the target site when bulged siRNAs were introduced but preferential binding of AGO2 on addition of perfectly matched siRNAs. These findings implicate sequence and thermodynamics of the small RNA duplex as determinates of the associated AGO protein, which has precedence in *Drosophila* studies.[96] AGO3 has even been implicated in miR-10a-mediated TGS of the homeobox D4 (HOXD4) gene.[75] More studies are needed to determine the rules that govern the use of different AGOs and the importance to downstream TGS events.

BOX 1

HOW TO INDUCE LONG-TERM TGS AT A MAMMALIAN GENE

1. Determine the transcription start site (TSS)

   To design a TGS effector, one first needs to know the major TSS. Ideally, this is obtained experimentally for the specific cell line and conditions to be used in TGS experiments by 5′ RACE. This may not be necessary for cell lines where TSSs are reported in databases such as the Database of Transcription Start Sites (http://dbtss.hgc.jp). If the major TSS is unknown, one can map the transcription start region (TSR).[147] Both human and mouse TSRs may be viewed and downloaded from the SwissRegulon Portal (http://www.swissregulon.unibas.ch/cgi-bin/regulon).
2. Determine the presence of promoter-associated transcripts

   All long-term TGS has shown the requirement for promoter-associated transcripts. Perform PCR across a region upstream of the TSS or spanning the TSR to verify promoter-associated transcripts are detectable.
3. Map transcription factor (TF) binding sites

   Heterochromatin formation has accompanied TGS mediated by effectors targeting TF binding sites.[69,74,85,120,121] However, to be sure that silencing is a result of epigenetic changes that provide durable silencing, and not just steric hindrance of TF binding to cognate response elements, TF binding sites should be mapped upstream of the TSS/spanning the TSR.
4. Design target sites using an siRNA design algorithm

   Many algorithms are now available online for the selection of siRNA target sites. These can be utilized on sequences upstream of the TSS/spanning the TSR, providing the target sites obtained are screened for TF binding sites.
5. Screen siRNAs

   The quickest way to find TGS effectors is to screen transfected siRNAs for target mRNA suppression by qRT-PCR. Typically, RNA extractions are performed 72 h posttransfection.
6. Confirm TGS

   To confirm that the silencing effects of any siRNA "hits" are mediated by TGS, and not PTGS, a nuclear run-on experiment should be performed.
7. Confirm epigenetic changes

   Heterochromatin formation at the targeted promoter is associated with sustained TGS. Chromatin immunoprecipitation experiments may be used to confirm the enrichment of histone modifications associated with silent-state chromatin at the targeted promoter on treatment with the siRNA. Additionally, bisulfite sequencing may assess DNA methylation.
8. Generate an expressed small RNA effector

   To obtain long-term TGS, the small RNA effectors may need to be present for longer than possible by transfection. Therefore, expression cassettes for RNAi pathway mimics should be generated such that, following processing, the same guide strand present in the siRNA is produced. To examine the kinetics of TGS, the expression cassettes should be driven by an inducible promoter. Expression cassettes may then be transduced into cells to give stable effector expression.

## D. Transcriptional Gene Activation

The apparent ability of promoter-targeting small duplex RNAs to associate with different proteins may underlie the observation that they can also direct TGA. The initial report of TGA showed increased expression of E-cadherin, p21, and VEGF on the transfection of promoter-targeting siRNAs into cultured cells,[97] which could be prevented by suppressing AGO2 expression. However, these findings require further validation, since gene activation may also be the result of an off-target PTGS effect on a transcriptional repressor, a phenomenon first described by Weinberg *et al.*[71] In this study, microarray analysis demonstrated that transcriptional activation with an antisense RNA targeting the HIV-1 long terminal repeat (LTR) promoter was accompanied by decreased expression of the C10orf76 transcript, which contained 3′UTR target sites recognized by the targeting guide strand. Further experiments revealed that the transcript functioned indirectly in regulating the HIV-1 LTR.

A different PTGS off-target effect was shown to cause previously observed TGA of the p21 gene.[97] In this case, a suppressive antisense transcript overlapping the promoter was the unintended target of the TGA-inducing siRNAs.[68] Despite these controversies, reports of TGA continued, including identification of endogenous effectors; transcriptional activation of E-cadherin and cold-shock domain-containing protein 2 (CSDC2) was observed following overexpression of miR-373.[98] However, there was no reciprocal E-cadherin and CSDC2 silencing on endogenous miR-373 knockdown. In addition, computational analysis revealed miR-373 targets within 1 kb of the TSS in more than 372 genes, but not all responded to miR-373 overexpression.[98] Although the miR-373-mediated TGA may function through a direct interaction of the miRNA with the promoter (Fig. 3A), further experimentation is required to exclude off-target effects on a regulatory ncRNA (Fig. 3B). Recently, miR-205 was shown to induce TGA of interleukin-24 (IL24) and IL32.[99] In this case, a luciferase reporter assay suggested that the miRNA-induced TGA was mediated by direct promoter targeting.

TGA has been induced using small RNAs targeted to distant enhancer regions of the IL10 gene.[100] Again, inhibition of a transcriptional suppressor was involved but, rather than mediating PTGS of a regulatory ncRNA, the siRNAs sterically prevented transcriptional repressor docking with its cognate binding site. In this case, the guide strand was able to induce transient TGA by blocking the association of a repressor protein at an enhancer region (Fig. 3C).

The most thorough examination of TGA mechanisms has emerged from the Corey group. agRNAs targeting the PGR promoter induced TGA in MCF7 cells, which express PGR at low levels.[101] Interestingly, these were similar effectors to those that induced TGS in T47D cells that express high levels of PGR.[79,87] Both silencing and activating agRNAs resulted in an enrichment of

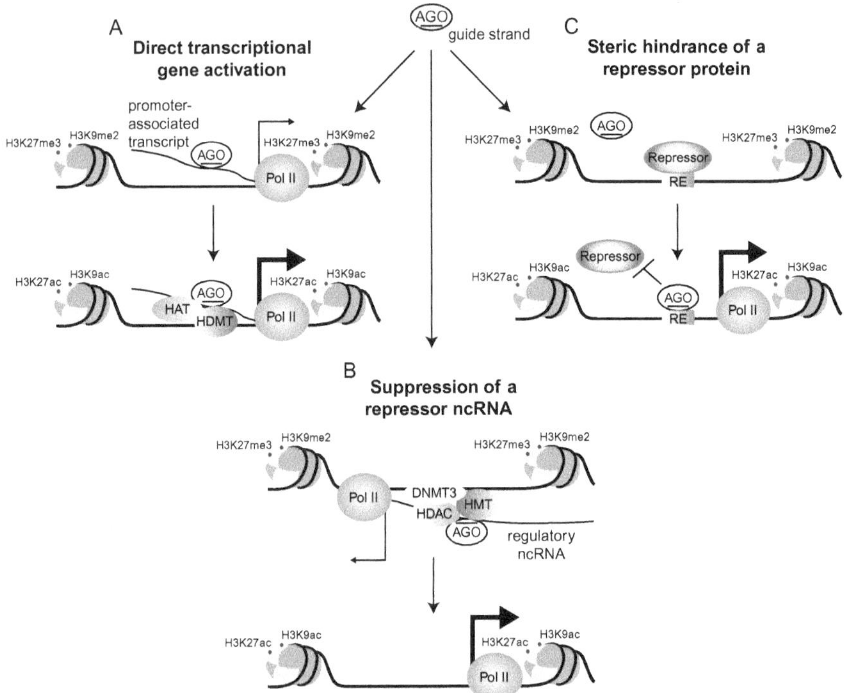


FIG. 3. *Proposed mechanisms of transcriptional gene activation.* (A) Direct transcriptional gene activation. The guide strand–AGO complex increases transcription at the targeted promoter on binding promoter-associated transcripts, perhaps through recruitment of epigenetic modifying enzymes, such as histone acetylases (HATs) and histone demethylases (HDMTs). (B) Suppression of a repressor ncRNA. The level of an endogenous, regulatory ncRNA at the targeted site is decreased by the guide strand–AGO complex. It is unclear whether this occurs by sequestration of the ncRNA in the nucleus or by PTGS of the ncRNA in the cytoplasm. (C) Steric hindrance of a repressor protein. Binding of a repressor protein to its response element (RE) in the targeted promoter is blocked by the guide strand–AGO complex.

AGO proteins at the target site.[88] Recently, the same group confirmed that AGO2, and not AGO1, 3, or 4, is necessary for small RNA-induced gene silencing and activation of PGR expression.[102] AGO2 was also shown to be involved in both shRNA-induced TGS and TGA at the VEGF-A gene.[76] Additional features in common between TGS and TGA include binding of agRNAs to promoter-associated transcripts and relocalization of hnRNPκ from DNA to the promoter-associated transcripts.[88] However, disruption of the transcript–agRNA association abrogated only TGA-mediated effects. In addition, no change in heterochromatin protein 1 gamma (HP1γ) at the targeted region was observed in TGS, but HP1γ levels were decreased in TGA, suggesting divergent mechanisms are involved. Nevertheless, TGA in particular requires further study to establish a consensus on the mechanism(s) involved.

Given the number of long ncRNA-mediated gene regulatory pathways already identified, it is perhaps not surprising that different small RNA-mediated transcription modulation pathways exist, particularly as the short RNA transcriptome is probably at least as complex, if not more so, than that of long RNAs.[103] However, it remains unclear how similar small RNAs can induce TGS in one context and TGA in another, particularly as Turunen *et al.* were able to induce TGS and TGA with VEGF-A promoter-targeting shRNAs within the same cell line.[76] Admittedly, a number of proteins are able to both activate and repress transcription, depending on context.[104–106] Thus, small RNAs are not the first transcriptional regulators to display opposing effects.

## IV. Therapeutic Application of TGS

The therapeutic application of small RNA-induced gene silencing has followed two distinct strategies: introduction of synthetic siRNAs/agRNAs, and the use of gene expression cassettes whose transcription produces RNAi precursors or antisense RNAs. Synthetic agRNAs were recently shown to retain their activity, either silencing or activating transcription, despite introduction of chemical modifications, which may facilitate *in vivo* applications.[107] siRNA precursors often comprise pre-miRNA or pri-miRNA mimics, the former also known as shRNAs. Following expression, these precursors are processed by the RNAi machinery to generate active guide strands in the cell. Each approach has its own benefits and shortcomings to therapeutic application, more specifically with respect to delivery, treatment duration, and possible toxic side effects (reviewed in Ref. 108). Whether synthetic siRNAs/agRNAs or expressed precursors are employed to regulate transcription or not, there are a number of advantages to this type of gene silencing over PTGS, including the potential for increased duration of gene suppression. Combined with the fact that a number of diseases are characterized by dysregulated heterochromatin, TGS is emerging as a valuable therapeutic approach.

### A. Duration and Potency of TGS

TGS accompanied by epigenetic changes may be sustained after the RNA effector has been removed. Hawkins *et al.* have shown that only 3 days of promoter-targeting shRNA induction in dividing cells and 2 days in serum-starved, nondividing cells are required to achieve long-term silencing of UBC.[64] The degree of permanency of TGS requires further examination, however. Although controversial, a number of studies suggest that transcription is pervasive,[109–118] which may lead to reactivation of the targeted gene or prevent complete silencing. In addition, given that the target gene is actively transcribed prior to TGS, other pathways in the cell may favor reactivation of the gene.

Duration of TGS may be influenced by long-range chromosomal interactions. Coregulated genes are preferentially associated within transcription factories.[119] As little is known about the regulation of transcription factory association, it is unclear whether long-term TGS requires a loss of target association with these specialized transcriptional hotspots. Hopefully, future studies will examine how 3D chromosomal interactions are affected by RNA-induced TGS.

Although the durability of TGS will vary between genes, the effector RNA will not require continual administration/expression if epigenetic changes are induced. Consequently, there is less likelihood of toxicity associated with saturation of the RNAi pathway, which is a concern with expressed PTGS strategies.[108] In combination with the fact that targets are in limited numbers, as opposed to many copies of mRNA, TGS may provide a more efficient silencing strategy than PTGS. In general, however, the silencing achieved by TGS has been less potent than that obtained by PTGS. Of course, incomplete silencing may at times be preferable, as complete knockdown of gene expression is not always desirable. For the most part, however, the low potency observed with classical TGS is seen as something to be improved upon, and may reflect the need for TGS effectors to access the nucleus.

Morris and coworkers used the MPG peptide to facilitate access of promoter-targeting siRNAs to the nucleus.[53] Subsequent studies have shown MPG is not necessary to obtain nuclear delivery of exogenous siRNAs when high concentrations of effector are used: up to 10 times that typically used to obtain PTGS.[63,64,79] Thus, despite the low number of TGS targets per cell, the siRNA-induced silencing may not be more efficient than PTGS because of the high concentration of effector needed to obtain sufficient nuclear delivery. In contrast to classical TGS, TI can be achieved with levels of siRNA comparable with those used to obtain PTGS.[79] The reason for this is unclear, particularly as both depend on nuclear delivery of siRNAs. Given this enhanced potency, TI may seem a more appealing silencing strategy than classical TGS. However, the lack of epigenetic changes accompanying TI bring into question the duration of the RNA-induced silencing. Additional studies are necessary to examine the long-term effects of TI.

## B. Expressed TGS Effectors

The problem of nuclear delivery may be overcome by using expression cassettes, such that the TGS effectors are produced in the nucleus. U6 promoter-driven expression of the antisense strand of a siRNA alone was sufficient to mediate TGS at the HIV-1 LTR.[72] Although the antisense effector was able to increase AGO1 at the target site,[120] dsRNAs that resemble RNAi pathway precursors may provide increased silencing because of more efficient incorporation into RITS than small antisense RNA alone.

Several groups have induced TGS through expression of RNAi pathway precursor mimics. shRNAs have successfully been used to transcriptionally silence human RASSF1A,[66,78] IL2,[85] UBC,[64] VEGF-A,[76] and the HIV-1 LTR.[121] It is currently unknown exactly how expressed shRNAs operate: if the precursors are processed by Dicer and the resulting guide loaded into RITS in the cytoplasm, the guide strand–AGO complex must undergo nuclear import. Although such import has been described in *Tetrahymena*[122] and *Caenorhabditis elegans*,[123] in humans the pathway has yet to be identified, although it may involve importin-8.[124] Alternatively, processing and loading may occur within the nucleus itself. AGO proteins are present in human nuclei,[124,125] and nuclear Dicer activity was proposed.[126] Dicer has been shown to shuttle between the nucleus and cytoplasm in fission yeast, and absence from the former prevents heterochromatin assembly.[127] Moreover, shRNA processing may not be required for TGS given that AGO2 has been shown to form an active RISC with pre-miRNAs in the nucleus.[125] Clarification on the cellular compartmentalization of RNAi pathway components will enhance our ability to design TGS effectors and understand their relative potencies. More information on the mechanisms that allow some miRNAs to modulate chromatin but restrict others to mRNA regulation in the cytoplasm will, hopefully, facilitate this.

## C. Disease Treatment

A number of diseases are characterized by dysregulated heterochromatin formation, which may be alleviated by small RNA-induced modulation of transcription. For example, reduced heterochromatin formation characterizes fascioscapulohumeral muscular dystrophy (FSHD), a dominant autosomal disease caused by a complex epigenetic mechanism and not by any mutation in a specific disease gene. Patients have reduced D4Z4 repeats, resulting in a number of nearby genes being inappropriately overexpressed.[128] The mechanism is unclear, but may involve an inability to establish heterochromatin at the repeat region and resultant failure of heterochromatin spreading to nearby genes.[129] RNA-induced TGS might be able to restore normal levels of gene expression, despite the lack of heterochromatin at the nearby repeat region.

The potential to silence genes transcriptionally has obvious application to oncogenes in cancer. TI was induced at the MYC gene, a key regulator of cell proliferation and death, resulting in decreased growth of prostate cancer cells.[80] Also, employing prostate cancer cells, Pulukrui *et al.* described epigenetic changes at the targeted urokinase plasminogen activator (PLAU) promoter.[81] The TGS was accompanied by decreased tumor cell invasion and angiogenesis *in vitro*, and inhibited tumor cell growth and lung metastasis in a mouse model,[81] a dramatic demonstration of the therapeutic potential of TGS. In another *in vivo* study, stable expression of shRNAs following local lentiviral gene transfer in an ischemic mouse hind-limb model was able to regulate transcription of the VEGF-A gene, accompanied by the same histone modifications observed in endothelial cells *in vitro.*[76] These studies demonstrate that small RNA-induced transcriptional regulation can occur *in vivo* with therapeutic benefit. The work also highlights the gene specificity imparted by effectors' base-pairing with target sequences, which is not possible with drugs that inhibit epigenetic modifying enzymes, such as HMTs, DNMTs, and HDACs.

Additional specificity arises with the possibility that TGS effectors have an inherent selectivity for cancer cells. This follows observations that the concentration of promoter-associated transcripts correlates with gene expression[89,117,130] and that TGS efficiency may correlate with the transcriptional rate of the target gene.[79,80] Oncogene overexpression that accompanies cancer may generate more promoter-associated transcripts that render the oncogenes more amenable to TGS in cancerous cells. The caveat is that levels of transcription may also result in unintended effects by TGS effectors. As previously mentioned, an agRNA targeting the PGR promoter induced TI in highly expressing T47D cells, but a similar agRNA caused TGA in low expressing MCF7 cells.[88] Thus, systemic administration of TGS effectors could have alternate effects, potentially causing disastrous activation of oncogenes in normal tissue.

TGA could also be used to combat cancer, providing the disparate mechanisms underlying TGS and TGA are understood. The obvious targets in this case are TSGs. For example, Majid *et al.* restored miR-205 expression in prostate cancer cells, resulting in TGA at the TSGs IL24 and IL32, which was accompanied by decreased migration and increased apoptosis.[99] Thus, specific transcriptional activation of TSGs by miRNAs, or their mimics, may present a novel therapeutic approach for the treatment of cancer.

Perhaps the most effective anticancer strategy will involve indirect TGA. Sense–antisense transcripts are widespread,[131] particularly at TSGs.[132] Moreover, perturbation of the antisense can alter expression of the sense gene.[133] For example, the antisense p15 transcript is upregulated in leukemia, resulting in heterochromatin formation and transcriptional silencing of the sense promoter.[132] It has already been demonstrated that PTGS of a regulatory antisense ncRNA spanning the p21 promoter induces TGA.[68] Thus, a similar strategy for indirect TGA induction could be employed at the p15 locus in the hope of restoring normal p15 expression levels in patients.

TGA would also be an attractive strategy for the treatment of disorders involving dysregulation of imprinting control regions. Prader–Wili syndrome and Angelman syndrome both involve aberrant imprinting of genes in the 15q11–13 region.[134] An ideal treatment would be to activate silenced alleles on the chromosome that has not undergone deletion or impairment. More detail on the mechanisms underlying establishment of epigenetic modifications at these regions is still required, however, to enable specific TGA treatment design.

Transcriptional regulation of viruses has therapeutic potential. TGS has been induced with siRNAs at both the human papilloma virus-16 (HPV-16) promoter[73] and enhancer,[74] leading to decreased production of the oncogenic E6 and E7 viral proteins[73,74] and increased cell death and senescence.[73] In addition, TGS has successfully been induced at both the SIV[67] and HIV-1[69,72,120,121] LTRs. However, similar to RNA-induced TGS at the viral LTR, latent provirus exhibits suppressed expression through heterochromatin formation and associations with heterochromatic chromosome regions in *trans*.[135] Thus, despite stable expression of LTR-targeted shRNA sustaining suppressed HIV-1 gene expression for more than 1 year in a CD4+ T cell line,[121] concerns persist over whether TGS of the HIV-1 promoter would in fact be detrimental to patients by increasing the latent reservoir of the virus. TGA may, therefore, offer greater therapeutic benefit than TGS at the HIV-1 LTR through reversing latency, rendering the replicating virus susceptible to antiviral drugs. Modulation of chromatin to eradicate latent viral reservoirs is not a new idea: an HDAC inhibitor, in combination with intensified antiviral drugs, has been used to deplete latent virus and accelerate clearance of HIV in patients.[136] Small RNA-induced TGA may prove more practical than an HDAC inhibitor since effects would be specific to the viral promoter. However, TGA of the viral promoter has not yet been reported, except when inadvertently induced by off-target PTGS.[71] Given these difficulties, transcriptional modulation may be better employed at promoters of host factors required for HIV-1 replication, such as the coreceptor CCR5,[63,66] rather than at the viral promoter itself.

## D. Targeting Nonpromoter Regions

Therapeutic application of TGS is not limited to targeting promoters. Recently, small RNA-targeting of sequences beyond the mRNA 3′ terminus of the PGR gene induced TGS.[77] Targeting the 3′ end may lead to transcriptional silencing through the association of 5′ and 3′ termini in gene loops. Indeed, the 3′ terminus and promoter of PGR were in close proximity.[77] Gene loops may be a general feature of transcription, already described in *Saccharomyces cerevisiae*[137] as well as in human cells at the BRCA1 gene[138] and the HIV-1 proviral DNA[139] in a manner dependent on transcription.[138,139] Interestingly, noncoding transcripts have been identified at the 3′ ends of genes,[112] including transcription initiation RNAs (tiRNAs—see below).[130] Targeting the 3′ regions associated with noncoding transcripts may lead to TGS if induced heterochromatin spreads to the associated promoter. It seems, therefore, that RNA-mediated regulation of transcription can occur over large genomic distances.[130]

As well as transcriptional regulation of an entire gene, TGS can modulate alternative splicing. siRNAs targeting sequences near an alternative exon in the fibronectin gene increased its inclusion in the mature mRNA.[140] The effect required HP1α and involved decreased Pol II processivity.[140] Transcriptional elongation rate had previously been linked to the regulation of alternative splicing,[141] but Alló *et al.* were the first to manipulate processivity and splicing by introduction of siRNAs. The siRNAs induced enrichment for AGO1 and H3K9 and H3K27 methylation at the target site, consistent with a TGS mechanism.[140] Alternative splicing affects the expression of ~70% of human genes and aberrant splicing is implicated in many diseases. Consequently, TGS could be used to treat a number of diseases where exon inclusion carries therapeutic benefit. This greatly expands the potential therapeutic application of small RNA-induced transcriptional regulation.

## V. The Scope of TGS Therapeutics

A number of genes have been subjected to small RNA-induced transcriptional regulation (Table I). Nevertheless, several barriers are present to the application of TGS genome-wide. These include a dearth of knowledge on (a) where in the promoter to target; (b) which genes to target; (c) when during differentiation/activation/the cell cycle to target; and (d) how to achieve tissue-specificity.

### A. Tissue Specificity

TGS effectors may have varying effects in different tissues, which may depend on the levels of target expression, as mentioned above. Further complications arise from variation in regulatory antisense RNA expression.[68,132] Methods for cell-type-specific delivery/local administration of effectors will go some way to resolving these problems. No doubt, the field of TGS will benefit from progress made in developing PTGS effector delivery strategies.[108] However, the other barriers are more specific to TGS.

### B. Promoter Target Design

At present, only one study has examined where in the promoter to target in any depth. A series of overlapping agRNAs were designed to target the open complex region, revealing that the resulting TI was extremely sensitive to single

TABLE I

Mammalian Genes Transcriptionally Regulated by AGO-Bound Small RNA Effectors

| Proposed mechanism | Gene(s) | Gene symbol | Effector | Cell line | Epigenetic changes | References |
|---|---|---|---|---|---|---|
| TGS | Eukaryotic translation elongation factor 1 alpha | EEF1A1 | siRNA | HEK293T | Yes | 53,63,72 |
| | E-cadherin | CDH1 | siRNA | HCT116; MCF7 | Yes | 54 |
| | INK4B/cyclin-dependent kinase inhibitor 2B/p15 + ARF + INK4A/cyclin-dependent kinase inhibitor 2A isoform 3/p16 | CDKN2B + CDKN2A | siRNA | HEK293T | Yes | 62 |
| | INK4A/cyclin-dependent kinase inhibitor 2A isoform 3/p16 | CDKN2A | siRNA | HEK293T | Yes | 70 |
| | Plasminogen activator, urokinase | PLAU | siRNA | PC3[a] | Yes | 81 |
| | Chemokine receptor 5 | CCR5 | siRNA | HEK293T | Yes | 63,66 |
| | Breast cancer-associated gene 1 | BRCA1 | siRNA | T47D | Not reported | 77 |
| | Interleukin 2 | IL2 | shRNA | Jurkat | Yes | 85 |
| | Ubiquitin C | UBC | siRNA; shRNA[b] | HEK293GT | Yes | 64 |
| | Transforming growth factor-β receptor II | TGFβRII | shRNA[b] | rat SBC10 | Yes | 82 |
| | Vascular endothelial growth factor | VEGF-A | shRNA[b] | mouse C166[a] | Yes | 76 |
| | Ras association domain family 1 | RASSF1A | shRNA[b] | HeLa | Yes | 66,78 |
| | Tubulin folding cofactor E-like | TBCEL/LRRC35 | miRNA | HCT116 | Yes | 62 |
| | Ras p21 protein activator 2 | RASA2 | miRNA | HCT116 | Yes | 62 |
| | Rhophilin, Rho GTPase binding protein 2 | RHPN2 | miRNA | HCT116 | Yes | 62 |
| | Wolf–Hirschhorn syndrome candidate 1 | WHSC1 | miRNA | HCT116 | Yes | 62 |
| | Homeobox D4 | HOXD4 | miRNA | MCF7; MDA-MB-231 | Yes | 75 |

(Continues)

TABLE I (*Continued*)

| Proposed mechanism | Gene(s) | Gene symbol | Effector | Cell line | Epigenetic changes | References |
|---|---|---|---|---|---|---|
| TI | Progesterone receptor | PGR | siRNA | T47D | Conflicting reports | 77,79,87,88,101,102 |
| | Huntington | HD | siRNA | T47D | No | 87 |
| | Androgen receptor | AR | siRNA | T47D | No | 87 |
| | c-Myc | MYC | siRNA | PC3 | No | 80 |
| TGA | E-cadherin | CDH1 | siRNA | PC3; DU145 | Yes | 97 |
| | Interleukin 10 | IL10 | siRNA | THP-1 | Not reported | 100 |
| | Cyclin-dependent kinase inhibitor 1A/p21 | CDKN1A | siRNA | HeLa; MCF7; PC3 | Not reported | 68,97 |
| | Progesterone receptor | PGR | siRNA | MCF7; T47D | Yes | 77,88,101,102 |
| | Vascular endothelial growth factor | VEGF-A | siRNA | HeLa | Not reported | 97 |
| | | VEGF-A | shRNA[b] | Mouse C166[a] | Yes | 76 |
| | E-cadherin | CDH1 | siRNA | PC3; DU145 | Yes | 97 |
| | | CDH1 | miRNA | PC3 | Not reported | 98 |
| | Cold-shock domain-containing protein 2 | CSDC2 | miRNA | PC3; HCT116 | Not reported | 98 |
| | Interleukin 24 | IL24 | | PC3 | Yes | 99 |
| | Interleukin 32 | IL32 | miRNA | PC3 | Yes | 99 |

[a] Also *in vivo* in mouse.
[b] Transduced, not transfected, effector.

base shifts.[79] The sensitivity was attributed to the accessibility of the target to the RNA effector. Less is known about where to target in a promoter to obtain classical TGS. Some studies have used pools of siRNA, and so provide little information as to what constitutes a good target site.[62,81] Most have focused on sequences up to 200 bp of the predicted TSS, the rationale being that this is most likely to correspond to the nucleosome-depleted region (NDR) and, therefore, the best place to focus chromatin remodeling. In contrast, Gonzalez *et al.* demonstrated TGS of four genes by miR-17-5p and miR-20a with the majority of target sites residing $>1$ kb upstream of the RefSeq TSS.[62] The same article also reported that exogenous siRNAs could induce TGS of the INK4b–ARF–INK4a locus by targeting the well-known regulatory domain $>1$ kb upstream. Thus, similar to TGA,[100] TGS may be induced by targeting distant regions. Furthermore, regulatory elements may be important in choosing target sites and, thus, the optimal distance from the TSS could vary substantially between genes.

Unfortunately, no studies have examined the promoter architecture of the targeted gene. Janowski *et al.* observed that agRNAs spanning TF Sp1 binding sites were better inducers of TI.[87] In this case, blocking the TF binding site did not result in chromatin remodeling, which is desirable to attain long-term silencing. However, other studies have designed effectors that span TF binding sites that did result in heterochromatin formation.[69,74,85,120,121] Insufficient data on negative target sites is available to enable a bioinformatic analysis on what makes a good TGS target site. Algorithms designed to predict PTGS target sites are often used, on the assumption that similar thermodynamic principles apply, as AGO proteins are involved in both small RNA-mediated gene silencing pathways. However, such algorithms do not take into account the context of the sequence within the promoter architecture.

## C. Promoter Architecture and Gene Susceptibility

Promoter architecture is important for selection of target genes. Gonzalez *et al.* searched for genes significantly downregulated in two different colon cancer cell lines on miR-17-5p and miR-20a expression and with a canonical seed-match hexamer in the forward strand of the promoter regions.[62] Of the 104 genes fulfilling this criteria, they focused on five genes at random, only four of which were susceptible to TGS upon introduction of pre-miRNAs or miRNA mimics.[62,142] Furthermore, in yeast, siRNA-initiated silencing was observed at only a subset of target loci, which correlated with antisense transcription and overexpression of Swi6, the yeast homologue of HP1.[142] Consequently, the target sequence alone is not sufficient for TGS, and other promoter characteristics are required for a TGS-permissive state.

Studying genes with contrasting promoter architectures may inform which promoter characteristics confer enhanced susceptibility to TGS. Cairns has suggested that some yeast promoters show extreme contrasts in architecture, classed as "open" or "covered."[143] Open promoters are typically TATA-less and associated with constitutively expressed genes. They are characterized by large (~ 150 bp) NDRs containing poly(dA:T) tracts that deter nucleosome formation,[144] maintaining exposure of *cis*-regulatory elements. In contrast, covered promoter *cis*-regulatory elements are fully or partially hidden by nucleosomes when the genes are in their repressed state.[143] Covered promoters are, therefore, more reliant on chromatin remodeling for activity and are associated with highly regulated genes.[143,145,146] Of course, the majority of genes possess promoters that are somewhere between these two extremes.

Mammalian promoters can also be classified into two types and, in this case, the two groups include the majority. Using data from the Functional Annotation of the Mammalian Genome (FANTOM) project,[145,146] promoters were classed on the basis of their C or G (CG) and CpG dinucleotide content.[147] The FANTOM research project is an international collaboration to characterize the mammalian transcriptome. The project led to the development of cap analysis of gene expression (CAGE) technology[148]: through a combination of cap trapping, cleavage by restriction enzymes, and deep sequencing, the 5′ 27-nt sequence of capped mRNAs are obtained.[112,113] These can be aligned to the genome providing a genome-wide map of TSSs.

Unlike *S. cerevisiae,*[149] human genes rarely have a defined TSS, with transcription initiation more commonly occurring over a region.[150] During the latest stage of the FANTOM project, the transcription regulatory network was examined by mapping TSSs in a differentiating macrophage-like cancer cell line, THP-1.[146] Interestingly, TSSs may be only a few base pairs apart with different expression patterns.[147] To accommodate this when constructing a "promoterome," only nearby coexpressed TSSs were grouped into transcription start clusters (TSCs). TSCs with overlapping proximal promoters were further clustered into transcription start regions (TSRs). Significantly, promoters of low CG and CpG content were associated with narrow TSRs ($<$ 10 bp), and high CG and CpG content with wide TSRs (25–250 bp).[147] Most importantly, more than 91% of promoters had a high probability of belonging to either the high- or low-CpG class. Thus, the method is suitable for unambiguous classification of human promoters into two classes, and different mechanisms may be involved in their regulation.

The two promoter classes are enriched in different core promoter elements, which may contribute to their differential regulation and focus the TSS. TATA boxes were strongly overrepresented in narrow TSR promoters, whereas consensus Sp1 sites were overrepresented in broad promoters.[113] This is consistent with the ability of Sp1 to recruit the TATA-binding protein in the absence of TATA boxes.[151] Other core promoter elements, such as initiator, BRE, MTE, DPE, and XCPE1, also specify different mechanisms of transcription and are differentially associated with promoters of different shapes.[150] Clearly, promoters can be classed on the basis of their architectures, which reflects differences in regulation and transcriptional plasticity.

Since, chromatin modifications and nucleosome remodeling are required for classical TGS, one could speculate that the two classes of human promoters are differentially susceptible to TGS. One hypothesis is that genes with broad TSRs are less conducive to TGS because of features within the promoter that deter nucleosome formation. It has been suggested that promoters have an intrinsic tendency to exclude nucleosomes.[152] Also, it is most common to find only a single TSC per TSR (i.e., several TSSs with the same expression profile). In this case, a stochastic process within the transcription machinery-accessible region may determine the TSS usage.[147] Hence, a broad TSR with one TSC implies a large accessible region, or NDR. In addition, broad TSRs and CpG islands are associated with ubiquitous transcripts and narrow TSRs and TATA-box promoters with tightly regulated transcripts.[113] However, whilst the majority of promoters are high-CpG class,[113,147] the majority of genes is not expressed in any given cell line at one time, indicating that broad TSR, high-CpG class promoters are not all constitutively expressed and undergo silencing.

A contrary hypothesis is that broad TSRs are more susceptible to TGS by favoring production of promoter-associated transcripts. CpG islands are associated with bidirectional promoter activity.[153] Further, divergent transcription and gene activity correlate with each other and with promoters containing a CpG island.[89] Promoter-associated transcript levels have also been shown to correlate generally, though not exclusively, with gene expression.[90,130] These include tiRNAs, which are ~ 18-nt long and preferentially associate with GC-rich promoters.[130] Recently, it was suggested that these tiRNAs play a role in epigenetic regulation of gene expression.[154]

Of course, things are not so simple. Broad TSRs do not always indicate single, dispersed promoters with large NDRs, but also overlapping proximal promoters with tissue- and temporal-specific expression (multiple TSCs). Indeed, distinct TSS usage is observed in different tissues within TSRs.[155] Broad TSRs can, therefore, be regions with "fuzzy" nucleosome architecture, with nucleosome position defining TSS usage.[156,157] Distinct TSS usage within TSRs may produce variation in promoter-associated transcripts, including the extent of sense–antisense pair formation. Whether a gene is susceptible to TGS may depend, therefore, on the distinct TSS active within a TSR. Consequently, TGS susceptibility may vary with, not just cell type, but also state of cell activation and differentiation. More generally, a large number of TSCs within a TSR may be associated with increased gene susceptibility to TGS, as it implies increased regulation of the region and a propensity for nucleosome remodeling.

In short, it is unknown whether small RNAs are capable of directing *de novo* heterochromatin formation at every gene promoter. There is a need to understand the context-dependent ability of small RNAs to induce TGS. RNA-mediated heterochromatin formation may require a permissive state at the promoter, which could depend on a number of factors, including levels of gene expression, presence of promoter-associated transcripts, and accessibility of *cis*-regulatory and nucleosome positioning elements. These, in turn, may depend on cell type and differentiation and activation states. Cooperativity is hardly a new concept in gene regulation: several TFs can be involved in regulating gene expression in a manner that is also dependent on chromatin architecture.[158] Requiring cooperativity amongst so many factors may underscore the low potency sometimes observed with classical TGS. If we can characterize the permissive state, it may be possible to enhance TGS susceptibility by inducing it.

## D. Inducing a TGS-Permissive State

Nascent Pol II transcripts are required for small RNA-induced transcriptional regulation (see above). Therefore, enhanced TGS susceptibility may be achieved by inhibiting the exosome, which has been shown to increase the life span of promoter-associated RNAs.[159]

Selective periods of the cell cycle may be permissive to TGS. In *S. pombe*, brief centromeric repeat transcription during the S phase is required to maintain heterochromatin at the region through successive cell divisions by producing the RNAi effectors.[157] Similarly, convergent genes in *S. pombe* employ a cell-cycle-regulated transcription termination mechanism that involves dsRNA production and heterochromatin formation.[160] In human cells, TGS-inducing rRNAs are increased in a cell-cycle-regulated manner.[161] Further, intrachromosomal interactions between enhancer regions and the TSS can be periodic, varying in a ligand- and cell-cycle-dependent fashion.[100] Such periodicity may affect target gene susceptibility if using enhancer-targeting effectors. These studies suggest that RNA-induced heterochromatin formation may be favored during specific periods of the cell cycle, through upregulation of pathway components and/or enhanced association of the targeted region with the proximal promoter.

Decreasing the elongation rate of Pol II may enhance siRNA-mediated heterochromatin formation. In *S. pombe*, 6-azauracil inhibits Pol II elongation and enhances RNA-induced gene silencing (S. I. S. Grewal, personal communication). Similarly, in human cells, decreased Pol II processivity has been associated with heterochromatin marks and recruitment of HP1 isoforms[140] (Muchardt, 2011).[163] Consequently, decreased Pol II processivity may create a TGS-permissive state.

Does decreasing Pol II elongation to enhance TGS susceptibility contradict the suggested requirement for high promoter activity? A model consistent with both involves high promoter activity to generate sufficient nascent promoter-associated transcript, which serves as a scaffold for the recruitment of the guide strand AGO complex. Initial binding is inefficient, but inhibition of the Pol II elongation rate by the small RNA increases the frequency of silent-state histone modifications, perhaps through Pol II-associated enzymes. Such heterochromatic modifications as H3K9 dimethylation stabilize the nascent transcript-guide strand–AGO complex via bridging chromodomain proteins, such as HP1. Recently, siRNAs have been shown to inhibit Pol II during elongation in *C. elegans*.[162] Thus, introducing a drug to inhibit Pol II processivity at the same time as introducing/expressing the TGS effector may facilitate TGS induction.

## VI. Conclusions

A number of biogenesis pathways exist for the generation of small RNAs that associate with AGO proteins. These complexes can mediate gene silencing, including at the transcriptional level. Although a model for the mechanism of classical TGS has gained general acceptance, a detailed description of the protagonists involved, as well as their relative importance, is still required. Furthermore, the ability of small RNAs to modulate transcription by other mechanisms, including TI and TGA, underscores the need to elucidate how these effectors regulate recruitment of different protein complexes, leading to different outcomes. Undoubtedly, better understanding of TGS, TI, and TGA will be facilitated by the identification of endogenous effectors for all pathways.

Expressed TGS effectors offer therapeutic potential, particularly when epigenetic changes are induced, resulting in longer lasting effects than other small RNA-based gene therapies. Although potency of silencing has not always matched that obtained by PTGS, the possible selectivity for cancer cells and the plethora of diseases characterized by heterochromatin dysregulation indicates that TGS will prove a valuable therapeutic tool. Furthermore, not only does RNA-mediated TGS have the capacity to silence genes, but it can also regulate alternative splicing, expanding the potential application of this strategy.

TGS therapeutic development will undoubtedly benefit from the structural similarity of the effectors to those involved in the canonical arm of the PTGS RNAi pathway. Several lessons have been learnt from research in the latter, including how to minimize toxicity, off-target effects, RNAi pathway saturation, and immune response activation as well as how to optimize expression systems and delivery. These findings will be applied to TGS therapeutics, which will, hopefully, speed up their progress to the clinic.

What is clear is that the potential of TGS as a therapy is currently limited by our understanding of the context-dependent effects small RNAs have on chromatin. Although at present this is cause for concern, context-dependent effects may ultimately provide great selectivity. Currently, there is no incentive to report instances where TGS has "failed," particularly as it is difficult to rule out that the right target site within the promoter was not found. This is compounded by the expense entailed in tiling a promoter with siRNAs. Nevertheless, resolving the barriers to TGS therapy would be assisted if researchers shared negative data and rigorously examined the context in which TGS is successfully induced. Given the correlations among TSR length, CpG content, core promoter element distribution, promoter-associated RNA presence, and gene expression level, it would be interesting to study the relative susceptibility of various extremes of promoter architecture to TGS, in the hope of identifying promoter characteristics associated with the TGS-permissive state. This would inform the scope of RNA-induced transcriptional regulation and the genome-wide applicability of this silencing strategy.

## Acknowledgments

The authors thank Kevin Morris and Patrick Arbuthnot for critical reading and comments on the manuscript. V. A. G. is the recipient of scholarship support from the Mellon Foundation and the Poliomyelitis Research Foundation (PRF). M. S. W. acknowledges grants from the National Research Foundation (NRF), Medical Research Council of South Africa (MRC), the Mellon Foundation, and the PRF.

## References

1. Fire A, Xu S, Montgomery MK, Kostas SA, Driver SE, Mello CC. Potent and specific genetic interference by double-stranded RNA in *Caenorhabditis elegans*. *Nature* 1998;**391**:806–11.
2. Lee Y, Ahn C, Han J, Choi H, Kim J, Yim J, et al. The nuclear RNase III Drosha initiates microRNA processing. *Nature* 2003;**425**:415–9.
3. Cai X, Hagedorn CH, Cullen BR. Human microRNAs are processed from capped, polyadenylated transcripts that can also function as mRNAs. *RNA* 2004;**10**:1957–66.
4. Denli AM, Tops BB, Plasterk RH, Ketting RF, Hannon GJ. Processing of primary microRNAs by the Microprocessor complex. *Nature* 2004;**432**:231–5.
5. Han J, Lee Y, Yeom KH, Nam JW, Heo I, Rhee JK, et al. Molecular basis for the recognition of primary microRNAs by the Drosha–DGCR8 complex. *Cell* 2006;**125**:887–901.
6. Zeng Y, Yi R, Cullen BR. Recognition and cleavage of primary microRNA precursors by the nuclear processing enzyme Drosha. *EMBO J* 2005;**24**:138–48.

7. Zhang X, Zeng Y. The terminal loop region controls microRNA processing by Drosha and Dicer. *Nucleic Acids Res* 2010;**38**(21):7689–97.
8. Morlando M, Ballarino M, Gromak N, Pagano F, Bozzoni I, Proudfoot NJ. Primary microRNA transcripts are processed co-transcriptionally. *Nat Struct Mol Biol* 2008;**15**(9):902–9.
9. Kim YK, Kim VN. Processing of intronic microRNAs. *EMBO J* 2007;**26**:775–83.
10. Kim VN, Han J, Siomi MC. Biogenesis of small RNAs in animals. *Nat Rev Mol Cell Biol* 2009;**10**:126–39.
11. Han J, Lee Y, Yeom KH, Kim YK, Jin H, Kim VN. The Drosha–DGCR8 complex in primary microRNA processing. *Genes Dev* 2004;**18**:3016–27.
12. Davis BN, Hilyard AC, Lagna G, Hata A. SMAD proteins control DROSHA-mediated microRNA maturation. *Nature* 2008;**454**:56–61.
13. Fukuda T, Yamagata K, Fujiyama S, Matsumoto T, Koshida I, Yoshimura K, et al. DEAD-box RNA helicase subunits of the Drosha complex are required for processing of rRNA and a subset of microRNAs. *Nat Cell Biol* 2007;**9**:604–11.
14. Guil S, Caceres JF. The multifunctional RNA-binding protein hnRNP A1 is required for processing of miR-18a. *Nat Struct Mol Biol* 2007;**14**:591–6.
15. Gregory RI, Yan KP, Amuthan G, Chendrimada T, Doratotaj B, Cooch N, et al. The Microprocessor complex mediates the genesis of microRNAs. *Nature* 2004;**432**:235–40.
16. Shiohama A, Sasaki T, Noda S, Minoshima S, Shimizu N. Nucleolar localization of DGCR8 and identification of eleven DGCR8-associated proteins. *Exp Cell Res* 2007;**313**:4196–207.
17. Siomi H, Siomi MC. Posttranscriptional regulation of microRNA biogenesis in animals. *Mol Cell* 2010;**38**:323–32.
18. Yi R, Qin Y, Macara IG, Cullen BR. Exportin-5 mediates the nuclear export of pre-microRNAs and short hairpin RNAs. *Genes Dev* 2003;**17**:3011–6.
19. Lund E, Guttinger S, Calado A, Dahlberg JE, Kutay U. Nuclear export of microRNA precursors. *Science* 2004;**303**:95–8.
20. Zeng Y, Cullen BR. Structural requirements for pre-microRNA binding and nuclear export by Exportin 5. *Nucleic Acids Res* 2004;**32**:4776–85.
21. Okada C, Yamashita E, Lee SJ, Shibata S, Katahira J, Nakagawa A, et al. A high-resolution structure of the pre-microRNA nuclear export machinery. *Science* 2009;**326**:1275–9.
22. Bernstein E, Caudy AA, Hammond SM, Hannon GJ. Role for a bidentate ribonuclease in the initiation step of RNA interference. *Nature* 2001;**409**:363–6.
23. Hutvagner G, McLachlan J, Pasquinelli AE, Balint E, Tuschl T, Zamore PD. A cellular function for the RNA-interference enzyme Dicer in the maturation of the let-7 small temporal RNA. *Science* 2001;**293**:834–8.
24. Chendrimada TP, Gregory RI, Kumaraswamy E, Norman J, Cooch N, Nishikura K, et al. TRBP recruits the Dicer complex to Ago2 for microRNA processing and gene silencing. *Nature* 2005;**436**:740–4.
25. Jinek M, Doudna JA. A three-dimensional view of the molecular machinery of RNA interference. *Nature* 2009;**457**:405–12.
26. Ma JB, Yuan YR, Meister G, Pei Y, Tuschl T, Patel DJ. Structural basis for 5′-end-specific recognition of guide RNA by the *A. fulgidus* Piwi protein. *Nature* 2005;**434**:666–70.
27. Parker JS, Roe SM, Barford D. Structural insights into mRNA recognition from a PIWI domain-siRNA guide complex. *Nature* 2005;**434**:663–6.
28. Meister G, Landthaler M, Patkaniowska A, Dorsett Y, Teng G, Tuschl T. Human Argonaute2 mediates RNA cleavage targeted by miRNAs and siRNAs. *Mol Cell* 2004;**15**:185–97.
29. Liu J, Carmell MA, Rivas FV, Marsden CG, Thomson JM, Song JJ, et al. Argonaute2 is the catalytic engine of mammalian RNAi. *Science* 2004;**305**:1437–41.
30. MacRae IJ, Ma E, Zhou M, Robinson CV, Doudna JA. In vitro reconstitution of the human RISC-loading complex. *Proc Natl Acad Sci USA* 2008;**105**:512–7.

31. Diederichs S, Haber DA. Dual role for argonautes in microRNA processing and posttranscriptional regulation of microRNA expression. *Cell* 2007;**131**:1097–108.
32. Matranga C, Tomari Y, Shin C, Bartel DP, Zamore PD. Passenger-strand cleavage facilitates assembly of siRNA into Ago2-containing RNAi enzyme complexes. *Cell* 2005;**123**:607–20.
33. Rand TA, Petersen S, Du F, Wang X. Argonaute2 cleaves the anti-guide strand of siRNA during RISC activation. *Cell* 2005;**123**:621–9.
34. Khvorova A, Reynolds A, Jayasena SD. Functional siRNAs and miRNAs exhibit strand bias. *Cell* 2003;**115**:209–16.
35. Schwarz DS, Hutvagner G, Du T, Xu Z, Aronin N, Zamore PD. Asymmetry in the assembly of the RNAi enzyme complex. *Cell* 2003;**115**:199–208.
36. Lewis BP, Shih IH, Jones-Rhoades MW, Bartel DP, Burge CB. Prediction of mammalian microRNA targets. *Cell* 2003;**115**:787–98.
37. Bartel DP. MicroRNAs: target recognition and regulatory functions. *Cell* 2009;**136**:215–33.
38. Guo H, Ingolia NT, Weissman JS, Bartel DP. Mammalian microRNAs predominantly act to decrease target mRNA levels. *Nature* 2010;**466**:835–40.
39. Okamura K, Hagen JW, Duan H, Tyler DM, Lai EC. The mirtron pathway generates microRNA-class regulatory RNAs in Drosophila. *Cell* 2007;**130**:89–100.
40. Ruby JG, Jan CH, Bartel DP. Intronic microRNA precursors that bypass Drosha processing. *Nature* 2007;**448**:83–6.
41. Ender C, Krek A, Friedlander MR, Beitzinger M, Weinmann L, Chen W, et al. A human snoRNA with microRNA-like functions. *Mol Cell* 2008;**32**:519–28.
42. Bogerd HP, Karnowski HW, Cai X, Shin J, Pohlers M, Cullen BR. A mammalian herpesvirus uses noncanonical expression and processing mechanisms to generate viral MicroRNAs. *Mol Cell* 2010;**37**:135–42.
43. Cole C, Sobala A, Lu C, Thatcher SR, Bowman A, Brown JW, et al. Filtering of deep sequencing data reveals the existence of abundant Dicer-dependent small RNAs derived from tRNAs. *RNA* 2009;**15**:2147–60.
44. Lee YS, Shibata Y, Malhotra A, Dutta A. A novel class of small RNAs: tRNA-derived RNA fragments (tRFs). *Genes Dev* 2009;**23**:2639–49.
45. Haussecker D, Huang Y, Lau A, Parameswaran P, Fire AZ, Kay MA. Human tRNA-derived small RNAs in the global regulation of RNA silencing. *RNA* 2010;**16**:673–95.
46. Tam OH, Aravin AA, Stein P, Girard A, Murchison EP, Cheloufi S, et al. Pseudogene-derived small interfering RNAs regulate gene expression in mouse oocytes. *Nature* 2008;**453**:534–8.
47. Watanabe T, Totoki Y, Toyoda A, Kaneda M, Kuramochi-Miyagawa S, Obata Y, et al. Endogenous siRNAs from naturally formed dsRNAs regulate transcripts in mouse oocytes. *Nature* 2008;**453**:539–43.
48. Babiarz JE, Ruby JG, Wang Y, Bartel DP, Blelloch R. Mouse ES cells express endogenous shRNAs, siRNAs, and other Microprocessor-independent, Dicer-dependent small RNAs. *Genes Dev* 2008;**22**:2773–85.
49. Buhler M, Moazed D. Transcription and RNAi in heterochromatic gene silencing. *Nat Struct Mol Biol* 2007;**14**:1041–8.
50. Grewal SI, Elgin SC. Transcription and RNA interference in the formation of heterochromatin. *Nature* 2007;**447**:399–406.
51. Matzke MA, Birchler JA. RNAi-mediated pathways in the nucleus. *Nat Rev Genet* 2005;**6**:24–35.
52. Verdel A, Jia S, Gerber S, Sugiyama T, Gygi S, Grewal SI, et al. RNAi-mediated targeting of heterochromatin by the RITS complex. *Science* 2004;**303**:672–6.
53. Morris KV, Chan SW, Jacobsen SE, Looney DJ. Small interfering RNA-induced transcriptional gene silencing in human cells. *Science* 2004;**305**:1289–92.

54. Ting AH, Schuebel KE, Herman JG, Baylin SB. Short double-stranded RNA induces transcriptional gene silencing in human cancer cells in the absence of DNA methylation. *Nat Genet* 2005;**37**:906–10.
55. Kornberg RD, Klug A. The nucleosome. *Sci Am* 1981;**244**:52–64.
56. Luger K, Mader AW, Richmond RK, Sargent DF, Richmond TJ. Crystal structure of the nucleosome core particle at 2.8 A resolution. *Nature* 1997;**389**:251–60.
57. Strahl BD, Allis CD. The language of covalent histone modifications. *Nature* 2000;**403**:41–5.
58. Sarma K, Reinberg D. Histone variants meet their match. *Nat Rev Mol Cell Biol* 2005;**6**:139–49.
59. Suzuki MM, Bird A. DNA methylation landscapes: provocative insights from epigenomics. *Nat Rev Genet* 2008;**9**:465–76.
60. Weber M, Hellmann I, Stadler MB, Ramos L, Paabo S, Rebhan M, et al. Distribution, silencing potential and evolutionary impact of promoter DNA methylation in the human genome. *Nat Genet* 2007;**39**:457–66.
61. Okano M, Bell DW, Haber DA, Li E. DNA methyltransferases Dnmt3a and Dnmt3b are essential for de novo methylation and mammalian development. *Cell* 1999;**99**:247–57.
62. Gonzalez S, Pisano DG, Serrano M. Mechanistic principles of chromatin remodeling guided by siRNAs and miRNAs. *Cell Cycle* 2008;**7**:2601–8.
63. Han J, Kim D, Morris KV. Promoter-associated RNA is required for RNA-directed transcriptional gene silencing in human cells. *Proc Natl Acad Sci USA* 2007;**104**:12422–7.
64. Hawkins PG, Santoso S, Adams C, Anest V, Morris KV. Promoter targeted small RNAs induce long-term transcriptional gene silencing in human cells. *Nucleic Acids Res* 2009;**37**:2984–95.
65. Kim DH, Saetrom P, Snove Jr. O, Rossi JJ. MicroRNA-directed transcriptional gene silencing in mammalian cells. *Proc Natl Acad Sci USA* 2008;**105**:16230–5.
66. Kim DH, Villeneuve LM, Morris KV, Rossi JJ. Argonaute-1 directs siRNA-mediated transcriptional gene silencing in human cells. *Nat Struct Mol Biol* 2006;**13**:793–7.
67. Lim HG, Suzuki K, Cooper DA, Kelleher AD. Promoter-targeted siRNAs induce gene silencing of simian immunodeficiency virus (SIV) infection in vitro. *Mol Ther* 2008;**16**:565–70.
68. Morris KV, Santoso S, Turner AM, Pastori C, Hawkins PG. Bidirectional transcription directs both transcriptional gene activation and suppression in human cells. *PLoS Genet* 2008;**4**:e1000258.
69. Suzuki K, Juelich T, Lim H, Ishida T, Watanebe T, Cooper DA, et al. Closed chromatin architecture is induced by an RNA duplex targeting the HIV-1 promoter region. *J Biol Chem* 2008;**283**:23353–63.
70. Wang X, Feng Y, Pan L, Wang Y, Xu X, Lu J, et al. The proximal GC-rich region of p16(INK4a) gene promoter plays a role in its transcriptional regulation. *Mol Cell Biochem* 2007;**301**:259–66.
71. Weinberg MS, Barichievy S, Schaffer L, Han J, Morris KV. An RNA targeted to the HIV-1 LTR promoter modulates indiscriminate off-target gene activation. *Nucleic Acids Res* 2007;**35**:7303–12.
72. Weinberg MS, Villeneuve LM, Ehsani A, Amarzguioui M, Aagaard L, Chen ZX, et al. The antisense strand of small interfering RNAs directs histone methylation and transcriptional gene silencing in human cells. *RNA* 2006;**12**:256–62.
73. Hong D, Lu W, Ye F, Hu Y, Xie X. Gene silencing of HPV16 E6/E7 induced by promoter-targeting siRNA in SiHa cells. *Br J Cancer* 2009;**101**:1798–804.
74. Palanichamy JK, Mehndiratta M, Bhagat M, Ramalingam P, Das B, Das P, et al. Silencing of integrated human papillomavirus-16 oncogenes by small interfering RNA-mediated heterochromatization. *Mol Cancer Ther* 2010;**9**:2114–22.
75. Tan Y, Zhang B, Wu T, Skogerbo G, Zhu X, Guo X, et al. Transcriptional inhibiton of Hoxd4 expression by miRNA-10a in human breast cancer cells. *BMC Mol Biol* 2009;**10**:12.

76. Turunen MP, Lehtola T, Heinonen SE, Assefa GS, Korpisalo P, Girnary R, et al. Efficient regulation of VEGF expression by promoter-targeted lentiviral shRNAs based on epigenetic mechanism: a novel example of epigenetherapy. *Circ Res* 2009;**105**:604–9.
77. Yue X, Schwartz JC, Chu Y, Younger ST, Gagnon KT, Elbashir S, et al. Transcriptional regulation by small RNAs at sequences downstream from 3′ gene termini. *Nat Chem Biol* 2010;**6**:621–9.
78. Castanotto D, Tommasi S, Li M, Li H, Yanow S, Pfeifer GP, et al. Short hairpin RNA-directed cytosine (CpG) methylation of the RASSF1A gene promoter in HeLa cells. *Mol Ther* 2005;**12**:179–83.
79. Janowski BA, Huffman KE, Schwartz JC, Ram R, Hardy D, Shames DS, et al. Inhibiting gene expression at transcription start sites in chromosomal DNA with antigene RNAs. *Nat Chem Biol* 2005;**1**:216–22.
80. Napoli S, Pastori C, Magistri M, Carbone GM, Catapano CV. Promoter-specific transcriptional interference and c-myc gene silencing by siRNAs in human cells. *EMBO J* 2009;**28**:1708–19.
81. Pulukuri SM, Rao JS. Small interfering RNA directed reversal of urokinase plasminogen activator demethylation inhibits prostate tumor growth and metastasis. *Cancer Res* 2007;**67**:6637–46.
82. Kim JW, Zhang YH, Zern MA, Rossi JJ, Wu J. Short hairpin RNA causes the methylation of transforming growth factor-beta receptor II promoter and silencing of the target gene in rat hepatic stellate cells. *Biochem Biophys Res Commun* 2007;**359**:292–7.
83. Park CW, Zeng Y, Steer CJ. Human DNA methyltransferase 3a does not associate with microRNAs in the regulation of DNA methylation. *J Cardiovasc Transl Res* 2010;**3**:290–5.
84. Metivier R, Gallais R, Tiffoche C, Le Peron C, Jurkowska RZ, Carmouche RP, et al. Cyclical DNA methylation of a transcriptionally active promoter. *Nature* 2008;**452**:45–50.
85. Murayama A, Sakura K, Nakama M, Yasuzawa-Tanaka K, Fujita E, Tateishi Y, et al. A specific CpG site demethylation in the human interleukin 2 gene promoter is an epigenetic memory. *EMBO J* 2006;**25**:1081–92.
86. Pruitt K, Zinn RL, Ohm JE, McGarvey KM, Kang SH, Watkins DN, et al. Inhibition of SIRT1 reactivates silenced cancer genes without loss of promoter DNA hypermethylation. *PLoS Genet* 2006;**2**:e40.
87. Janowski BA, Huffman KE, Schwartz JC, Ram R, Nordsell R, Shames DS, et al. Involvement of AGO1 and AGO2 in mammalian transcriptional silencing. *Nat Struct Mol Biol* 2006;**13**:787–92.
88. Schwartz JC, Younger ST, Nguyen NB, Hardy DB, Monia BP, Corey DR, et al. Antisense transcripts are targets for activating small RNAs. *Nat Struct Mol Biol* 2008;**15**:842–8.
89. Core LJ, Waterfall JJ, Lis JT. Nascent RNA sequencing reveals widespread pausing and divergent initiation at human promoters. *Science* 2008;**322**:1845–8.
90. Seila AC, Calabrese JM, Levine SS, Yeo GW, Rahl PB, Flynn RA, et al. Divergent transcription from active promoters. *Science* 2008;**322**:1849–51.
91. Buhler M, Verdel A, Moazed D. Tethering RITS to a nascent transcript initiates RNAi- and heterochromatin-dependent gene silencing. *Cell* 2006;**125**:873–86.
92. Milne L, Xu Y, Perrin DM, Sigman DS. An approach to gene-specific transcription inhibition using oligonucleotides complementary to the template strand of the open complex. *Proc Natl Acad Sci USA* 2000;**97**:3136–41.
93. Jain A, Magistri M, Napoli S, Carbone GM, Catapano CV. Mechanisms of triplex DNA-mediated inhibition of transcription initiation in cells. *Biochimie* 2010;**92**:317–20.
94. Martianov I, Ramadass A, Serra Barros A, Chow N, Akoulitchev A. Repression of the human dihydrofolate reductase gene by a non-coding interfering transcript. *Nature* 2007;**445**:666–70.

95. Giles KE, Ghirlando R, Felsenfeld G. Maintenance of a constitutive heterochromatin domain in vertebrates by a Dicer-dependent mechanism. *Nat Cell Biol* 2010;**12**:94–9 (sup pp. 91–96).
96. Tomari Y, Du T, Zamore PD. Sorting of Drosophila small silencing RNAs. *Cell* 2007;**130**:299–308.
97. Li LC, Okino ST, Zhao H, Pookot D, Place RF, Urakami S, et al. Small dsRNAs induce transcriptional activation in human cells. *Proc Natl Acad Sci USA* 2006;**103**:17337–42.
98. Place RF, Li LC, Pookot D, Noonan EJ, Dahiya R. MicroRNA-373 induces expression of genes with complementary promoter sequences. *Proc Natl Acad Sci USA* 2008;**105**:1608–13.
99. Majid S, Dar AA, Saini S, Yamamura S, Hirata H, Tanaka Y, et al. MicroRNA-205-directed transcriptional activation of tumor suppressor genes in prostate cancer. *Cancer* 2010;**116** (24):5637–49.
100. Matilainen JM, Husso T, Toropainen S, Seuter S, Turunen MP, Gynther P, et al. Primary effect of 1alpha,25(OH)(2)D(3) on IL-10 expression in monocytes is short-term down-regulation. *Biochim Biophys Acta* 2010;**1803**:1276–86.
101. Janowski BA, Younger ST, Hardy DB, Ram R, Huffman KE, Corey DR. Activating gene expression in mammalian cells with promoter-targeted duplex RNAs. *Nat Chem Biol* 2007;**3**:166–73.
102. Chu Y, Yue X, Younger ST, Janowski BA, Corey DR. Involvement of argonaute proteins in gene silencing and activation by RNAs complementary to a non-coding transcript at the progesterone receptor promoter. *Nucleic Acids Res* 2010;**38**(21):7736–48.
103. Kapranov P, Willingham AT, Gingeras TR. Genome-wide transcription and the implications for genomic organization. *Nat Rev Genet* 2007;**8**:413–23.
104. Chahrour M, Jung SY, Shaw C, Zhou X, Wong ST, Qin J, et al. MeCP2, a key contributor to neurological disease, activates and represses transcription. *Science* 2008;**320**:1224–9.
105. Scsucova S, Palacios D, Savignac M, Mellstrom B, Naranjo JR, Aranda A. The repressor DREAM acts as a transcriptional activator on Vitamin D and retinoic acid response elements. *Nucleic Acids Res* 2005;**33**:2269–79.
106. Willy PJ, Kobayashi R, Kadonaga JT. A basal transcription factor that activates or represses transcription. *Science* 2000;**290**:982–5.
107. Watts JK, Yu D, Charisse K, Montaillier C, Potier P, Manoharan M, et al. Effect of chemical modifications on modulation of gene expression by duplex antigene RNAs that are complementary to non-coding transcripts at gene promoters. *Nucleic Acids Res* 2010;**38**:5242–59.
108. Grimm D. Small silencing RNAs: state-of-the-art. *Adv Drug Deliv Rev* 2009;**61**:672–703.
109. van Bakel H, Nislow C, Blencowe BJ, Hughes TR. Most "dark matter" transcripts are associated with known genes. *PLoS Biol* 2010;**8**:e1000371.
110. Bertone P, Stolc V, Royce TE, Rozowsky JS, Urban AE, Zhu X, et al. Global identification of human transcribed sequences with genome tiling arrays. *Science* 2004;**306**:2242–6.
111. Birney E, Stamatoyannopoulos JA, Dutta A, Guigo R, Gingeras TR, Margulies EH, et al. Identification and analysis of functional elements in 1% of the human genome by the ENCODE pilot project. *Nature* 2007;**447**:799–816.
112. Carninci P, Kasukawa T, Katayama S, Gough J, Frith MC, Maeda N, et al. The transcriptional landscape of the mammalian genome. *Science* 2005;**309**:1559–63.
113. Carninci P, Sandelin A, Lenhard B, Katayama S, Shimokawa K, Ponjavic J, et al. Genome-wide analysis of mammalian promoter architecture and evolution. *Nat Genet* 2006;**38**:626–35.
114. Cheng J, Kapranov P, Drenkow J, Dike S, Brubaker S, Patel S, et al. Transcriptional maps of 10 human chromosomes at 5-nucleotide resolution. *Science* 2005;**308**:1149–54.
115. Kampa D, Cheng J, Kapranov P, Yamanaka M, Brubaker S, Cawley S, et al. Novel RNAs identified from an in-depth analysis of the transcriptome of human chromosomes 21 and 22. *Genome Res* 2004;**14**:331–42.

116. Kapranov P, Cawley SE, Drenkow J, Bekiranov S, Strausberg RL, Fodor SP, et al. Large-scale transcriptional activity in chromosomes 21 and 22. *Science* 2002;**296**:916–9.
117. Kapranov P, Cheng J, Dike S, Nix DA, Duttagupta R, Willingham AT, et al. RNA maps reveal new RNA classes and a possible function for pervasive transcription. *Science* 2007;**316**:1484–8.
118. Schadt EE, Edwards SW, GuhaThakurta D, Holder D, Ying L, Svetnik V, et al. A comprehensive transcript index of the human genome generated using microarrays and computational approaches. *Genome Biol* 2004;**5**:R73.
119. Schoenfelder S, Sexton T, Chakalova L, Cope NF, Horton A, Andrews S, et al. Preferential associations between co-regulated genes reveal a transcriptional interactome in erythroid cells. *Nat Genet* 2010;**42**:53–61.
120. Turner AM, De La Cruz J, Morris KV. Mobilization-competent lentiviral vector-mediated sustained transcriptional modulation of HIV-1 expression. *Mol Ther* 2009;**17**:360–8.
121. Yamagishi M, Ishida T, Miyake A, Cooper DA, Kelleher AD, Suzuki K, et al. Retroviral delivery of promoter-targeted shRNA induces long-term silencing of HIV-1 transcription. *Microbes Infect* 2009;**11**:500–8.
122. Noto T, Kurth HM, Kataoka K, Aronica L, DeSouza LV, Siu KW, et al. The Tetrahymena argonaute-binding protein Giw1p directs a mature argonaute-siRNA complex to the nucleus. *Cell* 2010;**140**:692–703.
123. Guang S, Bochner AF, Pavelec DM, Burkhart KB, Harding S, Lachowiec J, et al. An Argonaute transports siRNAs from the cytoplasm to the nucleus. *Science* 2008;**321**:537–41.
124. Weinmann L, Hock J, Ivacevic T, Ohrt T, Mutze J, Schwille P, et al. Importin 8 is a gene silencing factor that targets argonaute proteins to distinct mRNAs. *Cell* 2009;**136**:496–507.
125. Tan GS, Garchow BG, Liu X, Yeung J, Morris JPT, Cuellar TL, et al. Expanded RNA-binding activities of mammalian Argonaute 2. *Nucleic Acids Res* 2009;**37**:7533–45.
126. Lee NS, Kim DH, Alluin J, Robbins M, Gu S, Li H, et al. Functional and intracellular localization properties of U6 promoter-expressed siRNAs, shRNAs, and chimeric VA1 shRNAs in mammalian cells. *RNA* 2008;**14**:1823–33.
127. Emmerth S, Schober H, Gaidatzis D, Roloff T, Jacobeit K, Buhler M. Nuclear retention of fission yeast dicer is a prerequisite for RNAi-mediated heterochromatin assembly. *Dev Cell* 2010;**18**:102–13.
128. Gabellini D, Green MR, Tupler R. Inappropriate gene activation in FSHD: a repressor complex binds a chromosomal repeat deleted in dystrophic muscle. *Cell* 2002;**110**:339–48.
129. Hahn M, Dambacher S, Schotta G. Heterochromatin dysregulation in human diseases. *J Appl Physiol* 2010;**109**:232–42.
130. Taft RJ, Glazov EA, Cloonan N, Simons C, Stephen S, Faulkner GJ, et al. Tiny RNAs associated with transcription start sites in animals. *Nat Genet* 2009;**41**:572–8.
131. Rosok O, Sioud M. Systematic identification of sense–antisense transcripts in mammalian cells. *Nat Biotechnol* 2004;**22**:104–8.
132. Yu W, Gius D, Onyango P, Muldoon-Jacobs K, Karp J, Feinberg AP, et al. Epigenetic silencing of tumour suppressor gene p15 by its antisense RNA. *Nature* 2008;**451**:202–6.
133. Katayama S, Tomaru Y, Kasukawa T, Waki K, Nakanishi M, Nakamura M, et al. Antisense transcription in the mammalian transcriptome. *Science* 2005;**309**:1564–6.
134. Lee S, Wevrick R. Identification of novel imprinted transcripts in the Prader-Willi syndrome and Angelman syndrome deletion region: further evidence for regional imprinting control. *Am J Hum Genet* 2000;**66**:848–58.
135. Dieudonne M, Maiuri P, Biancotto C, Knezevich A, Kula A, Lusic M, et al. Transcriptional competence of the integrated HIV-1 provirus at the nuclear periphery. *EMBO J* 2009;**28**:2231–43.

136. Lehrman G, Hogue IB, Palmer S, Jennings C, Spina CA, Wiegand A, et al. Depletion of latent HIV-1 infection in vivo: a proof-of-concept study. *Lancet* 2005;**366**:549–55.
137. O'Sullivan JM, Tan-Wong SM, Morillon A, Lee B, Coles J, Mellor J, et al. Gene loops juxtapose promoters and terminators in yeast. *Nat Genet* 2004;**36**:1014–8.
138. Tan-Wong SM, French JD, Proudfoot NJ, Brown MA. Dynamic interactions between the promoter and terminator regions of the mammalian BRCA1 gene. *Proc Natl Acad Sci USA* 2008;**105**:5160–5.
139. Perkins KJ, Lusic M, Mitar I, Giacca M, Proudfoot NJ. Transcription-dependent gene looping of the HIV-1 provirus is dictated by recognition of pre-mRNA processing signals. *Mol Cell* 2008;**29**:56–68.
140. Allo M, Buggiano V, Fededa JP, Petrillo E, Schor I, de la Mata M, et al. Control of alternative splicing through siRNA-mediated transcriptional gene silencing. *Nat Struct Mol Biol* 2009;**16**:717–24.
141. de la Mata M, Alonso CR, Kadener S, Fededa JP, Blaustein M, Pelisch F, et al. A slow RNA polymerase II affects alternative splicing in vivo. *Mol Cell* 2003;**12**:525–32.
142. Iida T, Nakayama J, Moazed D. siRNA-mediated heterochromatin establishment requires HP1 and is associated with antisense transcription. *Mol Cell* 2008;**31**:178–89.
143. Cairns BR. The logic of chromatin architecture and remodelling at promoters. *Nature* 2009;**461**:193–8.
144. Struhl K. Naturally occurring poly(dA-dT) sequences are upstream promoter elements for constitutive transcription in yeast. *Proc Natl Acad Sci USA* 1985;**82**:8419–23.
145. Kawaji H, Kasukawa T, Fukuda S, Katayama S, Kai C, Kawai J, et al. CAGE basic/analysis databases: the CAGE resource for comprehensive promoter analysis. *Nucleic Acids Res* 2006;**34**:D632–6.
146. Kawaji H, Severin J, Lizio M, Waterhouse A, Katayama S, Irvine KM, et al. The FANTOM web resource: from mammalian transcriptional landscape to its dynamic regulation. *Genome Biol* 2009;**10**:R40.
147. Balwierz PJ, Carninci P, Daub CO, Kawai J, Hayashizaki Y, Van Belle W, et al. Methods for analyzing deep sequencing expression data: constructing the human and mouse promoterome with deepCAGE data. *Genome Biol* 2009;**10**:R79.
148. Maeda N, Nishiyori H, Nakamura M, Kawazu C, Murata M, Sano H, et al. Development of a DNA barcode tagging method for monitoring dynamic changes in gene expression by using an ultra high-throughput sequencer. *Biotechniques* 2008;**45**:95–7.
149. David L, Huber W, Granovskaia M, Toedling J, Palm CJ, Bofkin L, et al. A high-resolution map of transcription in the yeast genome. *Proc Natl Acad Sci USA* 2006;**103**:5320–5.
150. Juven-Gershon T, Hsu JY, Theisen JW, Kadonaga JT. The RNA polymerase II core promoter—the gateway to transcription. *Curr Opin Cell Biol* 2008;**20**:253–9.
151. Butler JE, Kadonaga JT. The RNA polymerase II core promoter: a key component in the regulation of gene expression. *Genes Dev* 2002;**16**:2583–92.
152. Sandelin A, Carninci P, Lenhard B, Ponjavic J, Hayashizaki Y, Hume DA. Mammalian RNA polymerase II core promoters: insights from genome-wide studies. *Nat Rev Genet* 2007;**8**:424–36.
153. Trinklein ND, Aldred SF, Hartman SJ, Schroeder DI, Otillar RP, Myers RM. An abundance of bidirectional promoters in the human genome. *Genome Res* 2004;**14**:62–6.
154. Taft RJ, Simons C, Nahkuri S, Oey H, Korbie DJ, Mercer TR, et al. Nuclear-localized tiny RNAs are associated with transcription initiation and splice sites in metazoans. *Nat Struct Mol Biol* 2010;**17**:1030–4.
155. Kawaji H, Frith MC, Katayama S, Sandelin A, Kai C, Kawai J, et al. Dynamic usage of transcription start sites within core promoters. *Genome Biol* 2006;**7**:R118.

156. Jiang C, Pugh BF. Nucleosome positioning and gene regulation: advances through genomics. *Nat Rev Genet* 2009;**10**:161–72.
157. Chen ES, Zhang K, Nicolas E, Cam HP, Zofall M, Grewal SI. Cell cycle control of centromeric repeat transcription and heterochromatin assembly. *Nature* 2008;**451**:734–7.
158. Moazed D. Small RNAs in transcriptional gene silencing and genome defence. *Nature* 2009;**457**:413–20.
159. Preker P, Nielsen J, Kammler S, Lykke-Andersen S, Christensen MS, Mapendano CK, et al. RNA exosome depletion reveals transcription upstream of active human promoters. *Science* 2008;**322**:1851–4.
160. Gullerova M, Proudfoot NJ. Cohesin complex promotes transcriptional termination between convergent genes in *S. pombe*. *Cell* 2008;**132**:983–95.
161. Santoro R, Schmitz KM, Sandoval J, Grummt I. Intergenic transcripts originating from a subclass of ribosomal DNA repeats silence ribosomal RNA genes in *trans*. *EMBO Rep* 2010;**11**:52–8.
162. Guang S, Bochner AF, Burkhart KB, Burton N, Pavelec DM, Kennedy S. Small regulatory RNAs inhibit RNA polymerase II during the elongation phase of transcription. *Nature* 2010;**465**:1097–101.
163. Saint-André V, Batsché E, Rachez C, Muchardt C. Histone H3 lysine 9 trimethylation and HP1 favor inclusion of alternative exons. *Nat Struct Mol Biol* 2011;**18**:337–44.

# MicroRNA Function in the Nervous System

Roberto Fiore,[1] Sharof Khudayberdiev,[1] Reuben Saba,[1] and Gerhard Schratt

*Interdisziplinäres Zentrum für Neurowissenschaften, SFB488 Junior Group, Universität Heidelberg, and Institut für Neuroanatomie, Universitätsklinikum Heidelberg, Im Neuenheimer Feld 345, Heidelberg, Germany*



MicroRNAs (miRNAs) are an extensive class of small noncoding RNAs that control posttranscriptional gene expression. miRNAs are highly expressed in neurons where they play key roles during neuronal differentiation, synaptogenesis, and plasticity. It is also becoming increasingly evident that miRNAs have a profound impact on higher cognitive functions and are involved in the etiology of several neurological diseases and disorders. In this chapter, we summarize our current knowledge of miRNA functions during neuronal development, physiology, and dysfunction.

[1]These authors contributed equally to this work.

Progress in Molecular Biology
and Translational Science, Vol. 102
DOI: 10.1016/B978-0-12-415795-8.00004-0

# I. Introduction

The brain is arguably the most complex organ system within our body, with billions of nerve cells forming intricate networks via trillions of specialized connections known as synapses. Nerve cells in the mammalian brain are not a homogenous entity, but come in different flavors (excitatory, inhibitory), and greatly vary in their morphological and functional properties depending on the specific brain area they are embedded in. Even within a distinct brain area, for example, the hippocampus, dozens of different neural subtypes are represented, each contributing in an essential yet distinct way to information processing. Although the intricate circuits formed by neurons undoubtedly lie at the heart of brain function, the functional relevance of the other major cell type in the brain, the glial cells, is increasingly being appreciated. Similar to neurons, glial cells also display a high morphological and functional divergence, adding another layer of complexity.

Considering all these facets, it is not surprising that the brain uses a nested set of sophisticated mechanisms to regulate gene expression to generate cellular diversity, maintain cellular identity, and specifically control cellular function. On top of this, gene regulation in the brain requires a series of special features that, for example, ensure the high degree of compartmentalization of nerve cells and their morphological and functional plasticity in response to changes in the environment. Therefore, gene expression research is particularly intense in the neural system, and a plethora of complex gene regulatory mechanisms, some of them unique to neurons, have been discovered in the mammalian brain. These operate at virtually all levels, including, but not limited to, epigenetic modifications of DNA, mRNA transcription, splicing, transport, translation, and stability, as well as posttranslational modifications of proteins. Less than 20 years ago, microRNAs (miRNAs) entered the stage as posttranscriptional regulators of gene expression, and they are now considered to regulate the expression of up to 50% of all genes at the level of mRNA translation and/or stability. In this chapter, we undertake the challenging task of summarizing our knowledge of miRNA function in the nervous system, gleaned mostly from research within the last 5 years, illustrating the wide variety of neural processes, including neurogenesis, synapse development, plasticity and neurological disorders, which are influenced in one way or the other by miRNAs. From these examples, it is easy to appreciate that the brain is, and will continue to be, a particularly rich source of miRNA-regulated mechanisms that will also be instrumental in our understanding of the gene regulatory processes underlying physiology and pathology in many other systems.

## II. miRNAs in Neural Differentiation

The nervous system of vertebrates is composed of diverse cell types, most of which are generated during embryonic development by a small number of cells, namely, neural stem cells (NSCs). NSCs have the potential to self-renew (proliferate) and differentiate into restricted lineages of cells (e.g., neurons and glia). There are several types of NSCs that are produced at different times of development in mammals: in the embryonic development, neuroepithelial cells and radial glial cells; in the adult, B cells of the subventricular zone (SVZ) and radial astrocytes of the dentate gyrus subgranular zone (SGZ) in the hippocampus.[1] All these cells have the potential to perform symmetrical divisions into two daughter cells that maintain progenitor cell fate. However, as neurogenesis proceeds, NSCs start to divide asymmetrically, and only one of the daughter cells remains as a progenitor whereas the other becomes an intermediate progenitor cell (IPC). IPCs, depending on the origin (from which type of NSC), developmental stage, and location, will perform one or more terminal divisions to form neurons or glial cells (oligodendrocytes, astrocytes). The precise spatiotemporal regulation of NSC function during pre- and postnatal development is achieved by an intricate network of gene regulatory events, involving epigenetical, transcriptional, and posttranscriptional machinery.

miRNAs, key players of posttranscriptional gene regulation, have been initially identified as regulators of developmental timing in *Caenorhabditis elegans*, underlying their important role in cell fate decisions.[2] Since then, it has become apparent that miRNAs are also critically involved in the multiple steps of neurogenesis and differentiation. In this section, we discuss (1) the effect of Dicer inactivation for neural development; (2) the role of specific miRNAs in neurogenesis; and (3) gliogenesis.

### A. The Role of Dicer in Neural Development

A function of miRNAs during nervous system development *in vivo* has initially been shown using constitutive Dicer knockout zebrafish. Owing to the lack of functional Dicer, these animals are unable to process hairpin precursor miRNAs into their mature counterparts. Dicer loss-of-function resulted, in addition to abnormal morphogenesis during gastrulation, somitogenesis, and heart development, in severe morphological malformations of the brain and spinal cord.[3] Intriguingly, the introduction of the miR-430 family of miRNAs suppressed the brain-associated phenotypes in Dicer-deficient mutants, implying specific and important roles of miRNAs in the development of the brain. Interestingly, early patterning and cell fate specification in the fish embryonic nervous system were largely unaffected by the lack of miRNAs. However, the normal neural differentiation, underlying brain morphogenesis, was dependent on Dicer activity.

Unlike in zebrafish, the constitutive loss of Dicer in mice led to early embryonic lethality, due to malfunctioning of stem cells. This suggests a role for the miRNA pathway in maintaining the stem cell population.[4] The Dicer mutant embryos died (embryonic day 7.5) before the onset of neurogenesis (E10), which made these animals unsuitable for investigating the role of miRNAs later in neural development.

In order to determine the role of the miRNA pathway early in central nervous system (CNS) development, particularly in NSC maintenance and differentiation, several brain-specific conditional Dicer knockout mouse models have subsequently been generated. Thus, the selective ablation of Dicer in neuroepithelial cells of the dorsal telencephalon—the embryonic brain structure that later develops into the cerebral cortex—at the onset of neurogenesis had been performed.[5,6] Although miRNA processing had been already impaired in neuroepithelial cells, these cells did not show detectable survival defects. However, as soon as the neuroepithelial cells differentiated into a fate-restricted NSC population, namely, radial glial cells, abundant cell death, which was also evident in the newly formed neurons, was observed. As a result, these mice displayed a dramatically hypotrophic cortex with defective layering, as well as smaller and massively disorganized hippocampi. Consistent with these studies, deletion of Dicer in the olfactory progenitor cells blocked their terminal differentiation into the mature olfactory neurons, ultimately resulting in cell death.[7] Taken together, these observations support a critical function of the miRNA pathway during neural differentiation, for example, the transition from NSCs to postmitotic neurons. This nicely fits with the theory that miRNAs mainly confer precision and robustness to developmental processes.[8]

Intriguingly, the observations from specific deletion of Dicer in postmitotic neurons led to conflicting results. Notably, the Dicer loss of function in excitatory forebrain neurons,[9] midbrain dopaminergic neurons,[10] and cerebellar Purkinje neurons[11] caused widespread apoptosis. In contrast, Dicer null striatal dopaminoceptive neurons[12] and olfactory neurons[7] did not show any signs of cell death. It should be noted that, although dopaminoceptive neurons survived over the life of the animal, their biological function had been perturbed. As a result, these animals showed several phenotypes, including ataxia, reduced brain size, and smaller neurons.[12] The discrepancy in the observed results may be due to a requirement of an miRNA pathway in the survival of some neuron types but not others. Alternatively, the timing of Dicer removal may determine whether neurons will undergo apoptosis or survive. Thus, early removal of Dicer may lead to cell death, whereas later removal may result in the other observed phenotypes, such as smaller neuron size,[12] reduced dendritic branch elaboration, and increased spine length.[9] These observations underscore the important role of Dicer and miRNAs for normal functioning of postmitotic neurons, which will be discussed in more depth in the next section.

On a quick note, several recent findings challenge whether all the observed phenotypes of Dicer null animals are due to depletion of mature miRNAs. It has been proposed that Dicer is also required for the generation of some small RNAs, such as snoRNAs,[13] endo-siRNAs,[14] noncanonical miRNAs, mirtrons, and short hairpin miRNAs.[15] Particularly, endo-siRNAs have been thought to regulate gene expression posttranscriptionally in murine oocytes and embryonic stem cells (ESCs). In agreement with this, deletion of DGCR8, an RNA-binding protein that is part of the pri-miRNA processing complex ("microprocessor"), in mouse oocytes and ESCs did not recapitulate the Dicer null phenotype.[16,17] However, when DGCR8 and its enzymatic partner, Drosha, have been deleted in different tissues, such as skin,[18] cardiomyocytes,[19] and T cells,[20] the phenotype appeared to be very similar to the corresponding Dicer null tissues. These findings suggest that endo-siRNAs and/or other classes of DGCR8/Drosha-independent, Dicer-dependent small RNAs may play less important (if any) roles in somatic tissues. Therefore, the phenotypes observed in mice with Dicer-deficient postmitotic neurons (or even neural progenitor cells) might, indeed, mainly originate from direct or indirect loss of miRNAs.

## B. The Role of Specific miRNAs in Neurogenesis

Despite the fact that knockout animals for miRNA biogenesis pathway components (Dicer, DGCR8, Drosha) can provide us with valuable general information *in vivo*, they are not useful to determine the impact of specific miRNAs in the developing nervous system. In order to assess specific functions of miRNAs, investigators first performed expression profiling of miRNAs during the different stages of neural development *in vitro* (neuronal stem cells–IPCs–mature neurons, and glial cells) and *in vivo* (during embryonic and postnatal development). Data on miRNA expression in NSCs during proliferation and maintenance have been reported.[21–23] However, the functional significance of these miRNAs for the maintenance of NSC identity awaits further studies.

Most of the findings concentrate on the changes in miRNA expression levels upon induction of NSC differentiation. With the onset of NSC differentiation, the levels of many mature miRNAs are upregulated, including miR-9, miR-124, miR-125, miR-128, and some of the let-7 family members.[24–27]

### 1. Global Regulators of Neurogenesis: let-7, miR-9, and mir-124

Among the miRNAs whose expression is upregulated upon NSC differentiation, let-7, miR-9, and miR-124 were identified to be expressed in all neuron types investigated so far. This underscores that these miRNAs are essential players to promote NSC differentiation into virtually all types of neurons (Table I; Fig. 1).

TABLE I

LIST OF MICRORNAS INVOLVED IN NEURAL DEVELOPMENT AND THEIR MRNA TARGETS

| MicroRNA | Organism | Function | Target (s) | References |
|---|---|---|---|---|
| lsy-6 | Worm | Chemosensory neuron differentiation | COG-1 | 28 |
| let-7 | Rodent | Promotes neuronal differentiation | LIN28, TLX, HMGA | 26,29,30 |
| | Rodent | Neural tube closure | MLIN41 | 31 |
| miR-7 | Fly | Promotes photoreceptor cell differentiation | YAN | 32 |
| miR-9 | Fly | Inhibits neuronal differentiation | SENS | 33 |
| | Zebrafish | Promotes neuronal differentiation | HER5, HER9 and FGF signaling pathway | 34 |
| | Rodent | Promotes neuronal differentiation | FOXG1, TLX, STAT3, REST, FGF8, FGFR1 | 34–37 |
| | Human | Promotes proliferation of early human ESC-derived hNSCs | STMN1 | 38 |
| miR-9* | Rodent | Promotes neuronal differentiation | BAF53a, | 39 |
| | Human cell line | ? | coREST | 37 |
| miR-17 | Human cell line | Inhibits neural differentiation | ? | 40 |
| miR-124 | Mouse | Promotes neuronal differentiation | SCP1, PTBP1, SOX9, DLX2, JAG1 BAF53a, | 39,41–43 |
| miR-125 | Mouse | Promotes neuronal differentiation | GLI1, SMO, LIN28 | 26,44 |
| miR-128 | Human cell line | Inhibits NSC proliferation | BMI1 | 45 |
| miR-133b | Mouse | Modulates maturation of dopaminergic neurons | PITX3 | 10 |
| miR-200 | Mouse | Promotes olfactory progenitor differentiation | ? | 7 |
| miR-206 | Human cell line | Promotes proliferation of oligodendrocyte progenitor cells | TPPP | 46 |
| miR-219 | Mouse | Promotes oligodendrocyte differentiation | SOX6, HES5, Zfp238/RP58 PDGFRα, FOXJ3 | 47,48 |
| miR-273 | Worm | Chemosensory neuron differentiation | DIE-1 | 28 |
| miR-279 | Fly | Promotes neuronal differentiation | Nerfin-1 | 49 |
| miR-324-5p | Mouse | Promotes neuronal differentiation | GLI1, SMO | 44 |
| miR-326 | Mouse | Promotes neuronal differentiation | GLI1, SMO | 44 |
| miR-338 | Mouse | Promotes oligodendrocyte differentiation | SOX6, HES5, Zfp238/RP58 | 47 |
| miR-430 | Zebrafish | Neuronal morphogenesis | ? | 3 |

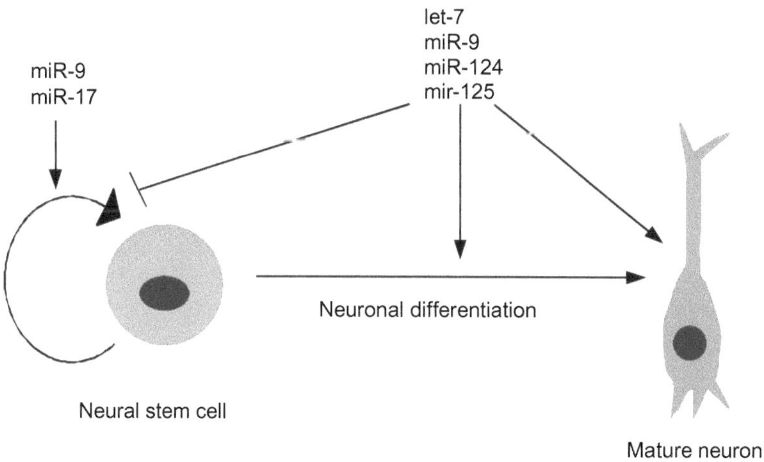


FIG. 1. *The role of miRNA at different stages of neuronal differentiation.* The proliferation and maintenance of NSCs are positively regulated by miR-9 and miR-17. On the other hand, neuronal differentiation and maintenance of the mature neural state are enhanced by the activity of let-7, miR-9, miR-124, and miR-125.

The let-7 family of miRNAs was among the first to be identified and is well conserved among the animal phyla. Currently, it is one of the most intensively studied miRNA families that have been implicated in suppression of proliferation of different cancer cells.[50] On the other hand, a role in NSC differentiation has been proposed for let-7 as well.[26,29–31,51,52] Although most of the let-7 family members are transcribed during early embryogenesis, no inhibitory activity of let-7 is observed at this stage due to the lack of mature let-7. LIN28, an RNA-binding pluripotency factor, has been identified as inhibitor of pri- and pre-let-7 processing, thereby limiting the generation of mature let-7.[26,53,54] In cancer cell lines, it has been proposed that LIN28 may induce urydilation and decay of pre-let7.[55] The block of let-7 processing by LIN28 is relieved at the onset of neurogenesis (E9.5), leading to the accumulation of mature let-7. Intriguingly, it has been found that LIN28 mRNA can be targeted by miR-125 and let-7. Therefore, let-7 and LIN28 are part of an autoinhibitory feedback loop: the increase in miR-125 level may reduce expression of LIN28, which in turn results in "leaky" expression of let-7, thereby completely inhibiting the expression of Lin28 at the posttranscriptional level.

In contrast to the inhibitory action of LIN28, the TRIM-NHL protein TRIM32 was found to enhance the activity of several miRNAs, including let-7, by interacting with AGO1. As NSCs proceed with asymmetric divisions, TRIM32 becomes polarized and accumulates in the daughter cell which will later differentiate into a neuron. On one hand, TRIM32 ubiquitinates and degrades the proliferative factor c-Myc; on the other, it increases the activity of let-7.[52] Further, Zhao *et al.*[30] and Nishino *et al.*[29] showed that let-7b regulates NSC proliferation and differentiation by targeting key stem cell regulators such as TLX (also known as NR2E1) and HMGA2, respectively. The transcription factors TLX and HMGA2 are essential for NSC proliferation and self-renewal, and their downregulation by let-7 contributes to the cell cycle arrest and subsequent differentiation into mature neurons.

miR-9 is another well-conserved miRNA that has been implicated in neural development. In the *Drosophila* peripheral nervous system, each external sensory organ develops from the division of a single sensory organ precursor (SOP) cell. miR-9a inhibited neuronal fate in non-SOP cells, including those that are adjacent to SOPs within proneural clusters, by downregulating the proneuronal zinc-finger transcription factor SENS. The overexpression of miR-9a resulted in severe loss of SOPs, whereas depletion of miR-9a led to production of extra SOPs.[33] This suggests that miR-9a may ensure the generation of a precise number of neuronal precursor cells during neural development.

Although miR-9 is also abundantly expressed in the developing nervous system of vertebrates, it seems that there, in contrast to the fly, it acts as a proneurogenic factor. For example, miR-9 in zebrafish has been proposed to promote neurogenesis by suppressing the antineurogenic factors HER5 and HER9. In addition, owing to inhibition of several components of the FGF signaling pathway, miR-9 defines the boundaries of midbrain–hindbrain boundary organizer activity in zebrafish.[34] In rodents, miR-9 has been implicated in differentiation of embryonic Cajal–Retzius cells, one of the first neuron types generated from NSCs. The transcription factor FOXG1, which promotes the proliferation of cortical progenitor cells and inhibits their premature differentiation into Cajal–Retzius cells, has been identified as a target of miR-9 in this scenario.[35] Interestingly, cortical progenitor cells from $FOXG1^{+/-}$ heterozygous mice have similar FOXG1 expression levels (i.e., 50% of that of the wild type) as the one from mice overexpressing miR-9, but these cells do not show increased production of Cajal–Retzius cells. This may indicate that miR-9 has additional targets, which in combination with FOXG1 give rise to the observed differentiation phenotype.

Indeed, Zhao and colleagues identified the nuclear receptor TLX, an essential regulator of NSC renewal, as a direct target of miR-9 in adult NSCs.[36] They showed that the overexpression of miR-9 in murine adult NSCs undergoing retinoic acid induced terminal differentiation suppressed TLX expression. These cells show reduced proliferation, which was accompanied by increased differentiation and migration of the newborn neurons. The observed phenotype was rescued by exogenous expression of miR-9-insensitive

TLX in NSCs. In addition to being a direct target of miR-9, TLX also transcriptionally inhibited miR-9 genes.[36] Thus, in analogy to the let-7/Lin28 interaction, miR-9 and nuclear receptor TLX form a feedback regulatory loop to regulate NSC proliferation and differentiation. It is important to note that investigators could not observe alterations in cellular differentiation upon knockdown of miR-9 using antisense RNA. Moreover, overexpression of miR-9 alone in cultured NSCs was not sufficient to induce neural differentiation. Therefore, this represents another example where miRNAs act primarily by fine-tuning gene expression programs.

In contrast to the findings suggesting an inhibitory role of miR-9 in NSC proliferation,[34–36] a recent study demonstrated that miR-9 promotes proliferation and impairs migration of early human ESC-derived human NSCs (hNSCs).[38] This effect was in part mediated by a direct target of miR-9, stathmin—a key regulatory protein of microtubule dynamics. The difference between this study and the one of Zhao *et al.*[36] might be attributed to differences in the species examined (human vs. mouse), in stages of hNSCs (early vs. adult neurogenesis), in origins of NSCs (ESCs vs. brain-derived), or in the experimental approaches (loss of function vs. gain of function). Indeed, in the study by Delaloy *et al.*[38] the authors report that TLX expression is induced only at the later stages of neural induction. Thus, the lack of TLX expression in early hNSCs may explain why these two studies came to different conclusions. However, the two studies raise the interesting possibility that miR-9 may act as a regulatory switch at different stages of neuronal differentiation by interacting with multiple target genes that are expressed in a temporal and region-specific manner.

Another well-studied CNS-specific miRNA during neuronal development is miR-124. The relatively extensive analysis of miR-124 function in brain development carried out by many research groups suggests that miR-124 is a key component in neuronal differentiation. Its cooperative action with other neuron-specific factors ensures the complete transition of NSC into a mature neuron.

Initial profiling studies demonstrate that the expression of this miRNA increases during mouse brain development.[24] Interestingly, overexpression of miR-124 in HeLa cells (cervical cancer cell line) downregulated about 100 mRNAs and shifted the mRNA expression profile toward that of a neuron.[56] Conversely, the inhibition of miR-124 activity with 2′-*O*-methyl oligonucleotides in mouse neurons resulted in the activation of nonneuronal mRNA expression.[57] In nonneuronal cells, including NSCs, neuron-specific gene expression is repressed by the transcriptional repressor REST (repressor element1 silencing transcription factor, also called NRSF). REST is a zinc finger protein that binds to a conserved 23-bp motif known as RE1 (repressor element 1, also called NRSE) and mediates active repression via recruitment of histone deacetylases by its corepressors mSin3 and coREST. In addition to repressing genes that are characteristic of the neuronal phenotype (such as ion channels, synaptic vesicle proteins, and others), REST also represses brain-specific miRNAs such as miR-9/9*, miR-132, and miR-124.[57] The downregulation of the REST protein by the proteasome pathway and repression of its transcription by retinoic acid receptor complex remove the inhibitory block from neuronal genes. Consequently, the level of miR-124 and miR-9/-9* increases, which in turn represses nonneuronal transcripts, thereby facilitating the switch from NSCs to differentiated neurons. It is interesting to note that miR-9 and-9* have been shown to target REST and coREST in nonneuronal cells (HEK293T), respectively.[37] Whether the same mechanism is employed in differentiating NSCs awaits further investigation.

To date, it has been shown that miR-124 targets several key factors involved in the maintenance of nonneuronal gene expression.[39,41–43] Among the factors recruited by REST to repress neuronal genes, small C-terminal domain phosphatase 1 (SCP1) has been shown to be crucial for the repression. Notably, the overexpression of a dominant negative form of SCP1 promoted neurogenesis in P19 cells by blocking the antineural function of REST. Interestingly, Visvanathan *et al.*[41] showed that miR-124 posttranscriptionally repressed SCP1 in the developing spinal cord of chicken as well as P19 cells. This attenuated the proliferation of neural progenitors and promoted their precocious neurogenesis.

Another important target of miR-124 is the RNA-binding protein PTBP1. PTBP1 is a global repressor of alternative pre-mRNA splicing and is expressed primarily in nonneuronal cells. It binds to pyrimidine-rich sequences of pre-mRNAs, which leads to the inhibition of splicing of nearby neuron-specific alternative exons. This results either in expression of the canonical form of the mRNA or in nonsense mediated decay of the incorrectly spliced neuronal mRNA.[42] Experiments carried out *in vitro* and *in vivo* have determined that direct posttranscriptional repression of PTBP1 by miR-124 is essential for the switch from nonneuronal mRNA splicing to neuronal lineage-specific splicing.

miR-124 (and miR-9*) have also been shown to regulate the components of the ATP-dependent chromatin-remodeling complex, such as BAF (also called mammalian SWI/SNF). The BAF chromatin remodeling complex has been implicated in several neural developmental processes, including neural progenitor proliferation and activity-dependent neurite outgrowth in postmitotic neurons. The ability to switch different subunits of the BAF complex gives the cell an opportunity to regulate the expression of specific genes at different stages of development. Thus, neural-progenitor-specific BAF (npBAF) complexes include the BAF53a subunit, which promotes neural progenitor proliferation. Upon commitment to the neuronal cell fate, the level of miR-124 and miR-9* increases, which in turn represses posttranscriptionally the BAF53a gene in postmitotic neurons.[39] In mature neurons, this results in the replacement of BAF53a by a homologous form—BAF53b—within neuron-specific BAF (nBAF) complexes. The exogeneous expression of the BAF53a isoform lacking miR-124 and miR-9* target sites disrupts dendritic morphogenesis of mature neurons. In addition, expression of REST in postmitotic neurons leads to derepression of BAF53a, indicating that REST-mediated repression of miRNAs directs the essential switch of chromatin regulatory complexes.[39]

Most of the work in determining the function of miR-124 has been carried out in cell cultures *in vitro* and during embryonal brain development *in vivo*. A recent finding suggested that miR-124 is also an important regulator of the temporal progression of adult neurogenesis in mice.[43] In the adult mammalian brain, the SVZ is the largest neurogenic niche. Interestingly, the knockdown of endogenous miR-124 maintained purified SVZ stem cells as dividing precursors. In agreement with previous reports, ectopic expression of miR-124 led to precocious and increased neuron formation. Furthermore, they identified several target genes (DLX1, JAG1, SOX9) of miR-124. Notably, the attenuating effect of miR-124 on adult stem cell proliferation and differentiation into neuroblasts was due to the downregulation of the transcription factor SOX9. Thus, miR-124-mediated repression of SOX9 is important for progression along the SVZ stem cell lineage to neurons.[43] In a different study, overexpression of miR-124 in the E12.5 mouse lateral cortex was performed using *in utero* electroporation. This procedure promoted neural progenitor transition from the apical to the basal compartment and stimulated direct differentiation of apical progenitors to postmitotic neurons, supporting most of the earlier reports.[58]

### 2. Specific Modulators of Neuronal Subtypes

As illustrated above, most miRNAs characterized so far have a more general impact on neural differentiation, for example, by promoting NSC differentiation, and/or maintenance of mature neurons in many neuronal cell types. To date, only a few reports on miRNA(s) involved in differentiation or maturation of specific neuronal types exist.[28,32,40,44,45,49,59] (Table I).

One of the most elegant mechanisms elucidated so far involves two specific miRNAs, lsy-6 and miR-273, which promote cell-fate specification of chemosensory neurons in *C. elegans*.[59] In this nematode, chemosensory progenitor neurons form two morphologically symmetric chemosensory neurons, ASE left (ASEL) and ASE right (ASER), on the left and right lateral sides of the animal, respectively. However, these neurons display different (asymmetric) expression profiles of chemosensory genes, which are maintained by lsy-6 in ASEL and miR-273 in ASER neurons. Notably, in ASEL neurons, lsy-6, whose expression is positively regulated by the ASEL-specific transcription factor DIE-1, represses translation of ASER-specific transcription factor COG-1. On the other hand, in ASER neurons, COG-1 positively regulates expression of miR-273, which in turn posttranscriptionally downregulates die-1. Thus, both miRNAs (lsy-6 and miR-273) and transcription factors (COG-1 and DIE-1) form a double-negative feedback loop that maintains different neuronal fates.[59]

Another example of miRNA-dependent differentiation of specific neuronal cells was described in *Drosophila*. It has been shown that, as photoreceptor progenitors differentiate into mature photoreceptor cells, the expression of progenitor-cell-specific transcription factor, YAN, decreases, whereas the levels of miR-7 increase. Interestingly, it turned out that YAN and miR-7 form a reciprocal negative feedback loop, which ensures their mutually exclusive expression. Thus, in mature photoreceptor cells, the expression of miR-7 led to downregulation of YAN, thereby stabilizing postmitotic cell phenotypes.[32]

Finally, in the vertebrate system, it has been proposed that miR-133b regulates the maturation of midbrain dopaminergic neurons by controlling the expression of paired-like homeodomain transcription factor PITX3 in mice.[10] In addition, mouse olfactory neurons were at least in part dependent on miR-200 family miRNAs for their maturation.[7]

## C. miRNAs in the Development of Glial Cells

The most abundant cell type in the mammalian CNS is the glial cell, which provides essential structural and functional support for postmitotic neurons. Although most of the initial work implicating miRNAs in the nervous system has been focused around NSCs and postmitotic neurons, recent findings indicate that miRNAs have in addition important roles in the distinct types of glial cells found within the CNS.

miRNA profiling studies determined that rat oligodendrocyte progenitor cells, which later differentiate onto mature oligodendrocytes, express the previously described abundant brain-enriched miRNA such as miR-9, miR-26a, miR-124, miR-125b, miR-181b, and let-7 family. However, during differentiation into mature oligodendrocytes, 23 miRNAs were downregulated, including the neuron-specific miRNAs miR-9 and miR-124. Interestingly, the expression level of 20 miRNAs such as miR-21, miR-152, miR-142–5p and -3p, mir-219, miR-338, miR-339, and miR-378 was increased.[60]

Several studies have investigated the functional role of the miRNA pathway in the oligodendrocyte differentiation using conditional Dicer inactivation in mice.[47,48,61,62] In the spinal cord, the development of oligodendrocytes occurs late in embryonic development after the formation of motor neurons. The analysis of Nestin-promoter-driven cre Dicer knockout mice, where Dicer is removed starting from E17, provided initial *in vivo* evidence about the role of miRNAs in oligodendrocyte development. The ablation of Dicer caused not only a decreased expansion of the oligodendrocyte precursor pool but also attenuated differentiation into mature oligodendrocytes in the E18.5

spinal cord.[61] Since these mutant mice are not viable owing to the loss of dicer in the entire CNS, these effects are difficult to interpret. However, a recent study supports some of the defects observed by Kawase-Koga *et al.*[61] in oligodendrocyte development. Specifically, the targeted ablation of dicer in ventral spinal cord oligodendrocyte progenitors, using Olig-1-Cre mice, resulted in suppression of oligodendrocyte expansion.[63]

In contrast to the above findings, two independent groups reported that mice lacking Dicer1 (Olig-1, Olig-2, or CNP cre) specifically in oligodendrocyte lineage cells display severe defects in oligodendrocyte maturation without any detrimental effects on the expansion of the oligodendrocyte progenitor pool *per se*.[47,48] In other words, oligodendrocyte neural progenitor cells (oNPCs) with Dicer1 deletion undergo extensive proliferation without further differentiation into mature oligodendrocytes. Differences between these two studies and the former by Zheng *et al.*[63] might be explained by the regional differences in Olig-Cre gene expression and the different developmental stages examined.

Zhao *et al.*[47] identified that miR-219 and miR-338, which are induced in differentiating oligodendrocytes, are essential for oligodendrocyte maturation. Inhibition of these miRNAs *in vitro* (oNPC cultures) and *in vivo* (zebrafish) resulted in a significant reduction of oligodendrocyte formation. On the other hand, overexpression of miR-219 and mir-338 in embryonal chick spinal cord and mouse cortex promoted precocious differentiation and maturation of oligodendrocytes. In addition, transfection of miR-219 and miR-338 mimics into oNPCs, derived from dicer knockout mice (Olig-1 cre), at least in part rescued the differentiation defect. Two inhibitors of oNPC differentiation, the transcription factors SOX6 and HES5, have been identified as targets of miR-219 and miR-338. Moreover, Zfp238/RP58, which is required for cortical neurogenesis, is also repressed by both miRNAs.[47] In addition to miR-219 and miR-338, Dugas *et al.*[48] found upregulation of miR-138 during oNPC differentiation. Furthermore, they determined that the translation of PDGFR$\alpha$ (positive regulator of cell proliferation) and FOXJ3 is inhibited by miR-219. In summary, these two studies suggest that miR-219 and miR-338 control oligodendrocyte maturation by negatively regulating both oligodendrocyte differentiation inhibitors and proneuronal differentiation factors.

A different study suggests that miRNAs under some circumstances inhibit rather than promote the differentiation of oNPC. Particularly, miR-206, which is expressed in the oNPCs but not in mature oligodendrocytes, inhibits the tubulin polymerization-promoting protein (TPPP). Since CG4 cells (an oligodendrocyte progenitor cell line), overexpressing miR-206, showed reduced differentiation into mature oligodendrocytes,[46] this suggests that miR-206 inhibits oNPC differentiation. Although TPPP mRNA is present in neurons and oligodendrocytes, TPPP protein accumulates only in glia cells. Concerning the underlying mechanism, it was suggested that, upon induction of oNPC differentiation, the expression of miR-206 is downregulated, leading to a removal of the translational block of the TPPP mRNA. This could at least in part contribute to the acquisition of a mature oligodendrocyte phenotype.

The recent finding that small RNAs, particularly miRNAs, are important regulators of posttranscriptional gene expression has added a new layer of complexity in CNS development. miRNAs have been implicated in different stages of neuronal development, including the maintenance of neuronal progenitors, their differentiation into postmitotic neurons, and, subsequently, the integration of postmitotic neurons into preexisting neural circuits. Based on these studies, a main theme emerges whereby miRNAs mostly function in cooperation with transcriptional and epigenetic factors to fine-tune gene expression programs that allow cells (neural progenitor and differentiating neurons) to switch from one developmental stage to another. So far, most of the work has been performed either *in vitro* or in Dicer mutant mice. It will be interesting to determine whether the knockout of specific miRNA(s) in a specific spatiotemporal manner will validate most of these *in vitro* results. In addition, revealing the key aspects of NSC renewal might give us insight for future therapeutic interventions against neuropathological conditions.

## III. miRNAs in Postmitotic Neurons

A postmitotic differentiated neuron has to confront the challenge of embedding itself correctly in the neuronal network. Proper wiring and functioning of this network is achieved thanks to the formation of trillions of specialized junctions between axons and neuronal cell bodies and dendrites, known as synapses. Thus synaptic function is essential for cognition, and synaptic dysfunction is the hallmark of several neurological disorders.

During neuronal development, synapse formation occurs through a stereotypical multistage program. However, synaptic connections are not static structures. During both postnatal development and adult life, neuronal networks are continuously rewired in response to sensory input. This process is known as experience-dependent plasticity, and the morphological and functional changes that occur during plasticity represent the basis of the brain's capability to store information while adapting to a changing environment. At the molecular level, synaptic plasticity is carried out via the constant interplay between neuronal activity and sophisticated programs of gene expression, which convert external stimuli into long-lasting changes in synapse structure and function. These changes often occur in a synapse-specific manner and are tightly spatially regulated within the neuron. Spatial regulation of synaptic plasticity is, at least in part, achieved because of the dendritic localization of a subset of mRNAs that code for important synaptic proteins. Thus, local control of mRNA translation offers the neuron the possibility to react to external stimuli with subcellular resolution, thereby enabling each synapse to respond autonomously to neuronal activity.

miRNAs are highly expressed in postmitotic neurons both at times of synapse development as well as in fully mature neurons. Their function as posttranscriptional regulators of gene expression makes them excellent candidates for controlling key aspects of synapse development and plasticity. So it comes as no surprise that intense research efforts have been focused on clarifying the role of miRNAs in compartmentalizing mRNA translation in neurons, as well as on the regulation of miRNA expression and function by neuronal activity. The importance of both these aspects of miRNA biology for synaptic function is now supported by solid experimental evidence that will be described in the next two sections. We will then summarize the current knowledge of the involvement of specific miRNAs in the different stages of synapse development and function (Table II).

TABLE II
List of MicroRNAs Involved in Functions of Postmitotic Neurons and Their mRNA Targets

| Process | MicroRNA | Target gene and pathway | References |
|---|---|---|---|
| Dendritic growth and arborization | miR-132 | p250RhoGap, Rho signaling | 64,65 |
| | miR-124 | Rho signaling, indirect regulation of CDC42 | 66 |
| | miR-125b | ? | 67 |
| | miR-134 | PUM2, mRNA metabolism | 68 |
| | miR-137 | MIB1, proteasome-dependent degradation | 69 |
| | bantam | ?, indirect regulation of AKT1 signaling | 70 |
| Synapse formation | Let-7 | Abrupt | 71 |
| | miR-134 | LIMK1, actin polymerization | 72 |
| | miR-138 | APT-1, palmitoylation | 73 |
| | miR-132 | p250RhoGap, rho signaling | 67,74 |
| | miR-137 | MIB1, proteasome-dependent degradation | 69 |
| | miR-125b | NR2A, neurotransmitter receptor composition | 67 |
| Synapse physiology and plasticity | miR-134 | CREB, activity-dependent transcription | 75 |
| | miR-132 | RFX4, circadian rhythm | 76 |
| | miR-284 | GLURA and GLURB, neurotransmitter receptor function | 77 |
| | miR-219 | SCOP, circadian rhythm | 76 |
| | miR-1 | UNC-29, UNC-63, MEF2, neurotransmitter receptor function and activity dependent transcription | 78 |

## A. Unique Features of miRNA in Postmitotic Neurons

### 1. Compartmentalized Expression

Because of the established importance for dendritic protein synthesis in some form of synaptic plasticity,[79] the isolation of several miRNAs associated with the translational machinery of postmitotic neurons[80] immediately suggested that the miRNAs pathway might operate locally within the neuron to control the expression of key synaptic proteins at a subset of synapses. This hypothesis found the first experimental support with the identification of several proteins involved in the miRNA pathway in the synapto-dendritic compartment in both vertebrate and invertebrate nervous systems. Specifically, Dicer, the argonaute homologue EIF2C,[81] Armitage and its mammalian homologue MOV10,[82,83] and fragile X mental retardation protein (FMRP) [84] have been all shown to be localized in dendrites. miR-134 has been the first functionally characterized dendritic miRNA. miR-134 localizes in granules in close proximity to dendritic spines in dissociated hippocampal neurons.[72] Spines are specialized actin-rich protrusions along the dendrites where the majority of excitatory synapses are located; the effect of miR-134 on dendritic spines will be described in further detail later. The repertoire of dendritic miRNAs has been greatly expanded by profiling studies. These investigations analyzed the small RNA content in preparations enriched in the synaptodendritic compartment by either laser capture microdissection of dendrites isolated from primary rat hippocampal neurons[85] or biochemical isolation of synaptosomes, a biological fraction of purified synaptic terminals from murine brain.[73,86] So far the list of dendritically enriched miRNAs includes more than 20 members. However, a function within the dendritic compartment has been identified for very few of them, namely miR-134, miR-138, and miR-132 that all regulate spine size in hippocampal neurons (see below). One key unanswered issue in the understanding of dendritic miRNA biology is how they reach their final destination in the dendrites. Several models are conceivable (for review, see Refs. 87,88), but none has been experimentally proven so far. Both miRNAs and RISC proteins show a granular pattern within the neuronal processes and share some components with the other ribonucleoprotein particles associated with mRNA transport and repression in dendrites. Furthermore, neuronal activity can alter the motility and composition of the miRNA-containing granules,[81,84,89] suggesting that their transport and dynamics represent an important additional layer of regulation of miRNA function in neurons.

The majority of miRNAs that show polarized expression in neurons have been identified in dendrites. However, experimental evidence for miRNAs that are either enriched in axons or exert their effect presynaptically is emerging. Genetic models with disrupted miRNA biogenesis are consistent with a role of miRNAs in axon. Both in *Drosophila m.* and mice, disruption of Dicer in olfactory neurons and retinal ganglion cells, respectively, leads to axonal pathfinding abnormalities.[90,91] However, it remains to be determined whether these defects arise from impaired biogenesis of specific axonal miRNAs. In the invertebrate *Aplysia*, miR-124 is localized in the presynaptic neuron of the sensory-motor synapse, where it regulates long-term facilitation (LTF), a form of synaptic plasticity (see below).[92] More recently, a number of miRNAs specifically enriched in the distal axon of rat sympathetic neurons have been isolated.[93] Between these, miR-338 regulates energy metabolism in the axon by controlling local translation of cytochrome c oxydase IV.[94] It is very likely that the list of miRNAs playing an important role in axonal development and function will increase in the near future, as it has been the case for dendritic miRNAs.

### 2. Regulation by Neuronal Activity

Neuronal activity plays an essential role in shaping synaptic strength and connectivity at all stages of neuronal development and function. A neuron reacts to changes in synaptic transmission by modulating gene expression programs, which in turn regulate important synaptic factors. Thus, it is not surprising that neuronal activity has a profound impact on all aspects of miRNA biogenesis and function (Fig. 2). A first layer of regulation occurs at the transcriptional level. Increased synaptic signaling induces calcium influx into the postsynaptic cells, which ultimately results in new gene transcription regulated by activity-dependent transcription factors.[95] Membrane depolarization and application of the neurotrophin BDNF (brain-derived neurotrophic factor) activate transcription of the miR-132 gene in dissociated neurons in a manner that depends on the well-known activity-regulated transcription factor CREB (cAMP response element binding protein).[64] Also, the transcription of miR379–410 cluster—a conserved, large mammalian miRNAs cluster encompassing more than 40 miRNAs, among them miR-134—is induced by neuronal activity upon activation of the transcription factor MEF2.[68]

Posttranscriptional regulation can further increase the complexity of the mechanisms connecting neuronal activity to miRNA function. The molecular mechanisms of regulated miRNA biogenesis have begun to be clarified in nonneuronal tissues and cell lines and are described in detail elsewhere.[96] It is likely that similar mechanisms are operative also in neurons, and the effect of neuronal activity on miRNAs biogenesis will surely be a field of exciting research. It has already been shown that brain-specific adenosine-to-inosine editing of the miR-376 family members can alter their silencing specificity.[97] Two experimental observations suggest that processing of miRNA precursors is regulated in neurons. The dendritic miR-138 is brain-specific, while its precursor pre-miR-138 is ubiquitously expressed.[98] This suggests that dicer-mediated cleavage of the precursor is the regulated step in the biogenesis of this specific

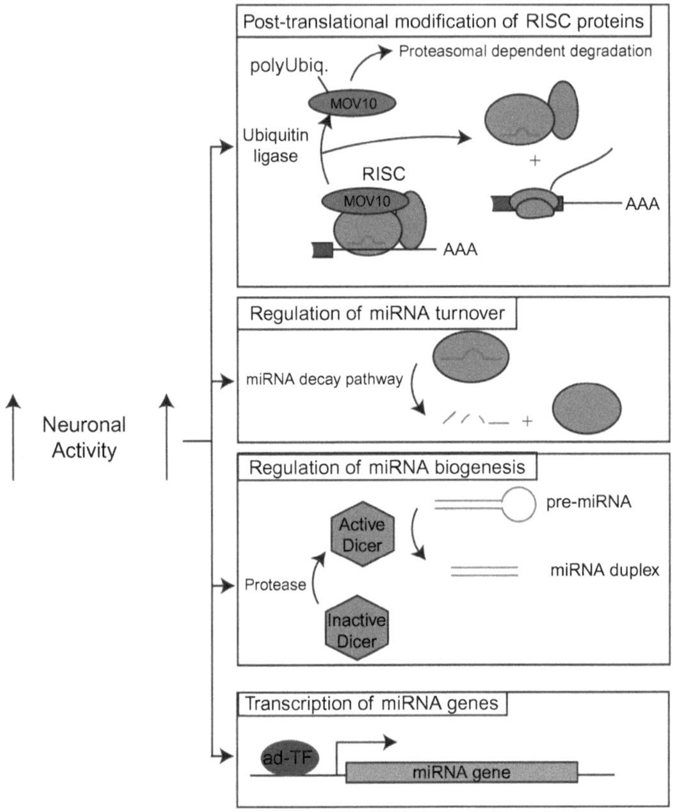


FIG. 2. *Mechanisms of miRNA regulation by neuronal activity.* The expression and activity of several neuronal miRNAs are controlled by neuronal activity at multiple levels, see Section 2 for details. ad-TF, activity-dependent transcription factor.

miRNA. Furthermore, in synaptosomes, activity-dependent calpain cleavage induces release of Dicer from the postsynaptic density (an electron-dense region at the membrane of a postsynaptic neuron where most neurotransmitter receptor localizes) and stimulates its RNase activity.[81] It remains to be experimentally demonstrated that increased Dicer activity is associated with the specific processing of dendritic miRNA precursors. Overall, it is becoming clear that neurons are able to dynamically regulate the levels of mature miRNAs. This is achieved not only by controlling their transcription but also their stability. In the retina and in dissociated neurons, increased neuronal activity accelerates turnover of several miRNAs, suggesting that miRNA levels are controlled by a regulated decay mechanism in neuron.[99]

Protein components of the RISC complex are also subject to regulation by neuronal activity.

In *Drosophila*, increased synaptic transmission induced by an olfactory learning paradigm leads to proteasomal-dependent degradation of the critical RISC component Armitage. This in turn leads to increased translation of the important plasticity gene CAMKII,[82] a calcium/camodulin-dependent kinase. A similar mechanism has also been observed in mammalian hippocampal neurons where MOV10 is degraded after activation of the *N*-methyl D-aspartate (NMDA) neurotransmitter receptors in an ubiquitin- and proteasome-dependent fashion, which is followed by the translation of the two known dendritic miRNA targets APT-1 and LIMK1 (see below).[83] Importantly, MOV10 degradation appeared to be restricted to activated synapses, in both studies. Thus neurons appear to be able to regulate individual RISC complexes in response to neuronal activity in a temporally and spatially restricted manner, a capacity that is well suited to control the long-lasting potentiation of specific synaptic connections that occurs during plasticity.

## B. Functions of miRNA in Postmitotic Neurons

### 1. miRNA's Roles During Dendritogenesis

Inhibitory and excitatory inputs from neighboring neurons converge on dendrites, which in turn transmit them to the soma. Thus the processing of information transmitted by different synapses established by a neuron is strongly influenced by the geometry of the dendritic tree. Unraveling the mechanisms regulating growth and remodeling of the dendritic arbor is essential to understand the functioning of neuronal circuits.

Many examples illustrate the importance of miRNAs at all stages of dendritic development. The initial stage of dendritogenesis is controlled by intrinsic programs of gene expression that govern the sprouting and branching of neurites. In dissociated, immature neuronal cultures, three miRNAs have been shown to be necessary at this stage. miR-132 and -124 are both necessary and sufficient to induce neurite outgrowth from primary cortical neurons and differentiating P19 cells, respectively. Both miRNAs exert their effect by decreasing the protein levels of members of the small rho GTPases family, a class of important regulators of actin cytoskeleton dynamics. Specifically, p250RhoGAP is downregulated by miR-132 and CDC42 by miR-124.[64,66] In contrast, miR-137 inhibits dendritic morphogenesis in primary neurons whereby the biologically significant target of miR-137 appears to be the ubiquitin ligase MIB1.[69] The relevance of the miRNA pathway for dendritic development has been confirmed *in vivo* by the use of different genetic models. In *Drosophila m.*, ectopic expression of the RISC associated proteins FMR1 and ME31B reduces the complexity of the dendritic tree of sensory neurons.[84] The same phenotype is observed when miR-124 is overexpressed, suggesting that *in vivo* the levels of this miRNA must be accurately regulated to ensure proper development of neuronal processes. Interestingly, FMR1 and miR-124 genetically interact, since the miR-124 gain-of-function phenotype is partially rescued in FMR1 null mutants. The rescue is associated with decreased levels of mature miR-124 and accumulation of its precursor, suggesting that FMR1 controls miR-124 activity by regulating its biogenesis.[100] In mice, conditional inactivation of Dicer in postmitotic neurons results in a variety of abnormalities that include a severe reduction of the complexity of the dendritic tree in the hippocampus.[9] However, owing to the pleiotropic effects of Dicer inactivation, it is difficult to determine whether the dendritic phenotypes are a primary or secondary consequence of the absence of miRNAs in neurons.

After the initial outgrowth and branching of the neuronal processes, the further remodeling of the dendritic tree is shaped by neuronal activity. Increased neuronal activity induces CREB-dependent expression of miR-132, which in turn is necessary for activity-dependent dendritogenesis in dissociated and organotypic hippocampal cultures. Thus miR-132 is necessary for dendritic outgrowth at multiple stages of dendritic development.[65] Increased neuronal activity boosts the expression also of the of miR379–410 cluster. Independent inhibition of multiple members of the cluster, including miR-134, prevents activity-dependent dendritogenesis but has no effect on the dendritic complexity of unstimulated neurons. The RNA-binding protein PUM2 is the key target of miR-134 in the context of activity-dependent dendritogenesis.[68] It is still unknown if other miRNAs from the miR379–410 cluster regulate activity-dependent dendritogenesis by targeting different components of the PUM2 pathway or independent signaling cascades. The analysis of mice that are haplo-insufficient for DCGR8, an essential component of the microprocessor complex, is consistent with the finding that expression of miRNAs belonging to the miR379–410 cluster is important for proper dendrite development *in vivo*. Maturation of several members of the cluster is impaired in DCGR8 mutant mice, and this correlates with reduced complexity of the dendritic tree of pyramidal neurons in the hippocampus.[101]

The final steps of dendritic development require their correct targeting to the synaptic partners and their maintenance in the receptive field of the neuron (the area within a neuron where stimulation excites the cell) during organogenesis and plasticity. It has been shown that both processes require miRNA activity. In *Drosophila m.*, the highly stereotyped projections of the dendrites of olfactory projection neurons are mistargeted in null mutants for the two miRNA biogenesis enzymes, Pasha and Dicer. Interestingly, just the dendrites of specific subsets of olfactory neurons were affected,[90] suggesting the existence of an miRNA code contributing to the wiring of the nervous system. Genetic analysis in *Drosophila m.* has shown that miRNAs are necessary not only for correct targeting but also to maintain coverage of the target area of sensory neuron dendrites. The miRNA bantam expressed in epithelial cells ensures that, during larval growth and development, sensory neuron dendrites that target the epithelium grow proportionally to maintain their receptive field.[70]

## 2. miRNA Roles in Synapse Formation and Maturation

A combination of studies in both invertebrate and vertebrate model systems has begun to shed some light on the role of the miRNA pathway during synaptogenesis. In *Drosophila*, the RISC components AGO1 and FMR1 are necessary for proper development of the neuromuscular junction (NMJ). FMR1 null mutants and FMR1/AGO1 double heterozygous show ectopic formation of synaptic boutons.[102] Subsequently, a role for the LET-7 miRNA family in the maturation of the fly NMJ has been revealed. Development of the NMJ is severely delayed in LET-7 mutants, suggesting that these miRNAs do not control the formation of the NMJ *per se*, but regulate its growth and maturation.[71] The discrepancy between the AGO1/FMR1 and the LET-7 phenotypes illustrates the difficulty of interpreting the results of genetic approaches in which the function of the whole miRNA pathway is impaired. First, most of the proteins involved in miRNA biogenesis and function have also other roles that do not depend on miRNAs themselves. Second, since miRNAs can control multiple aspects of the same biological process, complete suppression of this pathway will only uncover the dominant phenotype which could also be due to the synergistic effect of multiple miRNAs. Thus, to dissect the role of miRNAs in any biological process, it is imperative to identify the functions of individual miRNAs. Much progress in this direction has been achieved in mammals using primary neuronal cultures and organotypic slices. In dendrites of rat hippocampal neurons, miR-134 colocalizes with and suppresses the translation of the mRNA coding for the cytoskeleton regulator LIMK1. LIMK1 is a kinase that promotes actin polymerization, and the miR-134/LimK1 interaction leads to a decrease in dendritic spine size.[72] Interestingly, miR-134 repression of LimK1 translation is reversible. BDNF, which is released in response to synaptic transmission, can relieve the inhibition of LimK1 expression by miR-134. Thus this interaction might be subjected to regulation by neuronal activity and be relevant for plasticity since dendritic spine size is a good indicator of synaptic strength.

The control of different signaling pathways that modulate actin dynamics in spines seems to be a shared feature of the dendritic miRNAs that have been functionally characterized so far. Using a functional screening approach, miR-138 and miR-132 were identified as a positive and negative regulator of spine growth, respectively, in cultured hippocampal neurons. miR-138 inhibition led to increased spine volume associated with enhanced amplitude of the miniature excitatory postsynaptic currents (mEPSC). miR-138 downregulates expression of APT-1, an enzyme that catalyzes depalmitoylation of important synaptic proteins, including the actin cytoskeleton regulator Gα12/13, an RhoA regulator.[73] Similar to its growth-promoting effect on dendrites, miR-132 is necessary and sufficient for spinogenesis. Both in dissociated hippocampal neuron and organotypic hippocampal slices, miR-132 inhibition decreased spine density with a concomitant reduction in the frequency of mEPSC.[67,74] Thus, similar to miR-134, miR-132 is necessary at multiple stages of postmitotic neuronal development. However, in contrast to miR-134, which seems to have different key targets depending on the biological process that it regulates, miR-132 function in spinogenesis is again mediated by downregulation of p250GAP, a GTPase-activating protein that regulates the actin cytoskeleton via the Rac/PAK1 pathway.

Actin dynamics is not the only aspect of synapse formation and maturation controlled by miRNAs.

miR-137 modulates the activity of the proteasome pathway by repressing MIB1, which in turns leads to inhibition of spine development.[69] Spine development and mEPSC frequency are also impaired in hippocampal neurons overexpressing miR-125b. One target of miR-125b is NR2A, a subunit of the NMDA neurotransmitter receptors.[67] However, it remains to be experimentally demonstrated that NR2A is the key miR-125b target in the spinogenesis pathway.

### 3. miRNA's Roles in Synapse Physiology and Plasticity

miRNAs continue to play an important role in synapse physiology in the adult nervous system, where their roles range from the control of the efficiency of synaptic transmission to several forms of plasticity. The abundance and subunit composition of neurotransmitter receptors is one of the major determinants of synapse strength. Both of these aspects seem to be under the control of miRNAs in invertebrate and vertebrate synapses. The regulation of NR2A by miR-125b has already been mentioned above. At the fly NMJ, miR-284 controls the abundance of the glutamate receptor subunits GLURA and GLURB.[77] In *C. elegans*, miR-1 fine-tunes the efficacy of synaptic transmission at the NMJ through two mechanisms. First, miR-1 directly suppresses expression of two subunits of the acetylcholine receptors in the muscle, thus modulating the sensitivity of muscles to acetylcholine. Second, miR-1 also reduces expression of the transcription factor MEF2. The miR-1/MEF2 interaction is necessary to modulate a still-unknown retrograde signal that regulates neurotransmitter release on the presynaptic element of the NMJ.[78]

Since the long-lasting form of synaptic plasticity is dependent on protein synthesis, it was just a matter of time before experimental data on miRNAs regulating plasticity started to accumulate. The activity-dependent regulation of Armitage stability in *Drosophila m.* olfactory neurons has already been described above. Importantly, mutation in both Armitage and Dicer leads to impaired memory formation during olfactory learning due to the misregulated translation of CaMKII.[82] Successively, miR-124 was the first miRNA that was shown to be necessary for a specific form of plasticity: LTF at the sensory-motor synapse in the *Aplysia*. LTF is a form of synapse strengthening induced by repeated application of the neurotransmitter serotonin on the sensory neuron. miR-124 is expressed in the presynaptic neuron where it inhibits expression of CREB1. miR-124 seems to act as an inhibitory signal for plasticity at the sensory motor synapse, since the decrease of miR-124 activity is necessary and sufficient for LTF induced by repeated application of serotonin.[92]

Regulation of the critical plasticity gene CREB by miRNAs is necessary for some forms of synaptic plasticity also in mammalian CNS. CREB translation is inhibited by miR-134, and, *in vivo*, overexpression of miR-134 impairs both long-term potentiation (LTP; a form of synaptic plasticity) and memory formation in fear-related tests. A similar phenotype and decreased levels of the CREB protein are also observed in brain-specific knockdown of SIRT-1. SIRT-1 is a histone deacetylase that binds to a consensus sequence upstream of the miR-134 gene and suppresses its expression, suggesting that SIRT-1 indirectly controls CREB levels via miR-134. Consistently, viral-mediated inhibition of miR-134 rescues the phenotype observed upon SIRT-1 loss of function, restoring CREB levels.[75]

Finally, two miRNAs have been shown to be necessary for proper function of the circadian clock. Inhibition of miR-132 and miR-219 in the suprachiasmatic nucleus in the hypothalamus lengthens the circadian period and attenuates the entraining effects of light, presumably by disregulating the expression of two key clock genes.[76]

The development of a method to ectopically perturb *in vivo* the function of a specific miRNA has been instrumental in this study. Once this approach is applied to other miRNAs in parallel with gene targeting of brain specific miRNAs, it is very likely that our knowledge about the regulation of plasticity and cognitive functions by small RNAs will increase dramatically.

## IV. The Roles of miRNA in Neurological Diseases

The possible involvement of miRNAs in initiating and/or mediating the effects of neuropathological diseases and disorders is quite plausible because of their omnipresent role in gene regulation. At present, numerous studies have emerged illustrating the differential expression of miRNAs in several regions of the CNS in a wide assortment of neuropathological conditions. In this section, the neuropathological roles for miRNAs in neurodegenerative, neuropsychiatric, neurodevelopmental, and addictive conditions will be covered. Upon examination of these studies, a general consideration on whether this dysfunction in miRNA expression is the cause or consequence of the diseased state is greatly required.

## A. The Role of miRNAs in Neurodegenerative Diseases

### 1. Alzheimer's Disease

Alzheimer's disease (AD) is caused by the series of proteolytic cleavages to the cellular amyloid precursor protein (APP), first by the β-site APP-cleaving enzyme 1 (BACE1) and then by gamma secretase (γ secretase), to form amyloid β (Aβ). The accumulation of Aβ is highly neurotoxic to healthy neurons because of the aggregation of this molecule to form plaques that eventually signal neuronal cell death. Common symptoms of AD include the loss of memory, inability to think and concentrate, and other general cognitive disabilities including the failure to recognize people and objects. AD most often contributes to the death of the subject. A number of studies have reported on the aberrant expression of miRNAs in various brain tissues affected by the disease. Moreover, the differentially expressed miRNAs have the potential to target key proteins and enzymes involved in the production of the Aβ load, notably BACE1 and APP.

miRNA expression studies on whole brains and specific subregions of the brain affected by AD have been performed. One of the first studies to do so showed the upregulation of miR-9 and miR-128 in AD hippocampus upon comparison to age-matched control subjects.[103] The potential for these two miRNAs to target a wide array of target genes was suggested to contribute to the neurodegenerative process. miR-107, on the other hand, has been shown to be downregulated in the brains of patients with early stages of AD.[104] Coincidentally, BACE1 mRNA possesses at least one functional site for miR-107 in its 3′UTR, suggesting a reciprocal relationship between the two. Expression analysis of miRNAs in sporadic AD patients identified the miR-29 family as being downregulated in the disease. Once again, the gene encoding BACE1 was a target.[105] Indeed, upon further analysis of this relationship using *in vitro* reporter assays, a significant correlation between the downregulation of miR-29a/b and overexpression of BACE1 was identified. This reciprocal relationship between the miRNA and target gene has the potential to increase the production of APP in AD. Other miRNAs shown to target BACE1 include miR-298 and miR-328, which were found to display increased expression in the hippocampus in a mouse model of the disease.[106] miR-101, which recognizes one of two sites predicted for it in the 3′UTR of APP, has been shown to regulate the expression of this enzyme in both embryonic and postnatal primary hippocampal cultures.[107] Apart from directly targeting APP, miR-101 can also target the cyclooxygenase 2 (COX2) gene, the product of which has been implicated in the synthesis of APP by contributing to the synthesis of prostaglandin E2 (PGE2). A concerted downregulation in the expression of APP and COX2 (and indirectly PGE2) by miR-101 has the potential to reduce the Aβ load in AD. Similarly, miRNAs of the miR-20a family (miR-20a, miR-17-5p, and miR-106b) have also been shown to target APP.[108] Nevertheless, only miR-106b was significantly downregulated in sporadic cases of AD, but this change, however, did not correspond to an increased Aβ load.

Aberrantly expressed miRNAs can also contribute to more complex regulatory mechanisms involved in the pathobiology of the disease apart from directly targeting the central proteins and enzymes involved in the etiology of the disease. Many of these mechanisms maybe secondary in nature but, nevertheless, their dysfunction can contribute to exacerbate the disease process. In the cerebral cortex of an AD mouse model, miR-34a expression was found to be highly elevated.[109] A putative target of miR-34a is the Bcl2 gene which serves a protective role in AD brains. Downregulation of BCL2 in AD, potentially by the upregulation of miR-34a, can provide a mechanism by which the ensuing pathobiology can progress unhindered. TGF-β's (TGF-β1-3) are cytokines that play important roles in the growth and survival of neurons. Inhibiting the TGF-β pathway can increase the neurotoxicity induced by Aβ. The upregulated miR-106b in AD can target for repression TGF-β type II receptor (TβR II), a high-affinity receptor for TGF-βs, which is also decreased in the neurons of AD subjects.[110] The reduced expression of TβR II correlates with impairments in TGF-β signaling via SMAD proteins, together favoring neuronal degeneration.[110] The role of miR-146a expression in AD has been noted.[111,112] The expression of miR-146a in AD correlates with the decrease in the expression of complement factor H (CFH) which is involved in suppressing the immune and inflammatory responses in the brain. Unrestrained neuroinflammation during AD can accelerate the degenerative state through the amplification of the effects of proinflammatory molecules (cytokines and chemokines). The identification of aberrant miR-146a expression in AD is quite interesting, since this miRNA is a downstream effector molecule of the transcription factor NF-κB that plays a central role in orchestrating inflammatory responses and cell fate decisions.[113] Additionally, miR-146a has also been shown to be involved in inhibiting the expression of NF-κB through a negative feedback mechanism, suggesting that this miRNA can be involved in dampening the innate immune response.[113] The apparent dual role of miR-146a both as a positive and negative regulator of the innate immune response not only reinforces the general pleiotrophic nature of miRNA–mRNA interactions but also provides further insight into distinct molecular mechanisms of the brain that are perturbed in AD and how they may contribute to the pathobiology of the disease.

Prion diseases are similar to AD in that both are associated with the accumulation of a pathogenic fibrillar aggregate of their respective neurotoxic forms. In prion diseases, neuronal cell death is caused by the misfolding of the normal host encoded prion protein ($PrP^{c}$) to an aberrant neurotoxic isoform ($PrP^{Sc}$), which possesses the potential to self-replicate. Amongst neurodegenerative diseases, it maintains a unique status since it can possess sporadic, genetic, and infectious etiologies. Differential expression of miRNAs has been documented in brains of mouse and nonhuman primate models of the disease.[114,115] Although both studies identified unique subsets of differentially expressed miRNAs and numerous disease-specific target genes for future investigation, the upregulation of miR-342-3p was common to both. Additionally, this miRNA was also upregulated in a human manifestation of the disease, suggesting the potential for utilizing this miRNA as a biomarker of the disease pathobiology. One miRNA that was upregulated in prion disease and also consistently shown to be upregulated in the study of AD was miR-128.[114] Aberrant expression of this miRNA and the cognate target genes that it regulates may represent a general response to a neurodegenerative state.

## 2. Parkinson's Disease

Parkinson's disease (PD) is characterized by the progressive neurodegeneration of dopaminergic neurons in the substantia nigra, and manifests in such symptoms as tremors, rigidity, and bradykinesia. Initial establishment of possible role(s) for miRNAs in dopaminergic neurons was delineated through the inability of embryonic stem cells (ESCs) to differentiate into midbrain dopaminergic neurons upon Dicer ablation, suggesting a general role for miRNAs in the differentiation and survival of these cell types.[10] Furthermore, a subtractive microarray analysis which compared and contrasted miRNA expression in normal human adult brain versus that of PD patients revealed the impaired expression of miR-133b in diseased individuals. The reduced abundance of miR-133b was also confirmed in *Aphakia* mice that are often used as model organisms in the study of PD neuronal loss. A unique feature of this model system is the genomic scarceness of the paired-like homeodomain 3 (PITX3) gene, a transcription factor required for neuronal survival and normal motor activity and, coincidentally, a target gene of miR-133b. Based on overexpression and inhibitory tests *in vitro* and *in vivo*, miR-133b was shown to regulate Pitx3 and vice versa, essentially in a negative feedback loop consisting of the miRNA and the target gene. A disruption of this feedback mechanism may promote the multitude of symptoms associated with PD.[10] It is worth noting that, in another Dicer ablation study, specifically ablation in mouse dopaminoceptive neurons of the striatum, neuronal cell loss was not observed.[12] Nevertheless, a wide range of phenotypes were evident, including ataxia, front and hind limb clasping, reduced brain size, and smaller neurons, suggesting that probably neuronal dysfunction, and not just loss, contributes to disease pathobiology. A polymorphism in the fibroblast growth factor (FGF20) gene at the 8p21.3-22 locus was previously identified as a risk factor for PD, which in turn correlated with increased α-synuclein expression and the onset of PD.[116] Moreover, upon careful inspection of the risk-factor locus on chromosome 8, a single nucleotide polymorphism was identified within the 3′UTR of FGF20 that specifically disrupted the binding potential of miR-433 and in the process might permit unrestrained expression of the FGF20 gene.[117]

Specific mutations in the gene leucine-rich repeat kinase 2 (LRKK2) can give rise to some cases of PD in humans, though the molecular mechanisms that are involved are not yet clearly defined. A role for miRNAs in mediating the molecular effects of the pathogenic LRRK2 has been identified and a subsequent role in the PD neurodegenerative mechanism proposed.[118] Specifically, pathogenic LRRK2, which possesses increased kinase activity, was shown to reduce miRNA expression. This reduction was primarily due to the association of the pathogenic isoform with the RISC protein AGO1 as illustrated through coimmunoprecipitation studies. The association is believed to destabilize AGO1 interaction in the RISC. Furthermore, overexpression and downregulation of DCR1, another RISC component, suppressed and exacerbated the neurodegenerative effects of the pathogenic isoform, respectively. Upon examination of mRNA transcripts that are upregulated in response to the general decrease in miRNA production, the prominence of the transcription factor E2F1 was noted, as it has been previously implicated in PD. E2F1, in general, interacts with differentiation-regulated transcription factor proteins, such as DP. Together, their role in postmitotic neurons is to abort cell division leading to cell death. In *Drosophila* model of PD (expressing the pathogenic isoform of LRRK2), the expression of both proteins is upregulated; however, suppressing the expression of these proteins preserved dopaminergic neurons during the course of the disease. miRNA binding sites have been identified within E2f1 (miR-184*) and Dp (let-7) 3′UTRs.[118] Overexpression of these miRNAs in dopaminergic neurons of *Drosophila* partly rescued the phenotype associated with the pathogenic LRRK2, whereas the contrary was true upon knockdown of these miRNAs in wild-type flies. Taken together, the results provide compelling evidence for the role of miRNAs in mutated LRKK2 mediated PD.

### 3. Huntington's Disease

Phenotypic symptoms of Huntington's disease (HD) include random and uncontrollable movements as well as diminished cognitive and psychiatric abilities. At the molecular level, HD is believed to be caused by the expansion of the "CAG" trinucleotide in the Huntingtin (Htt) gene, resulting in the degeneration of neurons, particularly those in the striatum. Moreover, the trinucleotide repeat in Htt prevents the sequestration of the REST factor in the cytoplasm of neurons by the translated HTT, in the process enabling the translocation of the protein to the nucleus where it binds to RE1 repressor sequences upstream of target genes and promotes an overall downregulation in gene expression. Several studies have drawn attention to the deregulation of miRNAs in HD due to the aberrant function of REST. The expression of miRNAs miR-9/9*, miR-29a/b, miR-124a, miR-132, miR-330, miR-17, miR-196a, miR-222, miR-485, and miR-486 have all been shown to be downregulated in HD patients and mouse models of the disease, in the process leading to the upregulation of cognate target genes.[37,119] For example, upregulation of the Rho family GTPase-activating protein, p250GAP, due to a decrease in miR-132 levels, can negatively modulate dendritic plasticity.[65] miR-9 and miR-9* also target the REST complex (REST and coREST, respectively), suggesting the potential for the existence of a double negative feedback mechanism on miRNA expression.[65] Intriguingly, apart from sequestering REST, wild-type HTT has also been found to bind and stabilize interactions within the RISC complex as it copurified with AGO2.[120] Furthermore, wild-type HTT also colocalized with AGO2 in P bodies. In contrast, the mutant version inhibits the formation of such structures. Therefore, in HD, it appears that both the expression and also the function of miRNAs may be compromised, leading to aberrant target gene regulation.

### 4. Motor Neuron Diseases (Amyotrophic Lateral Sclerosis and Spinal Motor Neuron Disease)

Amyotrophic lateral sclerosis (ALS) is characterized by the progressive degeneration of motor neurons that are involved in the control of muscle movements. The symptoms of this disease include muscle atrophy, paralysis, and subsequently death. In ALS, the skeletal muscle specific miR-206 has been shown to sense motor neuron injury and/or loss and, in the process, ameliorate the resulting muscle damage by promoting regeneration of neuromuscular synapses.[121] The expression of miR-206 was highly upregulated in a mouse model of ALS, and upon knockdown of this miRNA, the symptoms of the disease progressed at a much faster rate leading to death of the animal. Similarly, in normal mice that have undergone sciatic nerve injury, leading to the deinnervation of muscle, increase in the expression of miR-206 in muscle fibers was apparent. Furthermore, by inhibiting the action of miR-206 in normal mice, the response of the neuromuscular junction (NMJ) to nerve injury was adversely affected via the formation of morphologically defective NMJs. Some of the possible *in vivo* effects of miR-206 in muscle cells appear to be mediated through the inhibition of histone deacetylase 4 (HDAC4). HDAC4 is known to inhibit nerve reinnervation by blocking the expression of fibroblast growth factor binding protein 1(FGFBP1), which is involved in potentiating the actions of a subset of fibroblast growth factor proteins (FGF7,10,22) on distal motor neuron terminals. The suppression of HDAC4, by miR-206, leads to the release of these fibroblast growth factors that promote the interactions between nerve and muscle. Moreover, the role of miR-206 at the injured NMJ appears to be very specific to adults only, as the loss of this miRNA had no impact on uninjured embryos and also on young mice. In general, miR-206 appears to possess a protective role in ALS where it preserves the integrity and plasticity of the adult neuromuscular synapse.[121]

Another miRNA implicated in motor neuron diseases is miR-9, which has been shown to be involved in the loss of spinal motor neurons (SMNs) leading to spinal muscular atrophy.[122] miR-9 was shown to be downregulated in SMNs that had undergone targeted ablation of Dicer. Although a general deletion of Dicer affects the biogenesis of a plethora of miRNAs, the decrease in the expression of miR-9 is especially interesting, as it targets for posttranscriptional gene silencing the heavy neurofilament (NEFH). NEFH has been previously implicated in motor neuron diseases where perturbations in the coordinated expression of this gene led to cytoskeletal defects of the axon. The increased expression of NEFH in SMNs perturbs the fine neurofilament subunit stoichiometry that exists between the light (NEFL), medium (NEFM), and NEFH.

## B. The Role of miRNAs in Neuropsychiatric Disorders

Neuropsychiatric disorders encompass a heterogeneous collection of cognitive and emotional disorders that have variable behavioral phenotypes and are challenging to diagnose and universally difficult to treat. Common disorders include schizophrenia (SZ), Tourette's syndrome (TS), autism spectrum disorders (ASDs), and various types of mood disorders. The exact molecular underpinnings of neuropsychiatric disorders has been elusive but the current hypothesis suggests that possible impairment of prominent biological pathways that contribute to neuronal synaptic plasticity, vulnerability to aberrant stimuli, and neurotrophic factor signaling may individually or concurrently contribute to the etiology of various types of disorders. The multiple roles mediated by miRNAs in gene regulation in the brain have rendered these molecules as possible candidates for investigation into the genesis and progression of neuropsychiatric disorders.

### 1. Schizophrenia

SZ is characterized by cognitive deficits that influence a patient's thoughts, perceptions, speech, mood, and behaviors. Several different types of molecular approaches have been used to identify and study the role of miRNAs in SZ. These can be divided into genetic variation studies and the subsequent extrapolation to larger subset(s) of the population and differential miRNA expression studies to document and to identify aberrantly expressed miRNAs in SZ.

One of the earliest genetic association studies involved the generation of a mouse model with a hemizygous 1.3-Mb deficiency of a genomic region that is syntenic to the 1.5-Mb region missing in the 22q11.2 microdeletion found in many SZ patients.[101] This region contains ~27 genes, including Di George syndrome (DGS) critical region 8 (DGCR8), a partner of Drosha. These animals showed behavioral alterations consistent with SZ, such as in locomotor hyperactivity, anxiety-like behaviors and deficits in prepulse inhibition, cued and contextual fear conditioning, and working memory. Microarray analysis of the hippocampus and prefrontal cortex of these animals showed alterations in the abundance levels of multiple genes, many of which have been previously linked to synaptic plasticity and function. Expression levels of DGCR8 were reduced, while no differences were seen in the expression levels of other proteins involved in the processing of miRNAs. Upon further investigation into the effect of this downregulation on miRNA processing, levels of pri-miRNAs were found to be increased while mature miRNAs had decreased levels, which was also confirmed in heterozygous knockout mice for Dgcr8. In general, this study was one of the first to suggest a potential link between a genetic alteration found in SZ patients, specifically the microdeletion, and an overall dysfunction of miRNA-dependent mechanisms.

miR-130b, which is located in the 22q11.2 susceptibility locus for SZ and targets the mRNA of MeCP2, has been investigated for possible genetic variation.[123] Albeit a sequence variation was identified in the 5′-upstream region of this gene, specifically at potential binding sites for transcription factors, it did not translate into statistically significant association with SZ upon extrapolation of the study to a much larger population. miR-346, which is located in the intron of glutamate receptor ionotrophic delta 1 (GRID1), has been the subject of a genetic association study since the host gene has been identified as an SZ susceptibility gene.[124] Additionally, miR-346 also targets for posttranscriptional regulation a large number of other SZ-related genes at a greater frequency than other brain-expressed miRNAs. Although no variation was apparent in the flanking regions surrounding miR-346 in a relatively small sample size, both the miRNA and the host gene were downregulated in the dorsolateral-prefrontal cortex (DL-PFC) of SZ patients. Upon examination of 59 miRNA genes on the X-chromosome, 9 were found to have genetic variation in either the pre-miRNA or mature miRNA sequence (let-7f-2, miR-188, miR-18b/18b*, miR-502, miR-510, miR-660, miR-325, miR-505, miR-509-3) in a set of SZ subjects.[125] Intriguingly, many of the target genes of these miRNAs are some of the defining genes of the disorder (e.g., neuregulin 1 (NRG1), disrupted in schizophrenia 1 (DISC1), regulator of G-protein signaling 4 (RGS4)). Analysis of genetic variants near or within 28 brain-expressed miRNAs identified a single nucleotide variation in miR-206 as being overrepresented, whereas a variation in miR-198 was underrepresented, in SZ subjects of European ancestry.[126] A single nucleotide variation in the pre miR 30e which can influence the liberation of the mature form was identified among SZ subjects of Chinese ancestry.[127,128] Lastly, a "hot-spot" for SZ on the 8p21-23 locus hosts six brain-expressed miRNAs (miR-124, miR-383, miR-320, miR-596, miR-597, and miR-598).[129] It is not difficult to fathom that structural variants at this locus, in the form of copy number variants (CNV), microdeletions, and/or microduplications, can possibly lead to dysfunctions in miRNA expression and/or function with the potential to influence the phenotypes often associated with SZ.

Large-scale miRNA expression studies have been performed on postmortem brain from human subjects afflicted with SZ.[40,130–132] In these particular studies, different regions of the affected brain were analyzed, different number of miRNAs were assayed, and the size of the sampling groups (affected vs. control) varied. The earliest of these studies examined the PFC (15 SZ cases vs. 21 control cases) and identified 15 downregulated and 1 upregulated miRNAs, of which 4 (miR-26b, miR-30b, miR-29b, and miR-106b) of the downregulated miRNAs were experimentally confirmed.[130] Putative target genes of the four confirmed downregulated miRNAs showed enrichment for signaling pathways related to synaptic function. Upon examination of the superior temporal gyrus (STG; seven SZ cases vs. seven control cases), two miRNAs (let-7g and miR-181b) were confirmed.[131] In particular, the target genes of miR-181b, upon assignment of gene ontology categories, also exhibited significant enrichment for synaptic function. Further analysis of the miRNA repertoire in SZ subjects (17 SZ cases vs. 17 control cases in STG and 15 SZ cases vs. 15 control cases in DL-PFC) showed differential expression in a large number of these molecules in both types of tissue. Nevertheless, a prominent upregulation of miR-181b, miR-219, and members of the miR-15 family was observed. Interestingly, members of the miR-15 family also possess enrichment for target genes associated with synaptic function.[40] In the largest set of miRNAs evaluated to date in SZ (PFC of 35 SZ subjects vs. 35 control cases), the aberrant expression of only seven miRNAs was determined.[132] Of the seven, three (miR-7, miR-132, miR-212) have been previously implicated in SZ. Taken together, these studies support a role for aberrantly expressed miRNAs in the regulation of genes with the potential to modulate the phenotypes often associated with SZ.

### 2. Tourette's Syndrome

TS is characterized by chronic vocal and motor tics which can manifest in behavioral abnormalities. A link between TS and miRNAs was proposed on the basis of the work that examined the Slit and Trk-like 1 (SLITRK1) gene, which is involved in neurite outgrowth and has also been implicated as a candidate gene in the disorder. Screening for mutations in the 3′UTR of this gene in TS patients identified a sequence variant in the binding site of miR-189 that was absent in the much larger subset of control subjects.[133] Moreover, the expression of SLITRK1 and miR-189 were found to overlap in many regions of the brain that have been associated with TS. It is interesting to note that several follow-up studies failed to segregate the genetic variation with TS patients, suggesting that further work, with a much larger sample size, is imperative to draw any further conclusions.

### 3. Autism Spectrum Disorders

ASDs are characterized by abnormalities in reciprocal social interactions, language development and/or usage, and restricted interests, as well as by repetitive behaviors. The aberrant expression of nine miRNAs was identified in postmortem cerebellar cortex of individuals afflicted with the disorder.[134] Some of the aberrantly expressed miRNAs target neurexin (NRX1) and Shank3, two genes that have been strongly implicated in autism. It is interesting to note that several other studies have also profiled the transcription of miRNAs in lymphoblasts derived from ASD subjects and have thereby identified differentially expressed miRNAs, many with neurologically relevant target genes.[135,136] In general, the latter set of studies support the use of lymphoblast cell lines as surrogate sources of miRNAs in the study of ASDs.

### 4. Mood Disorders

The potential role of miRNAs in the modulation of various mood disorders has been examined. Specifically, aberrant miRNA expression has been shown in the postmortem DL-PFC of individuals clinically diagnosed with bipolar disorder (BPD). Although many of the differentially expressed miRNAs were also shared among SZ subjects, there was a distinct subgroup, consisting of ~15 miRNAs, which were unique to the BPD subjects.[132] Even though miR-134 has not yet been identified as being significantly differentially expressed in BPD, it has nonetheless been investigated as a potential biomarker of the disorder.[137] Interestingly, the level of miR-134 in blood peripheral to the CNS was lower in BPD patients at both the drug-treated and drug-free stage than in healthy subjects. Furthermore, the level of miR-134 increased upon drug treatment in comparison to BPD subjects who were not subjected to treatment. From these results, it appears that the level of miR-134, in peripheral blood, has the potential to be a marker of the BPD, and that increases in plasma levels may be associated with the success of the treatments.

In the case of anxiety disorders, such as panic disorder and obsessive compulsive disorder (OCD), genetic association studies have been performed in order to suggest a possible molecular role for miRNAs in the etiology of the disorder. The neuronal plasticity gene, neurotrophin-3 receptor (NTRK1), which has been previously reported to play a key role in modulating anxiety-like behavior in mouse models, was shown to possess variations in several miRNA-binding sites in the 3′UTR.[138] Specifically, a nucleotide variation in the binding site for the brain-expressed miR-485-3p was found to be associated with the hording phenotype of OCD. Additionally, rare variants in the binding sites for miR-765, miR-509, and miR-128 have also been identified in the 3′UTR of this gene. In the case of the latter two miRNAs, nucleotide variations in their respective binding sites within the 3′UTR led to a significant recovery of NTRK1 expression. Within the serotonin receptor 1b (HTR1B), a common single nucleotide variation (A → G) was shown to attenuate the regulatory potential of miR-96.[139] Given that Htr1b knockout mice display an aggressive phenotype, it was postulated that this particular polymorphism would be related to a history of aggression-related behavior in a study of human subjects. Upon examination of a cohort of college students, it was apparent that those individuals homozygous for the A-element displayed more aggression-related conduct disorder behavior than those possessing one or two copies of the genetic variation. Therefore, it is entirely possible that this common polymorphism that attenuates the regulatory potential of an miRNA may also reduce the prevalence of a behavioral phenotype.

## 5. The Effect of Antipsychotics and Psychotomimetics on Neuronal miRNA Expression

Pharmacological agents that aid in the study and/or treatment of psychiatric diseases have profound effects on miRNA expression, thus further suggesting a role for miRNAs in modulating the molecular underpinnings of these disorders. Chronic treatment of primary hippocampal cells with mood stabilizers, lithium and valproic acid (VPA), resulted in 37 and 31 responsive miRNAs, respectively, with nine miRNAs in common between the two regimens. mir-34a, in particular, was downregulated in the presence of both types of agents. A putative target of miR-34a is the metabotropic glutamate receptor 7 (GRM7), which showed elevated levels in the hippocampal regions of the treated animals.[140] GRM7 is an interesting candidate in the context of mood stabilizers since mice lacking this gene show more risk-taking behavior compared to control mice.[141] Taken together, the results suggest that mood stabilizers mediate some of their molecular effects through the regulation of miRNAs, in this particular instance suppression of miR-34a expression. A decrease in miR-34a abundance leads to an increase in GRM7 levels that affect behavioral phenotypes, specifically dampening excessive behavioral excitement that is associated with risk-taking behavior. In an alternate study, mice that were acutely treated with the psychotomimetic dizolcipine, an NMDA-receptor antagonist, displayed decreased expression of miR-219 in the prefrontal cortex.[142] Additionally, reduced expression of miR-219 can also be observed in mice in which the expression of the critical NR1 subunit of the

NMDA receptor has been ablated. The decrease in miR-219 expression by dizolcipine was ablated upon pretreatment of cells with the antipsychotics haloperidol and colzapine. A potential target gene of miR-219 is CAMKIIγ, a member of the calcium/calmodulin-dependent protein kinase family of NMDA-receptor signaling molecules. Therefore, the upregulation of miR-219 in several SZ miRNA expression studies suggest that the hypoactivity of NMDA receptors may, in part, be due to the overexpression of this miRNA and its effect on downstream target genes.

## C. The Role of miRNAs in Neurodevelopmental Disorders

Given the role for miRNAs in the differentiation and development of neurons and also in mediating the putative functions of postmitotic neurons, possible functions for miRNAs in mediating the etiology of neurodevelopmental disorders and the phenotypes associated with such disorders have been postulated. Indeed, a role for miRNAs has been investigated in Down syndrome (DS), fragile X syndrome (FXS), Rett syndrome (RS), and Di George Syndrome (DGS).

### 1. Down Syndrome

Trisomy 21, commonly known as DS, is caused by a chromosomal abnormality in which there is a triplication of all, or a portion, of the human chromosome 21, leading to an increase in the expression of this region by ~50% in many tissues. How the general increase in the expression of human chromosome 21 leads to the manifestation of DS phenotypes is still under intense scrutiny. Nevertheless, DS is characterized by variable phenotypes, among them cognitive impairment(s), congenital defects of the heart, craniofacial abnormalities, gastrointestinal anomalies, leukemia, and AD. In humans, chromosome 21 contains genes for five miRNAs (miR-99a, let-7c, miR-125b-2, miR-155, and miR-802).[143] The expression of these miRNAs was found to be increased in DS fetal hippocampus samples as well as in fetal heart samples, consistent with the extra copy of chromosome 21. Two of the upregulated miRNAs, miR-155 and miR-802, were shown to target the gene encoding MeCP2, which is downregulated in DS brain specimens from both humans and mice.[144] In DS, the MeCP2 gene is a particularly promising candidate for investigation since it plays an important role in neurogenesis, a process that is dysfunctional in DS. Downstream targets of MeCP2, CREB1 and MEF2C, which are underexpressed and overexpressed by MeCP2 action, respectively, are also aberrantly expressed in the same samples. Furthermore, *in vivo* silencing of miR-155 or miR-802 in Ts65Dn mice (trisomic for miR-155 and miR-802) led to the reestablishment of the expression of miRNAs, target genes, and also the downstream effector genes to nontrisomic

21 levels. Taken together, these results suggest that perturbations in the cascade of molecular signals, initiated as a result of chromosome 21 trisomy, may be mediated by several of the upregulated miRNAs found within the vicinity of chromosome 21.[144]

### 2. Fragile X Syndrome

FXS is caused by the loss-of-function mutations in the gene encoding FMR1, specifically by the expansion of the trinucleotide CGG in the 5′UTR region of the gene. The Fmr1 gene encodes the FMRP. FMRP, in general, is an RNA-binding protein that acts as translational repressor in neurons. Phenotypes of FXS are manifested in the forms of severe cognitive and intellectual disabilities. Association of FMRP with the miRNA pathway was first illustrated in extensive work done in *Drosophila*.[66,84,100,145,146] In mammals, FMRP, Dicer, and eIF2c2 have been found to be in association with each other in postsynaptic densities.[81] At these sites, it is suspected that the association operates in an activity-dependent manner to regulate the proper translation of target mRNAs. Specifically, FMRP association with its target genes may provide the initial framework for targeting of the miRNA–RISC complex to the 3′UTR and subsequent translational repression. Additionally, the phosphorylation status of FMRP appears to regulate the association with the miRNA machinery.[147]

Specific miRNAs have been found to be associated with FMRP in immunoprecipitates prepared from mouse brain.[67] Interestingly, two of these miRNAs (miR-125b and miR-132) had diverging effects on dendritic spines. In dissociated hippocampal neurons, the overexpression of miR-125b generated long narrow spines that resulted in reduced mEPSC amplitude, whereas the knockdown of this miRNA increased the average width of the dendritic protrusions. The NMDA-receptor subunit, NR2A, was identified to be a potential target gene of miR-125b. NR2A mRNA also interacts with FMRP, suggesting that a complex of FMRP, miR-125b, and AGO1 might be involved in the regulation of NR2A mRNA. In contrast to the effects of miR-125b on dendritic spines, overexpression of miR-132 increased dendritic protrusion width and increased mEPSC amplitude, whereas the knockdown led to a significant reduction of dendritic complexity. Taken together, these results suggest that miRNAs may play central role(s) in the silencing of FMRP target mRNA molecules. Moreover, these results also provide an intriguing possibility for altering miRNA abundance as a means of challenging the abnormal spine phenotype associated with FXS.

### 3. Rett Syndrome

RS is an X-linked neurodevelopmental disorder that primarily occurs in females, with a prevalence of 1:12,500 births being affected by the age of 12. Mutations in the gene encoding MeCP2 have been linked to symptoms associated with RS, thus underlining the importance of this gene in the syndrome. Additionally, tight regulation of this gene in the brain is essential, as both increases and decreases in its expression have been shown to lead to marked neurodevelopmental defects associated with RS.[148,149] Interestingly, the 3′UTR of MeCP2 possesses an miRNA recognition element (MRE) for miR-132, a CREB pathway induced neuronal miRNA that is highly responsive to BDNF treatment.[64,65,74] Moreover, BDNF has emerged as a target of MeCP2, especially during neuronal development.[150] Taken together, these observations suggest the existence of a possible homeostatic mechanism whereby, as MeCP2 levels increase, so also the levels of BDNF, which in turn leads to the induction of miR-132 that can repress the translation of MeCP2.[151] Overall, the specific role of miR-132 may be to fine-tune the expression status of MeCP2 so as to keep the transcriptional corepressor at physiologically relevant levels; perturbations in this fine balance may provide the impetus for the symptoms associated with RS.

#### 4. Di George Syndrome

DGS is believed to be caused by a 3-Mb hemizygous deletion on chromosome 22 (22q11.2), which results in the loss of a subset of genes. Individuals with DGS suffer from both behavioral and cognitive deficits including attention-deficit hyperactivity disorder (ADHD), OCD, and ASD. A putative gene found within the deletion is Dgcr8, which is known for its intimate involvement with miRNA processing by partnering with Drosha. Knockout animals lacking DGCR8 have been shown to have a decreased abundance of mature miRNAs[101] (also refer to the section on Schizophrenia). Based on this work, it is tempting speculate that aberrant processing of miRNAs leading to altered regulation of host target genes may provide an explanation for some of the behavioral and cognitive phenotypes associated with DGS.

## D. The Role of miRNAs in Addiction to Drugs of Abuse

The designation of addiction to drugs of abuse, such as alcohol, cocaine, and nicotine, as a neurological disorder and/or disease is debatable. Nevertheless, the presence of characteristic signs and symptoms, as well as a progressive course of affliction, clearly suggests that addiction has the potential to be considered a pathology of the CNS. Long-term intake of various drugs of abuse can lead to changes in neuroplasticity that manifests in an individual's decreased sensitivity and/or increased propensity for the abused drug. However, the exact molecular mechanisms underlying these neuroadaptations are far from clear. Therefore, the potential role of miRNAs in governing the intricate molecular mechanisms of addiction must be considered. Although relatively new to the field of miRNA research, several pioneering works advocate an essential role for miRNAs in governing the molecular mechanisms of addiction to drugs of abuse.

1. Cocaine Addiction

The earliest evidence of a role for miRNAs in the modulation of neuroplasticty in response to the psychostimulant cocaine documented the upregulation of miR-181a and the downregulation of two other miRNAs, let-7d and miR-124a, in addiction-relevant regions of the brain.[152] Many of the target genes of these miRNAs were recognized as playing central roles in synaptic plasticity and also serving as putative genes in addiction-related processes, suggesting the possibility that aberrant miRNA expression and subsequent differential expression of target genes may contribute to the altered neuroadaptations. Further evidence for potential miRNA involvement in cocine addiction can also be gleaned from AGO2 deficiency in DRD2-expressing neurons in rat models.[153] Specific ablation of the Ago2 gene in these neurons leads to a decrease in the motivation of the animal to self-administer cocaine. The decreased motivation, in part, may be attributed to a loss in a subset of miRNAs (approximately 23 miRNAs) that show explicit expression patterns in DRD2 expressing neurons and are also highly responsive to cocaine administration. Many of the target genes of these miRNAs have the potential to regulate various facets of neuronal plasticity that has the potential to define addiction to the drug. Although large-scale identification of miRNAs with potential involvement in addiction-related processes is imperative to narrow down the search for a subset of miRNAs, these studies, however, do not identify a specific miRNA or set of miRNAs and the mechanism(s) by which they may operate. Even so, a handful of studies have emerged that do, in fact, identify probable mechanisms mediated by miRNAs in addiction. One study, in particular, has identified a differentially expressed miRNA that appears to alter the behavioral response to cocaine, primarily by serving as a conduit between controlled and uncontrolled drug-seeking behaviors. Specifically, the induction of miR-212 (and closely related miR-132 which is located on the same genomic region) in the dorsal striatum of rats that were given extended access to the drug via intravenous cocaine self-administration, a model that closely mimics the binge use of cocaine by humans, was found to limit the development of compulsive cocaine-seeking behavioral response in such rats.[154] In contrast, among animals with extended access to the drug, however with blocked miR-212 expression, cocaine seeking behavior flourished. Furthermore, among rats whose access to the drug was restricted throughout the study, miR-212 induction was not at all apparent in the dorsal striatum.[154] A possible explanation that takes into consideration all these unique observations may lie in the finding that miR-212 amplifies the activity of CREB and its crucial cofactor, Transducer of regulated CREB activity (TORC).[154] CREB activity, in turn, has been attributed to decrease the motivational responses to cocaine. The specific molecular mechanisms by which this may occur involves the miR-212-mediated suppression of several putative target genes, among them the gene encoding SPRED1. In turn, SPRED1 is known to limit the activity of the small GTPase, RAF1, a central protein recognized in the generation of the second messenger cyclic AMP by adenylyl cyclase. An increase in RAF1 activity (due to the miRNA-mediated repression of the SPRED1 gene) leads to greater production of cAMP, and ultimately to increased phosphorylation and activity of CREB1. Interestingly, CREB is also a known transcriptional activator of miR-212.[64,65] Taken together, miR-212 and CREB–TORC are likely to operate via a positive feedback loop, leading to a cessation in cocaine uptake as a consequence of decreased motivational cues for the drug, most notably during extended access conditions.[154]

An additional role for miR-212 and miR-132 in regulating the vulnerability to cocaine addiction has also been identified, specifically in conjunction with MeCP2.[155] MeCP2 expression in the dorsal striatum is increased upon cocaine administration in rats with extended access but not restricted access. MeCP2 expression, in turn, has been shown to negatively regulate the expression of miR-212 and miR-132, which alleviates the miRNA-mediated repression of BDNF.[155] Increased BDNF expression in the dorsal striatum promotes the increased sensitivity to the motivational properties of cocaine self-administration in rat models. Moreover, miR-212 and miR-132 also repress MeCP2 expression by a negative feedback loop.[151,155] Nevertheless, it is unclear at this time how cocaine stimulates MeCP2 expression, how MeCP2 regulates the expression of these miRNAs, and whether there is a recognition element in the MeCP2 transcript for these miRNAs. Delineating the complex interplay between MeCP2 and miR-212/132 in the dorsal striatum, also taking into consideration how other factors influence this fine balance, is imperative prior to assigning a role for miRNAs in mediating the vulnerability to compulsive cocaine-seeking behaviors.

### 2. Nicotine Addiction

Nicotine exposure on miRNA expression has been investigated *in vitro* in a rodent neuronal cell line.[156] Acute nicotine exposure led to the specific upregulation of 11 miRNAs and the downregulation of 14 other miRNAs. Of the differentially expressed miRNAs, several of the target genes regulated by miR-140* are important for neuronal function. Of particular importance is the target gene Dynamin-1 (DNM1), which encodes a large GTPase that plays a central role in synaptic endocytosis. Additionally, by blocking the recruitment of DNM1, LTD in the nucleus accumbens can be prevented.[157]

The upregulation of miR-140* was shown to downregulate the expression of the DNM1 gene, as shown by both a luciferase reporter assay and the measurement of mRNA abundance. Overall, nicotine-induced miR-140*-mediated regulation of DNM1 suggests a potential role for this miRNA in regulating drug-induced neuronal plasticity.

### 3. Ethanol Addiction

miRNA-mediated mechanisms have been implicated in ethanol addiction. Specifically, miR-9 upregulation has been documented in primary neuronal cells in response to ethanol exposure.[158] miR-9, in turn, targets for repression different mRNA splice variants of the main pore-forming $\alpha$ subunit of the BK channel, a high-conductance calcium- and voltage-dependent potassium channel. The various isoforms of the BK channel, which show differing sensitivity to ethanol, enable a neuronal cell to be "plastic" and respond quickly to various environmental cues. Those BK channel splice variants with high ethanol sensitivity were identified to contain the MRE for miR-9 in their respective 3′UTRs. Therefore, they are more likely repressed by an increase in miR-9 expression. In contrast, those variants with decreased sensitivity to ethanol lacked a complementary recognition region for miR-9. Therefore, upon ethanol exposure, the ratio of ethanol-sensitive to ethanol-tolerant variants in neuronal cells shifts toward ethanol-tolerant. The loss of various ethanol-sensitive isoforms has the potential to produce a molecular tolerance for ethanol. Moreover, this mechanism also results in the regulation of additional miR-9 targets relevant to alcohol abuse such as those important for synaptic plasticity, circadian rhythm, and neurotransmitter release. Thus, one can envision a model consisting of miRNA master switches, activated by alcohol, coordinating an integrated and dynamic response pattern among a number of downstream targets.

### 4. Opioid Addiction

Opioid drugs, such as morphine, belong to a powerful class of analgesics and antinociceptives that are used for the treatment of many forms of acute and chronic pain. The $\mu$-opioid receptor (MOR) is the primary receptor that is involved in mediating the analgesic and antinociceptive effects of opioids. Several studies have emerged implicating a role for miRNAs in modulating the function of opioids. One of the earliest evidences showed that the 3′UTR of MOR contains a binding site for miR-23b. Subsequently, miR-23b was shown to bind to this region via a K-box motif, in the process excluding the association of the transcript with polysomes and arresting protein production, without any overt effect on transcript abundance.[159] Moreover, prolonged exposure to morphine was shown to increase the expression of miR-23b in neuronal cells, suggesting the possible occurrence of an autoregulatory feedback mechanism on MOR expression leading to a tolerance for the drug.[160] A similar mechanism for the posttranscriptional regulation of the MOR transcript by let-7 has also been postulated.[161] Let-7, like miR-23b, is positively regulated by morphine and mediates the translocation and sequestering of the MOR transcript to P bodies, leading to translational repression without affecting transcript abundance.[161] Specific agonists of MOR that are known to activate distinct signaling cascades within the realm of opioids have been used to identify differentially expressed miRNAs and their target genes.[162,163] One miRNA identified in this manner, miR-190, located in the intronic regions of the Talin2 gene, showed decreased expression in response to fentanyl and limited response to morphine. Although both opioids activate phosphorylated ERK, fentanyl does so in a β-arrestin 2-dependent manner, leading to different subcellular distribution of the phosphorylated product. The decrease in miR-190 expression, in response to fentanyl, was shown to alleviate the repression of one of its target genes, NeuroD. In turn, NeuroD has the potential to contribute to a wide array of neuronal processes by activating the transcription of several protein products important in adult neurogenesis, axon morphology, dendritic morphology, and synaptogenesis.

## E. Concluding Remarks

Although significant research into miRNAs and their role in gene regulation has been actively going on for several years, it is still considered very much a budding field with numerous unanswered questions. Within the context of neuropathological diseases and disorders, the primary conundrum lies in whether dysfunction of miRNA(s) is the cause or the consequence of the pathological state. Furthermore, another layer of complexity has been added as a result of the involvement of miRNAs in brain tumor development and progression, where they have been shown to be fully integrated into many of the key pathways of tumor pathogenesis. At this conjuncture, it is quite premature, and also naive, to impose only one set of restrictions to miRNAs. Nevertheless, emerging evidence linking miRNAs to diseases and disorders of the CNS advocates the targeting of these putative molecules as potential points of therapeutic intervention through targeted *in vivo* overexpression or knockdown/knockout studies. A summary of the miRNAs implicated in CNS diseases and disorders is presented in Table III. Undoubtedly, additional research within this field will provide further insights into how these essential molecules contribute to disease pathobiology in the CNS.

TABLE III

LIST OF MIRNAS IMPLICATED IN NEUROPATHOLOGICAL DISEASES AND THEIR MRNA TARGETS

| Neuropathological state | miRNA(s) Identified | Potential target gene(s) identified and other features that may contribute to disease/disorder etiology | References |
|---|---|---|---|
| Neurodegenerative diseases | | | |
| Alzheimer's disease (AD) | miR-9↑, miR-128↑ | Wide array of CNS genes are targets | 103 |
| | miR-107↓ | Identified in early stages of AD. BACE1 mRNA is a target gene. | 104 |
| | miR-29 family↓ | Identified in sporadic AD. BACE1 mRNA is a target gene | 105 |
| | miR-298↑, miR-328↑ | Expressed in hippocampus of mouse model of AD. BACE1 mRNA is a target gene | 106 |
| | miR-101 | APP and Cox2 mRNA are target genes | 107 |
| | miR-20a family | APP mRNA is a target gene. MiR-106b, from the family, downregulated in sporadic AD but does not correspond to increased Aβ load | 108 |
| | miR-34a↑ | Cerebral cortex of a mouse model of AD. BCL2 mRNA is a target gene | 109 |
| | miR-106b↑ | TβR II mRNA is a target gene | 110 |
| | miR-146a↑ | CHF mRNA is a target gene. Also, operates in a negative feedback-loop to repress NF-κB which has the potential to dampen the innate immune response | 103,112,113 |
| | miR-342-3p↑; miR-128↑ | MiR-342-3p common to several models of Prion disease. MiR-128↑ also upregulated in several Alzheimer's studies | 114,115 |
| Parkinson's disease (PD) | miR-133b↓ | Pitx3 mRNA is a target gene. PITX3 also operates in a negative feedback-loop on the expression of miR-133b | 10 |
| | miR-433 | A mutation in the 3′UTR of the Fgf20 gene disrupts the miRNA binding site | 117 |
| | miR-184*↓, let-7↓ | Target E2f1 and Dp genes. Overexpression of these two genes in postmitotic neurons leads to abortive cell division and eventually cell death | 118 |

(*Continues*)

TABLE III (*Continued*)

| Neuropathological state | miRNA(s) Identified | Potential target gene(s) identified and other features that may contribute to disease/disorder etiology | References |
|---|---|---|---|
| Huntington's disease (HD) | miR-9/9*↓, miR-29a/b↓, miR-124a↓, miR-132↓, miR-330↓, miR-17↓, miR-196a↓, miR-222↓, miR-485↓, miR-486↓ | Several putative target genes including p250GAP, REST, coREST | 37,119 |
| Amyotrophic lateral sclerosis (ALS) | miR-206↑ | Targets HDAC4 mRNA | 121 |
| Spinal motor neuron disease (SMN) | miR-9↓ | Targets NEFH mRNA | 122 |
| Neuropsychiatric disorders | | | |
| Schizophrenia (SZ) | miR-130b | Targets MeCP2 mRNA. Sequence variation identified in the 5′ upstream region of this gene which may affect transcription factor binding site | 123 |
| | miR-346↓ | Located in the intron of GRID1. Targets a large number of SZ related genes | 140 |
| | let-7f-2, miR-188, miR-18b/18b*, miR-502, miR-510, miR-660, miR-325, miR-505, miR-509-3 | Many of these miRNAs target the mRNA sequences of NRG1, DISC1 and RGS4 | 125 |
| | miR-206, miR-198 | Overrepresentation of a nucleotide variation in miR-206 and underrepresentation of a nucleotide variation in miR-198 in SZ subjects of European ancestry | 126 |
| | miR-30e | A nucleotide variation in the pre-miR form can affect the processing of this miRNA in SZ subjects of Chinese ancestry | 127,128 |
| | miR-124, miR-383, miR-320, miR-596, miR-597, miR-598 | These miRNAs are located on a SZ associated "hot-spot" on 8p21–23 locus | 129 |
| | miR-26b↓, miR-30b↓, miR-29b↓, miR-106b↓ | Targets genes are primarily involved in synaptic functions | 130 |
| | let-7g↑, miR-181b↑ | Targets genes are primarily involved in synaptic functions | 131 |
| | miR-181b, miR-219, members of the miR-15 family of miRNAs | Members of the miR-15a family of miRNAs have an enrichment for target genes associated with synaptic function | 40 |
| | miR-34a, miR-132*, miR-132, miR-212, miR-544, miR-7, miR-154* | One of the largest SZ case studies conducted. MiR-7, miR-132 and miR-212 have been previously implicated in the disease process | 132 |

| | | | |
|---|---|---|---|
| Tourette's syndrome (TS) | miR-189 | Targets SLITRK1 gene. Although, follow up studies have failed to segregate the mutation with TS patients | 133 |
| Autism spectrum disorders | ~9 miRNAs | Many of the deregulated miRNAs target the SHANK3 and NRXN1 mRNA | 134 |
| Mood disorders | miR-7, miR-154*, miR-504, miR-454*, miR-29a, miR-520c-3p, miR-140-3p, miR-145*, miR-767-5p, miR-22*, miR-145, miR-874, miR-133b, miR-154*, miR-32, miR-573, miR-889 | Many of the BPD-related miRNAs correlate with those miRNAs identified to be differentially expressed in SZ | 132 |
| | miR-134 | Levels of this miRNA are lower in BPD cases and increases upon drug treatment | 137 |
| | miR-485-3p, miR-765, miR-509, miR-128 | Variations in the binding site for these miRNAs in the 3′UTR of the mRNA of NTRK3 have been identified. Implications in OCD | 138 |
| | miR-96 | Targets HTR1B mRNA. Implicated in aggression-related behavior | 139 |
| Antipsychotics and psychotomimetics | miR-34a↓ | This miRNA is downregulated in response to lithium and valproic acid. Targets the GRM7 mRNA | 140 |
| | miR-219↓ | This miRNA is downregulated in response to an NMDA antagonist. Targets the CAMKIIγ mRNA | 142 |
| Neurodevelopmental disorders | | | |
| Down syndrome | miR-99a↓, let-7c↓, miR-125b-2↓, miR-155↓, miR-802↓ | Both miR-155 and miR-802 target the MeCP2 gene | 143,144 |
| Fragile X syndrome | miR-125b, miR-132 | Found to be associated with FMRP. MiR-125b can target the NR2A mRNA | 67 |
| Rett syndrome | miR-132 | Targets the gene for MeCP2 via a homeostatic mechanism involving BDNF | 151 |
| Di George syndrome | General miRNA abundance | Overall decrease in miRNA abundance as a consequence of the deletion of Dgcr8 | 101 |

(*Continues*)

TABLE III (*Continued*)

| Neuropathological state | miRNA(s) Identified | Potential target gene(s) identified and other features that may contribute to disease/disorder etiology | References |
|---|---|---|---|
| Addiction | | | |
| Cocaine addiction | miR-181a↑, miR-124↓ let-7d↓ | Regulate various target genes leading to alteration in neuroplasticity in response to cocaine | 152 |
| | miR-476a, miR-544, miR-154, miR-467b, miR-500, miR-186, miR-337-3p, miR-138, miR-369-3p, miR-136, miR-7a, miR-137, miR-324-5p, miR-665, miR-380-3p, miR-376a, miR-130a, miR-301b, miR-148b, miR-488, miR-384-5p, miR-376c, miR-181a | Downregulated in DRD2 expressing striatal neurons in which AGO2 is knocked out. These miRNAs are also enriched and/or upregulated in DRD2-expressing neurons in response to cocaine treatment | 153 |
| | miR-212↑ (miR-132↑) | Decreases motivational response to cocaine in rats with extended access to the drug by amplifying CREB-TORC activity through the targeting of the gene encoding SPRED1 | 154 |
| | miR-212↓ (miR-132↓) | Regulate and are also regulated by MeCP2. Decrease is believed to increase the motivational response to cocaine in rats with extended access | 155 |
| Nicotine addiction | miR-140*↑ | Targets DNM1 mRNA | 156 |
| Ethanol addiction | miR-9↑ | Targets the transcripts encoding BK channel splice variants with high ethanol sensitivity | 158 |
| Opioid addiction | miR-23b↑ | Targets the μ-opioid receptor (MOR) gene | 159,160 |
| | miR-23b↑, let-7↑ | Targets the MOR gene | 161 |
| | miR-190↓ | Located in the intron of Talin2, targets NeuroD mRNA. Responsive to MOR against fentanyl | 162,163 |

## References


1. Kriegstein A, Alvarez-Buylla A. The glial nature of embryonic and adult neural stem cells. *Annu Rev Neurosci* 2009;**32**:149–84. doi:10.1146/annurev.neuro.051508.135600.
2. Lee RC, Feinbaum RL, Ambros V. The *C. elegans* heterochronic gene lin-4 encodes small RNAs with antisense complementarity to lin-14. *Cell* 1993;**75**:843–54.
3. Giraldez AJ, Cinalli RM, Glasner ME, Enright AJ, Thomson JM, Baskerville S, et al. MicroRNAs regulate brain morphogenesis in zebrafish. *Science* 2005;**308**:833–8. doi:10.1126/science.1109020.
4. Bernstein E, Kim SY, Carmell MA, Murchison EP, Alcorn H, Li MZ, et al. Dicer is essential for mouse development. *Nat Genet* 2003;**35**:215–7. doi:10.1038/ng1253.
5. De Pietri Tonelli D, Pulvers JN, Haffner C, Murchison EP, Hannon GJ, Huttner WB. miRNAs are essential for survival and differentiation of newborn neurons but not for expansion of neural progenitors during early neurogenesis in the mouse embryonic neocortex. *Development* 2008;**135**:3911–21. doi:10.1242/dev.025080.
6. Kawase-Koga Y, Low R, Otaegi G, Pollock A, Deng H, Eisenhaber F, et al. RNAase-III enzyme Dicer maintains signaling pathways for differentiation and survival in mouse cortical neural stem cells. *J Cell Sci* 2010;**123**:586–94. doi:10.1242/jcs.059659.
7. Choi PS, Zakhary L, Choi WY, Caron S, Alvarez-Saavedra E, Miska EA, et al. Members of the miRNA-200 family regulate olfactory neurogenesis. *Neuron* 2008;**57**:41–55. doi:10.1016/j.neuron.2007.11.018.
8. Bartel DP. MicroRNAs: genomics, biogenesis, mechanism, and function. *Cell* 2004;**116**:281–97.
9. Davis TH, Cuellar TL, Koch SM, Barker AJ, Harfe BD, McManus MT, et al. Conditional loss of Dicer disrupts cellular and tissue morphogenesis in the cortex and hippocampus. *J Neurosci* 2008;**28**:4322–30. doi:10.1523/JNEUROSCI.4815-07.2008.
10. Kim J, Inoue K, Ishii J, Vanti WB, Voronov SV, Murchison E, et al. A microRNA feedback circuit in midbrain dopamine neurons. *Science* 2007;**317**:1220–4. doi:10.1126/science.1140481.
11. Schaefer A, O'Carroll D, Tan CL, Hillman D, Sugimori M, Llinas R, et al. Cerebellar neurodegeneration in the absence of microRNAs. *J Exp Med* 2007;**204**:1553–8. doi:10.1084/jem.20070823.
12. Cuellar TL, Davis TH, Nelson PT, Loeb GB, Harfe BD, Ullian E, et al. Dicer loss in striatal neurons produces behavioral and neuroanatomical phenotypes in the absence of neurodegeneration. *Proc Natl Acad Sci USA* 2008;**105**:5614–9. doi:10.1073/pnas.0801689105.
13. Taft RJ, Glazov EA, Lassmann T, Hayashizaki Y, Carninci P, Mattick JS. Small RNAs derived from snoRNAs. *RNA* 2009;**15**:1233–40. doi:10.1261/rna.1528909.
14. Okamura K, Lai EC. Endogenous small interfering RNAs in animals. *Nat Rev Mol Cell Biol* 2008;**9**:673–8. doi:10.1038/nrm2479.
15. Babiarz JE, Ruby JG, Wang Y, Bartel DP, Blelloch R. Mouse ES cells express endogenous shRNAs, siRNAs, and other microprocessor-independent, Dicer-dependent small RNAs. *Genes Dev* 2008;**22**:2773–85. doi:10.1101/gad.1705308.
16. Wang Y, Medvid R, Melton C, Jaenisch R, Blelloch R. DGCR8 is essential for microRNA biogenesis and silencing of embryonic stem cell self-renewal. *Nat Genet* 2007;**39**:380–5. doi:10.1038/ng1969.
17. Suh N, Baehner L, Moltzahn F, Melton C, Shenoy A, Chen J, et al. MicroRNA function is globally suppressed in mouse oocytes and early embryos. *Curr Biol* 2010;**20**:271–7. doi:10.1016/j.cub.2009.12.044.

18. Yi R, Pasolli HA, Landthaler M, Hafner M, Ojo T, Sheridan R, et al. DGCR8-dependent microRNA biogenesis is essential for skin development. *Proc Natl Acad Sci USA* 2009;**106**:498–502. doi:10.1073/pnas.0810766105.
19. Rao PK, Toyama Y, Chiang HR, Gupta S, Bauer M, Medvid R, et al. Loss of cardiac microRNA-mediated regulation leads to dilated cardiomyopathy and heart failure. *Circ Res* 2009;**105**:585–94. doi:10.1161/CIRCRESAHA.109.200451.
20. Chong MM, Rasmussen JP, Rudensky AY, Littman DR. The RNAseIII enzyme Drosha is critical in T cells for preventing lethal inflammatory disease. *J Exp Med* 2008;**205**:2005–17. doi:10.1084/jem.20081219.
21. Kapsimali M, Kloosterman WP, de Bruijn E, Rosa F, Plasterk RH, Wilson SW. MicroRNAs show a wide diversity of expression profiles in the developing and mature central nervous system. *Genome Biol* 2007;**8**:R173. doi:10.1186/gb-2007-8-8-r173.
22. Nielsen JA, Lau P, Maric D, Barker JL, Hudson LD. Integrating microRNA and mRNA expression profiles of neuronal progenitors to identify regulatory networks underlying the onset of cortical neurogenesis. *BMC Neurosci* 2009;**10**:98. doi:10.1186/1471-2202-10-98.
23. Goff LA, Davila J, Swerdel MR, Moore JC, Cohen RI, Wu H, et al. Ago2 immunoprecipitation identifies predicted microRNAs in human embryonic stem cells and neural precursors. *PLoS ONE* 2009;**4**:e7192. doi:10.1371/journal.pone.0007192.
24. Krichevsky AM, King KS, Donahue CP, Khrapko K, Kosik KS. A microRNA array reveals extensive regulation of microRNAs during brain development. *RNA* 2003;**9**:1274–81.
25. Sempere LF, Freemantle S, Pitha-Rowe I, Moss E, Dmitrovsky E, Ambros V. Expression profiling of mammalian microRNAs uncovers a subset of brain-expressed microRNAs with possible roles in murine and human neuronal differentiation. *Genome Biol* 2004;**5**:R13. doi:10.1186/gb-2004-5-3-r13.
26. Rybak A, Fuchs H, Smirnova L, Brandt C, Pohl EE, Nitsch R, et al. A feedback loop comprising lin-28 and let-7 controls pre-let-7 maturation during neural stem-cell commitment. *Nat Cell Biol* 2008;**10**:987–93. doi:10.1038/ncb1759.
27. Smirnova L, Grafe A, Seiler A, Schumacher S, Nitsch R, Wulczyn FG. Regulation of miRNA expression during neural cell specification. *Eur J Neurosci* 2005;**21**:1469–77. doi:10.1111/j.1460-9568.2005.03978.x.
28. Johnston RJ, Hobert O. A microRNA controlling left/right neuronal asymmetry in *Caenorhabditis elegans*. *Nature* 2003;**426**:845–9. doi:10.1038/nature02255.
29. Nishino J, Kim I, Chada K, Morrison SJ. Hmga2 promotes neural stem cell self-renewal in young but not old mice by reducing p16Ink4a and p19Arf expression. *Cell* 2008;**135**:227–39. doi:10.1016/j.cell.2008.09.017.
30. Zhao C, Sun G, Li S, Lang MF, Yang S, Li W, et al. MicroRNA let-7b regulates neural stem cell proliferation and differentiation by targeting nuclear receptor TLX signaling. *Proc Natl Acad Sci USA* 2010;**107**:1876–81. doi:10.1073/pnas.0908750107.
31. Liang X, Stahlhut C, DelConte C, Stefani G, Slack FJ. The let-7 microRNA target gene, Mlin41/Trim71 is required for mouse embryonic survival and neural tube closure. *Cell Cycle* 2008;**7**:3935–42.
32. Li X, Carthew RW. A microRNA mediates EGF receptor signaling and promotes photoreceptor differentiation in the Drosophila eye. *Cell* 2005;**123**:1267–77. doi:10.1016/j.cell.2005.10.040.
33. Li Y, Wang F, Lee JA, Gao FB. MicroRNA-9a ensures the precise specification of sensory organ precursors in Drosophila. *Genes Dev* 2006;**20**:2793–805. doi:10.1101/gad.1466306.
34. Leucht C, Stigloher C, Wizenmann A, Klafke R, Folchert A, Bally-Cuif L. MicroRNA-9 directs late organizer activity of the midbrain–hindbrain boundary. *Nat Neurosci* 2008;**11**:641–8. doi:10.1038/nn.2115.

35. Shibata M, Kurokawa D, Nakao H, Ohmura T, Aizawa S. MicroRNA-9 modulates Cajal-Retzius cell differentiation by suppressing Foxg1 expression in mouse medial pallium. *J Neurosci* 2008;**28**:10415–21. doi:10.1523/JNEUROSCI.3219-08.2008.
36. Zhao C, Sun G, Li S, Shi Y. A feedback regulatory loop involving microRNA-9 and nuclear receptor TLX in neural stem cell fate determination. *Nat Struct Mol Biol* 2009;**16**:365–71. doi:10.1038/nsmb.1576.
37. Packer AN, Xing Y, Harper SQ, Jones L, Davidson BL. The bifunctional microRNA miR-9/miR-9* regulates REST and CoREST and is downregulated in Huntington's disease. *J Neurosci* 2008;**28**:14341–6. doi:10.1523/JNEUROSCI.2390-08.2008.
38. Delaloy C, Liu L, Lee JA, Su H, Shen F, Yang GY, et al. MicroRNA-9 coordinates proliferation and migration of human embryonic stem cell-derived neural progenitors. *Cell Stem Cell* 2010;**6**:323–35. doi:10.1016/j.stem.2010.02.015.
39. Yoo AS, Staahl BT, Chen L, Crabtree GR. MicroRNA-mediated switching of chromatin-remodelling complexes in neural development. *Nature* 2009;**460**:642–6. doi:10.1038/nature08139.
40. Beveridge NJ, Tooney PA, Carroll AP, Tran N, Cairns MJ. Down-regulation of miR-17 family expression in response to retinoic acid induced neuronal differentiation. *Cell Signal* 2009;**21**:1837–45. doi:10.1016/j.cellsig.2009.07.019.
41. Visvanathan J, Lee S, Lee B, Lee JW, Lee SK. The microRNA miR-124 antagonizes the anti-neural REST/SCP1 pathway during embryonic CNS development. *Genes Dev* 2007;**21**:744–9. doi:10.1101/gad.1519107.
42. Makeyev EV, Zhang J, Carrasco MA, Maniatis T. The MicroRNA miR-124 promotes neuronal differentiation by triggering brain-specific alternative pre-mRNA splicing. *Mol Cell* 2007;**27**:435–48. doi:10.1016/j.molcel.2007.07.015.
43. Cheng LC, Pastrana E, Tavazoie M, Doetsch F. miR-124 regulates adult neurogenesis in the subventricular zone stem cell niche. *Nat Neurosci* 2009;**12**:399–408. doi:10.1038/nn.2294.
44. Ferretti E, De Smaele E, Miele E, Laneve P, Po A, Pelloni M, et al. Concerted microRNA control of Hedgehog signalling in cerebellar neuronal progenitor and tumour cells. *EMBO J* 2008;**27**:2616–27. doi:10.1038/emboj.2008.172.
45. Godlewski J, Nowicki MO, Bronisz A, Williams S, Otsuki A, Nuovo G, et al. Targeting of the Bmi-1 oncogene/stem cell renewal factor by microRNA-128 inhibits glioma proliferation and self-renewal. *Cancer Res* 2008;**68**:9125–30. doi:10.1158/0008-5472.CAN-08-2629.
46. Lehotzky A, Lau P, Tokesi N, Muja N, Hudson LD, Ovadi J. Tubulin polymerization-promoting protein (TPPP/p25) is critical for oligodendrocyte differentiation. *Glia* 2010;**58**:157–68. doi:10.1002/glia.20909.
47. Zhao X, He X, Han X, Yu Y, Ye F, Chen Y, et al. MicroRNA-mediated control of oligodendrocyte differentiation. *Neuron* 2010;**65**:612–26. doi:10.1016/j.neuron.2010.02.018.
48. Dugas JC, Cuellar TL, Scholze A, Ason B, Ibrahim A, Emery B, et al. Dicer1 and miR-219 Are required for normal oligodendrocyte differentiation and myelination. *Neuron* 2010;**65**:597–611. doi:10.1016/j.neuron.2010.01.027.
49. Cayirlioglu P, Kadow IG, Zhan X, Okamura K, Suh GS, Gunning D, et al. Hybrid neurons in a microRNA mutant are putative evolutionary intermediates in insect CO2 sensory systems. *Science* 2008;**319**:1256–60. doi:10.1126/science.1149483.
50. Roush S, Slack FJ. The let-7 family of microRNAs. *Trends Cell Biol* 2008;**18**:505–16. doi:10.1016/j.tcb.2008.07.007.
51. Wulczyn FG, Smirnova L, Rybak A, Brandt C, Kwidzinski E, Ninnemann O, et al. Post-transcriptional regulation of the let-7 microRNA during neural cell specification. *FASEB J* 2007;**21**:415–26. doi:10.1096/fj.06-6130com.

52. Schwamborn JC, Berezikov E, Knoblich JA. The TRIM-NHL protein TRIM32 activates microRNAs and prevents self-renewal in mouse neural progenitors. *Cell* 2009;**136**:913–25. doi:10.1016/j.cell.2008.12.024.
53. Newman MA, Thomson JM, Hammond SM. Lin-28 interaction with the Let-7 precursor loop mediates regulated microRNA processing. *RNA* 2008;**14**:1539–49. doi:10.1261/rna.1155108.
54. Viswanathan SR, Daley GQ, Gregory RI. Selective blockade of microRNA processing by Lin28. *Science* 2008;**320**:97–100. doi:10.1126/science.1154040.
55. Heo I, Joo C, Cho J, Ha M, Han J, Kim VN. Lin28 mediates the terminal uridylation of let-7 precursor MicroRNA. *Mol Cell* 2008;**32**:276–84. doi:10.1016/j.molcel.2008.09.014.
56. Lim LP, Lau NC, Garrett-Engele P, Grimson A, Schelter JM, Castle J, et al. Microarray analysis shows that some microRNAs downregulate large numbers of target mRNAs. *Nature* 2005;**433**:769–73. doi:10.1038/nature03315.
57. Conaco C, Otto S, Han JJ, Mandel G. Reciprocal actions of REST and a microRNA promote neuronal identity. *Proc Natl Acad Sci USA* 2006;**103**:2422–7. doi:10.1073/pnas.0511041103.
58. Maiorano NA, Mallamaci A. Promotion of embryonic cortico-cerebral neuronogenesis by miR-124. *Neural Dev* 2009;**4**:40. doi:10.1186/1749-8104-4-40.
59. Chang S, Johnston Jr. RJ, Frokjaer-Jensen C, Lockery S, Hobert O. MicroRNAs act sequentially and asymmetrically to control chemosensory laterality in the nematode. *Nature* 2004;**430**:785–9. doi:10.1038/nature02752.
60. Lau P, Verrier JD, Nielsen JA, Johnson KR, Notterpek L, Hudson LD. Identification of dynamically regulated microRNA and mRNA networks in developing oligodendrocytes. *J Neurosci* 2008;**28**:11720–30. doi:10.1523/JNEUROSCI.1932-08.2008.
61. Kawase-Koga Y, Otaegi G, Sun T. Different timings of Dicer deletion affect neurogenesis and gliogenesis in the developing mouse central nervous system. *Dev Dyn* 2009;**238**:2800–12. doi:10.1002/dvdy.22109.
62. Shin D, Shin JY, McManus MT, Ptacek LJ, Fu YH. Dicer ablation in oligodendrocytes provokes neuronal impairment in mice. *Ann Neurol* 2009;**66**:843–57. doi:10.1002/ana.21927.
63. Zheng K, Li H, Zhu Y, Zhu Q, Qiu M. MicroRNAs are essential for the developmental switch from neurogenesis to gliogenesis in the developing spinal cord. *J Neurosci* 2010;**30**:8245–50. doi:10.1523/JNEUROSCI.1169-10.2010.
64. Vo N, Klein ME, Varlamova O, Keller DM, Yamamoto T, Goodman RH, et al. A cAMP-response element binding protein-induced microRNA regulates neuronal morphogenesis. *Proc Natl Acad Sci USA* 2005;**102**:16426–31.
65. Wayman GA, Davare M, Ando H, Fortin D, Varlamova O, Cheng HY, et al. An activity-regulated microRNA controls dendritic plasticity by down-regulating p250GAP. *Proc Natl Acad Sci USA* 2008;**105**:9093–8. doi:10.1073/pnas.0803072105.
66. Yu JY, Chung KH, Deo M, Thompson RC, Turner DL. MicroRNA miR-124 regulates neurite outgrowth during neuronal differentiation. *Exp Cell Res* 2008;**314**:2618–33. doi:10.1016/j.yexcr.2008.06.002.
67. Edbauer D, Neilson JR, Foster KA, Wang CF, Seeburg DP, Batterton MN, et al. Regulation of synaptic structure and function by FMRP-associated microRNAs miR-125b and miR-132. *Neuron* 2010;**65**:373–84. doi:10.1016/j.neuron.2010.01.005.
68. Fiore R, Khudayberdiev S, Christensen M, Siegel G, Flavell SW, Kim TK, et al. Mef2-mediated transcription of the miR379-410 cluster regulates activity-dependent dendritogenesis by fine-tuning Pumilio2 protein levels. *EMBO J* 2009;**28**:697–710. doi:10.1038/emboj.2009.10.
69. Smrt RD, Szulwach KE, Pfeiffer RL, Li X, Guo W, Pathania M, et al. MicroRNA miR-137 regulates neuronal maturation by targeting ubiquitin ligase mind bomb-1. *Stem Cells* 2010;**28**:1060–70. doi:10.1002/stem.431.

70. Parrish JZ, Xu P, Kim CC, Jan LY, Jan YN. The microRNA bantam functions in epithelial cells to regulate scaling growth of dendrite arbors in drosophila sensory neurons. *Neuron* 2009;**63**:788–802. doi:10.1016/j.neuron.2009.08.006.
71. Caygill EE, Johnston LA. Temporal regulation of metamorphic processes in Drosophila by the let-7 and miR-125 heterochronic microRNAs. *Curr Biol* 2008;**18**:943–50. doi:10.1016/j.cub.2008.06.020.
72. Schratt GM, Tuebing F, Nigh EA, Kane CG, Sabatini ME, Kiebler M, et al. A brain-specific microRNA regulates dendritic spine development. *Nature* 2006;**439**:283–9. doi:10.1038/nature04367.
73. Siegel G, Obernosterer G, Fiore R, Oehmen M, Bicker S, Christensen M, et al. A functional screen implicates microRNA-138-dependent regulation of the depalmitoylation enzyme APT1 in dendritic spine morphogenesis. *Nat Cell Biol* 2009;**11**:705–16. doi:10.1038/ncb1876.
74. Impey S, Davare M, Lasiek A, Fortin D, Ando H, Varlamova O, et al. An activity-induced microRNA controls dendritic spine formation by regulating Rac1-PAK signaling. *Mol Cell Neurosci* 2010;**43**:146–56. doi:10.1016/j.mcn.2009.10.005.
75. Gao J, Wang WY, Mao YW, Graff J, Guan JS, Pan L, et al. A novel pathway regulates memory and plasticity via SIRT1 and miR-134. *Nature* 2010;**466**:1105–9. doi:10.1038/nature09271.
76. Cheng HY, Papp JW, Varlamova O, Dziema H, Russell B, Curfman JP, et al. microRNA modulation of circadian-clock period and entrainment. *Neuron* 2007;**54**:813–29. doi:10.1016/j.neuron.2007.05.017.
77. Karr J, Vagin V, Chen K, Ganesan S, Olenkina O, Gvozdev V, et al. Regulation of glutamate receptor subunit availability by microRNAs. *J Cell Biol* 2009;**185**:685–97. doi:10.1083/jcb.200902062.
78. Simon DJ, Madison JM, Conery AL, Thompson-Peer KL, Soskis M, Ruvkun GB, et al. The microRNA miR-1 regulates a MEF-2-dependent retrograde signal at neuromuscular junctions. *Cell* 2008;**133**:903–15. doi:10.1016/j.cell.2008.04.035.
79. Sutton MA, Schuman EM. Dendritic protein synthesis, synaptic plasticity, and memory. *Cell* 2006;**127**:49–58. doi:10.1016/j.cell.2006.09.014.
80. Kim J, Krichevsky A, Grad Y, Hayes GD, Kosik KS, Church GM, et al. Identification of many microRNAs that copurify with polyribosomes in mammalian neurons. *Proc Natl Acad Sci USA* 2004;**101**:360–5. doi:10.1073/pnas.2333854100.
81. Lugli G, Larson J, Martone ME, Jones Y, Smalheiser NR. Dicer and eIF2c are enriched at postsynaptic densities in adult mouse brain and are modified by neuronal activity in a calpain-dependent manner. *J Neurochem* 2005;**94**:896–905. doi:10.1111/j.1471-4159.2005.03224.x.
82. Ashraf SI, McLoon AL, Sclarsic SM, Kunes S. Synaptic protein synthesis associated with memory is regulated by the RISC pathway in Drosophila. *Cell* 2006;**124**:191–205. doi:10.1016/j.cell.2005.12.017.
83. Banerjee S, Neveu P, Kosik KS. A coordinated local translational control point at the synapse involving relief from silencing and MOV10 degradation. *Neuron* 2009;**64**:871–84. doi:10.1016/j.neuron.2009.11.023.
84. Barbee SA, Estes PS, Cziko AM, Hillebrand J, Luedeman RA, Coller JM, et al. Staufen- and FMRP-containing neuronal RNPs are structurally and functionally related to somatic P bodies. *Neuron* 2006;**52**:997–1009. doi:10.1016/j.neuron.2006.10.028.
85. Kye MJ, Liu T, Levy SF, Xu NL, Groves BB, Bonneau R, et al. Somatodendritic microRNAs identified by laser capture and multiplex RT-PCR. *RNA* 2007;**13**:1224–34. doi:10.1261/rna.480407.
86. Lugli G, Torvik VI, Larson J, Smalheiser NR. Expression of microRNAs and their precursors in synaptic fractions of adult mouse forebrain. *J Neurochem* 2008;**106**:650–61. doi:10.1111/j.1471-4159.2008.05413.x.

87. Bicker S, Schratt G. microRNAs: tiny regulators of synapse function in development and disease. *J Cell Mol Med* 2008;**12**:1466–76. doi:10.1111/j.1582-4934.2008.00400.x.
88. Kosik KS. The neuronal microRNA system. *Nat Rev Neurosci* 2006;**7**:911–20. doi:10.1038/nrn2037.
89. Cougot N, Bhattacharyya SN, Tapia-Arancibia L, Bordonne R, Filipowicz W, Bertrand E, et al. Dendrites of mammalian neurons contain specialized P-body-like structures that respond to neuronal activation. *J Neurosci* 2008;**28**:13793–804. doi:10.1523/JNEUROSCI.4155-08.2008.
90. Berdnik D, Fan AP, Potter CJ, Luo L. MicroRNA processing pathway regulates olfactory neuron morphogenesis. *Curr Biol* 2008;**18**:1754–9. doi:10.1016/j.cub.2008.09.045.
91. Pinter R, Hindges R. Perturbations of microRNA function in mouse dicer mutants produce retinal defects and lead to aberrant axon pathfinding at the optic chiasm. *PLoS ONE* 2010;**5**: e10021. doi:10.1371/journal.pone.0010021.
92. Rajasethupathy P, Fiumara F, Sheridan R, Betel D, Puthanveettil SV, Russo JJ, et al. Characterization of small RNAs in aplysia reveals a role for miR-124 in constraining synaptic plasticity through CREB. *Neuron* 2009;**63**:803–17. doi:10.1016/j.neuron.2009.05.029.
93. Natera-Naranjo O, Aschrafi A, Gioio AE, Kaplan BB. Identification and quantitative analyses of microRNAs located in the distal axons of sympathetic neurons. *RNA* 2010;**16**:1516–29. doi:10.1261/rna.1833310.
94. Aschrafi A, Schwechter AD, Mameza MG, Natera-Naranjo O, Gioio AE, Kaplan BB. MicroRNA-338 regulates local cytochrome c oxidase IV mRNA levels and oxidative phosphorylation in the axons of sympathetic neurons. *J Neurosci* 2008;**28**:12581–90. doi:10.1523/JNEUROSCI.3338-08.2008.
95. Flavell SW, Greenberg ME. Signaling mechanisms linking neuronal activity to gene expression and plasticity of the nervous system. *Annu Rev Neurosci* 2008;**31**:563–90. doi:10.1146/annurev.neuro.31.060407.125631.
96. Siomi H, Siomi MC. Posttranscriptional regulation of microRNA biogenesis in animals. *Mol Cell* 2010;**38**:323–32. doi:10.1016/j.molcel.2010.03.013.
97. Kawahara Y, Zinshteyn B, Sethupathy P, Iizasa H, Hatzigeorgiou AG, Nishikura K. Redirection of silencing targets by adenosine-to-inosine editing of miRNAs. *Science* 2007;**315**:1137–40. doi:10.1126/science.1138050.
98. Obernosterer G, Leuschner PJ, Alenius M, Martinez J. Post-transcriptional regulation of microRNA expression. *RNA* 2006;**12**:1161–7. doi:10.1261/rna.2322506.
99. Krol J, Busskamp V, Markiewicz I, Stadler MB, Ribi S, Richter J, et al. Characterizing light-regulated retinal microRNAs reveals rapid turnover as a common property of neuronal microRNAs. *Cell* 2010;**141**:618–31. doi:10.1016/j.cell.2010.03.039.
100. Xu XL, Li Y, Wang F, Gao FB. The steady-state level of the nervous-system-specific microRNA-124a is regulated by dFMR1 in Drosophila. *J Neurosci* 2008;**28**:11883–9. doi:10.1523/JNEUROSCI.4114-08.2008.
101. Stark KL, Xu B, Bagchi A, Lai WS, Liu H, Hsu R, et al. Altered brain microRNA biogenesis contributes to phenotypic deficits in a 22q11-deletion mouse model. *Nat Genet* 2008;**40**:751–60. doi:10.1038/ng.138.
102. Jin P, Zarnescu DC, Ceman S, Nakamoto M, Mowrey J, Jongens TA, et al. Biochemical and genetic interaction between the fragile X mental retardation protein and the microRNA pathway. *Nat Neurosci* 2004;**7**:113–7. doi:10.1038/nn1174.
103. Lukiw WJ. Micro-RNA speciation in fetal, adult and Alzheimer's disease hippocampus. *Neuroreport* 2007;**18**:297–300. doi:10.1097/WNR.0b013e3280148e8b.
104. Wang WX, Rajeev BW, Stromberg AJ, Ren N, Tang G, Huang Q, et al. The expression of microRNA miR-107 decreases early in Alzheimer's disease and may accelerate disease

progression through regulation of beta-site amyloid precursor protein-cleaving enzyme 1. *J Neurosci* 2008;**28**:1213–23. doi:10.1523/JNEUROSCI.5065-07.2008.
105. Hebert SS, Horre K, Nicolai L, Papadopoulou AS, Mandemakers W, Silahtaroglu AN, et al. Loss of microRNA cluster miR-29a/b-1 in sporadic Alzheimer's disease correlates with increased BACE1/beta-secretase expression. *Proc Natl Acad Sci USA* 2008;**105**:6415–20. doi:10.1073/pnas.0710263105.
106. Boissonneault V, Plante I, Rivest S, Provost P. MicroRNA-298 and microRNA-328 regulate expression of mouse beta-amyloid precursor protein-converting enzyme 1. *J Biol Chem* 2009;**284**:1971–81. doi:10.1074/jbc.M807530200.
107. Vilardo E, Barbato C, Ciotti M, Cogoni C, Ruberti F. MicroRNA-101 regulates amyloid precursor protein expression in hippocampal neurons. *J Biol Chem* 2010;**285**:18344–51. doi:10.1074/jbc.M110.112664.
108. Hebert SS, Horre K, Nicolai L, Bergmans B, Papadopoulou AS, Delacourte A, et al. MicroRNA regulation of Alzheimer's Amyloid precursor protein expression. *Neurobiol Dis* 2009;**33**:422–8. doi:10.1016/j.nbd.2008.11.009.
109. Wang X, Liu P, Zhu H, Xu Y, Ma C, Dai X, et al. miR-34a, a microRNA up-regulated in a double transgenic mouse model of Alzheimer's disease, inhibits bcl2 translation. *Brain Res Bull* 2009;**80**:268–73. doi:10.1016/j.brainresbull.2009.08.006.
110. Wang H, Liu J, Zong Y, Xu Y, Deng W, Zhu H, et al. miR-106b Aberrantly expressed in a double transgenic mouse model for Alzheimer's disease targets TGF-beta type II receptor. *Brain Res* 2010;doi:10.1016/j.brainres.2010.08.023.
111. Lukiw WJ, Zhao Y, Cui JG. An NF-kappaB-sensitive micro RNA-146a-mediated inflammatory circuit in Alzheimer disease and in stressed human brain cells. *J Biol Chem* 2008;**283**:31315–22. doi:10.1074/jbc.M805371200.
112. Pogue AI, Li YY, Cui JG, Zhao Y, Kruck TP, Percy ME, et al. Characterization of an NF-kappaB-regulated, miRNA-146a-mediated down-regulation of complement factor H (CFH) in metal-sulfate-stressed human brain cells. *J Inorg Biochem* 2009;**103**:1591–5. doi:10.1016/j.jinorgbio.2009.05.012.
113. Taganov KD, Boldin MP, Chang KJ, Baltimore D. NF-kappaB-dependent induction of microRNA miR-146, an inhibitor targeted to signaling proteins of innate immune responses. *Proc Natl Acad Sci USA* 2006;**103**:12481–6. doi:10.1073/pnas.0605298103.
114. Saba R, Goodman CD, Huzarewich RL, Robertson C, Booth SA. A miRNA signature of prion induced neurodegeneration. *PLoS ONE* 2008;**3**:e3652. doi:10.1371/journal.pone.0003652.
115. Montag J, Hitt R, Opitz L, Schulz-Schaeffer WJ, Hunsmann G, Motzkus D. Upregulation of miRNA hsa-miR-342-3p in experimental and idiopathic prion disease. *Mol. Neurodegener* 2009;**4**:36. doi:10.1186/1750-1326-4-36.
116. van der Walt JM, Noureddine MA, Kittappa R, Hauser MA, Scott WK, McKay R, et al. Fibroblast growth factor 20 polymorphisms and haplotypes strongly influence risk of Parkinson disease. *Am J Hum Genet* 2004;**74**:1121–7. doi:10.1086/421052.
117. Wang G, van der Walt JM, Mayhew G, Li YJ, Zuchner S, Scott WK, et al. Variation in the miRNA-433 binding site of FGF20 confers risk for Parkinson disease by overexpression of alpha-synuclein. *Am J Hum Genet* 2008;**82**:283–9. doi:10.1016/j.ajhg.2007.09.021.
118. Gehrke S, Imai Y, Sokol N, Lu B. Pathogenic LRRK2 negatively regulates microRNA-mediated translational repression. *Nature* 2010;**466**:637–41. doi:10.1038/nature09191.
119. Johnson R, Zuccato C, Belyaev ND, Guest DJ, Cattaneo E, Buckley NJ. A microRNA-based gene dysregulation pathway in Huntington's disease. *Neurobiol Dis* 2008;**29**:438–45. doi:10.1016/j.nbd.2007.11.001.
120. Savas JN, Makusky A, Ottosen S, Baillat D, Then F, Krainc D, et al. Huntington's disease protein contributes to RNA-mediated gene silencing through association with Argonaute and P bodies. *Proc Natl Acad Sci USA* 2008;**105**:10820–5. doi:10.1073/pnas.0800658105.

121. Williams AH, Valdez G, Moresi V, Qi X, McAnally J, Elliott JL, et al. MicroRNA-206 delays ALS progression and promotes regeneration of neuromuscular synapses in mice. *Science* 2009;**326**:1549–54. doi:10.1126/science.1181046.
122. Haramati S, Chapnik E, Sztainberg Y, Eilam R, Zwang R, Gershoni N, et al. miRNA malfunction causes spinal motor neuron disease. *Proc Natl Acad Sci USA* 2010;**107**:13111–6. doi:10.1073/pnas.1006151107.
123. Burmistrova OA, Goltsov AY, Abramova LI, Kaleda VG, Orlova VA, Rogaev EI. MicroRNA in schizophrenia: genetic and expression analysis of miR-130b (22q11). *Biochemistry (Mosc)* 2007;**72**:578–82.
124. Zhu Y, Kalbfleisch T, Brennan MD, Li Y. A MicroRNA gene is hosted in an intron of a schizophrenia-susceptibility gene. *Schizophr Res* 2009;**109**:86–9. doi:10.1016/j.schres.2009.01.022.
125. Feng J, Sun G, Yan J, Noltner K, Li W, Buzin CH, et al. Evidence for X-chromosomal schizophrenia associated with microRNA alterations. *PLoS ONE* 2009;**4**:e6121. doi:10.1371/journal.pone.0006121.
126. Hansen T, Olsen L, Lindow M, Jakobsen KD, Ullum H, Jonsson E, et al. Brain expressed microRNAs implicated in schizophrenia etiology. *PLoS ONE* 2007;**2**:e873. doi:10.1371/journal.pone.0000873.
127. Xu Y, Li F, Zhang B, Zhang K, Zhang F, Huang X, et al. MicroRNAs and target site screening reveals a pre-microRNA-30e variant associated with schizophrenia. *Schizophr Res* 2010;**119**:219–27. doi:10.1016/j.schres.2010.02.1070.
128. Xu Y, Liu H, Li F, Sun N, Ren Y, Liu Z, et al. A polymorphism in the microRNA-30e precursor associated with major depressive disorder risk and P300 waveform. *J Affect Disord* 2010; doi:10.1016/j.jad.2010.05.019.
129. Tabares-Seisdedos R, Rubenstein JL. Chromosome 8p as a potential hub for developmental neuropsychiatric disorders: implications for schizophrenia, autism and cancer. *Mol Psychiatry* 2009;**14**:563–89. doi:10.1038/mp. 2009.2.
130. Perkins DO, Jeffries CD, Jarskog LF, Thomson JM, Woods K, Newman MA, et al. microRNA expression in the prefrontal cortex of individuals with schizophrenia and schizoaffective disorder. *Genome Biol* 2007;**8**:R27. doi:10.1186/gb-2007-8-2-r27.
131. Beveridge NJ, Tooney PA, Carroll AP, Gardiner E, Bowden N, Scott RJ, et al. Dysregulation of miRNA 181b in the temporal cortex in schizophrenia. *Hum Mol Genet* 2008;**17**:1156–68. doi:10.1093/hmg/ddn005.
132. Kim AH, Reimers M, Maher B, Williamson V, McMichael O, McClay JL, et al. MicroRNA expression profiling in the prefrontal cortex of individuals affected with schizophrenia and bipolar disorders. *Schizophr Res* 2010;doi:10.1016/j.schres.2010.07.002.
133. Abelson JF, Kwan KY, O'Roak BJ, Baek DY, Stillman AA, Morgan TM, et al. Sequence variants in SLITRK1 are associated with Tourette's syndrome. *Science* 2005;**310**:317–20. doi:10.1126/science.1116502.
134. Abu-Elneel K, Liu T, Gazzaniga FS, Nishimura Y, Wall DP, Geschwind DH, et al. Heterogeneous dysregulation of microRNAs across the autism spectrum. *Neurogenetics* 2008;**9**:153–61. doi:10.1007/s10048-008-0133-5.
135. Talebizadeh Z, Butler MG, Theodoro MF. Feasibility and relevance of examining lymphoblastoid cell lines to study role of microRNAs in autism. *Autism Res* 2008;**1**:240–50. doi:10.1002/aur.33.
136. Sarachana T, Zhou R, Chen G, Manji HK, Hu VW. Investigation of post-transcriptional gene regulatory networks associated with autism spectrum disorders by microRNA expression profiling of lymphoblastoid cell lines. *Genome Med* 2010;**2**:23. doi:10.1186/gm144.

137. Rong H, Liu TB, Yang KJ, Yang HC, Wu DH, Liao CP, et al. MicroRNA-134 plasma levels before and after treatment for bipolar mania. *J Psychiatr Res* 2010;**45**:doi:10.1016/j.jpsychires.2010.04.028.
138. Muinos-Gimeno M, Guidi M, Kagerbauer B, Martin-Santos R, Navines R, Alonso P, et al. Allele variants in functional MicroRNA target sites of the neurotrophin-3 receptor gene (NTRK3) as susceptibility factors for anxiety disorders. *Hum Mutat* 2009;**30**:1062–71. doi:10.1002/humu.21005.
139. Jensen KP, Covault J, Conner TS, Tennen H, Kranzler HR, Furneaux HM. A common polymorphism in serotonin receptor 1B mRNA moderates regulation by miR-96 and associates with aggressive human behaviors. *Mol Psychiatry* 2009;**14**:381–9. doi:10.1038/mp.2008.15.
140. Zhou R, Yuan P, Wang Y, Hunsberger JG, Elkahloun A, Wei Y, et al. Evidence for selective microRNAs and their effectors as common long-term targets for the actions of mood stabilizers. *Neuropsychopharmacology* 2009;**34**:1395–405. doi:10.1038/npp. 2008.131.
141. Cryan JF, Kelly PH, Neijt HC, Sansig G, Flor PJ, van Der Putten H. Antidepressant and anxiolytic-like effects in mice lacking the group III metabotropic glutamate receptor mGluR7. *Eur J Neurosci* 2003;**17**:2409–17.
142. Kocerha J, Faghihi MA, Lopez-Toledano MA, Huang J, Ramsey AJ, Caron MG, et al. MicroRNA-219 modulates NMDA receptor-mediated neurobehavioral dysfunction. *Proc Natl Acad Sci USA* 2009;**106**:3507–12. doi:10.1073/pnas.0805854106.
143. Kuhn DE, Nuovo GJ, Martin MM, Malana GE, Pleister AP, Jiang J, et al. Human chromosome 21-derived miRNAs are overexpressed in down syndrome brains and hearts. *Biochem Biophys Res Commun* 2008;**370**:473–7. doi:10.1016/j.bbrc.2008.03.120.
144. Kuhn DE, Nuovo GJ, Terry Jr. AV, Martin MM, Malana GE, Sansom SE, et al. Chromosome 21-derived microRNAs provide an etiological basis for aberrant protein expression in human Down syndrome brains. *J Biol Chem* 2010;**285**:1529–43. doi:10.1074/jbc.M109.033407.
145. Xu K, Bogert BA, Li W, Su K, Lee A, Gao FB. The fragile X-related gene affects the crawling behavior of Drosophila larvae by regulating the mRNA level of the DEG/ENaC protein pickpocket1. *Curr Biol* 2004;**14**:1025–34. doi:10.1016/j.cub.2004.05.055.
146. Yang L, Duan R, Chen D, Wang J, Chen D, Jin P. Fragile X mental retardation protein modulates the fate of germline stem cells in Drosophila. *Hum Mol Genet* 2007;**16**:1814–20. doi:10.1093/hmg/ddm129.
147. Cheever A, Ceman S. Phosphorylation of FMRP inhibits association with Dicer. *RNA* 2009;**15**:362–6. doi:10.1261/rna.1500809.
148. Fukuda T, Itoh M, Ichikawa T, Washiyama K, Goto Y. Delayed maturation of neuronal architecture and synaptogenesis in cerebral cortex of Mecp2-deficient mice. *J Neuropathol Exp Neurol* 2005;**64**:537–44.
149. Jugloff DG, Jung BP, Purushotham D, Logan R, Eubanks JH. Increased dendritic complexity and axonal length in cultured mouse cortical neurons overexpressing methyl-CpG-binding protein MeCP2. *Neurobiol Dis* 2005;**19**:18–27.
150. Chang Q, Khare G, Dani V, Nelson S, Jaenisch R. The disease progression of Mecp2 mutant mice is affected by the level of BDNF expression. *Neuron* 2006;**49**:341–8.
151. Klein ME, Lioy DT, Ma L, Impey S, Mandel G, Goodman RH. Homeostatic regulation of MeCP2 expression by a CREB-induced microRNA. *Nat Neurosci* 2007;**10**:1513–4.
152. Chandrasekar V, Dreyer JL. microRNAs miR-124, let-7d and miR-181a regulate cocaine-induced plasticity. *Mol Cell Neurosci* 2009;**42**:350–62. doi:10.1016/j.mcn.2009.08.009.
153. Schaefer A, Im HI, Veno MT, Fowler CD, Min A, Intrator A, et al. Argonaute 2 in dopamine 2 receptor-expressing neurons regulates cocaine addiction. *J Exp Med* 2010;**207**:1843–51. doi:10.1084/jem.20100451.

154. Hollander JA, Im HI, Amelio AL, Kocerha J, Bali P, Lu Q, et al. Striatal microRNA controls cocaine intake through CREB signalling. *Nature* 2010;**466**:197–202. doi:10.1038/nature09202.
155. Im HI, Hollander JA, Bali P, Kenny PJ. MeCP2 controls BDNF expression and cocaine intake through homeostatic interactions with microRNA-212. *Nat Neurosci* 2010;**13**:1120–7. doi:10.1038/nn.2615.
156. Huang W, Li MD. Nicotine modulates expression of miR-140*, which targets the 3′-untranslated region of dynamin 1 gene (Dnm1). *Int J Neuropsychopharmacol* 2009;**12**:537–46. doi:10.1017/S1461145708009528.
157. Brebner K, Wong TP, Liu L, Liu Y, Campsall P, Gray S, et al. Nucleus accumbens long-term depression and the expression of behavioral sensitization. *Science* 2005;**310**:1340–3. doi:10.1126/science.1116894.
158. Pietrzykowski AZ, Friesen RM, Martin GE, Puig SI, Nowak CL, Wynne PM, et al. Posttranscriptional regulation of BK channel splice variant stability by miR-9 underlies neuroadaptation to alcohol. *Neuron* 2008;**59**:274–87. doi:10.1016/j.neuron.2008.05.032.
159. Wu Q, Law PY, Wei LN, Loh HH. Post-transcriptional regulation of mouse mu opioid receptor (MOR1) via its 3′ untranslated region: a role for microRNA23b. *FASEB J* 2008;**22**:4085–95. doi:10.1096/fj.08-108175.
160. Wu Q, Zhang L, Law PY, Wei LN, Loh HH. Long-term morphine treatment decreases the association of mu-opioid receptor (MOR1) mRNA with polysomes through miRNA23b. *Mol Pharmacol* 2009;**75**:744–50. doi:10.1124/mol.108.053462.
161. He Y, Yang C, Kirkmire CM, Wang ZJ. Regulation of opioid tolerance by let-7 family microRNA targeting the mu opioid receptor. *J Neurosci* 2010;**30**:10251–8. doi:10.1523/JNEUROSCI.2419-10.2010.
162. Zheng H, Chu J, Zeng Y, Loh HH, Law PY. Yin Yang 1 phosphorylation contributes to the differential effects of mu-opioid receptor agonists on microRNA-190 expression. *J Biol Chem* 2010;**285**:21994–2002. doi:10.1074/jbc.M110.112607.
163. Zheng H, Zeng Y, Zhang X, Chu J, Loh HH, Law PY. mu-Opioid receptor agonists differentially regulate the expression of miR-190 and NeuroD. *Mol Pharmacol* 2010;**77**:102–9. doi:10.1124/mol.109.060848.

# The Role of MicroRNAs in Viral Infection

Ashley P.F. Roberts, Andrew P. Lewis, and Catherine L. Jopling

*School of Pharmacy, Centre for Biomolecular Sciences, University of Nottingham, Nottingham NG7 2RD, United Kingdom*



Progress in Molecular Biology
and Translational Science, Vol. 102
DOI: 10.1016/B978-0-12-415795-8.00002-7

MicroRNAs (miRNAs) are small non-coding RNA molecules that have emerged in recent years as central regulators of eukaryotic gene expression. In mammalian systems, miRNAs are associated with numerous pathological and physiological pathways. miRNAs are important in many viral infections, with different viral families expressing their own miRNAs, manipulating host miRNA expression, or showing direct or indirect regulation by host or viral miRNAs. In this chapter we will examine the current evidence for interplay between the miRNA pathway and viral infections in mammals.

## I. Introduction

MicroRNAs (miRNAs) are 21–23 nucleotide (nt) noncoding RNA molecules expressed by many eukaryotic species, and are an important component of the cellular RNA interference system.[1] miRNAs are encoded in hairpin precursor transcripts in the nucleus, but, following export to the cytoplasm, their biogenesis and function show much overlap with the short interfering RNA (siRNA) pathway. Both miRNAs and siRNAs are generated by processing of their double-stranded precursor by Dicer in the cytoplasm to yield the mature, single-stranded, functional molecule, and both function in association with a complex of proteins known as the RNAi-induced silencing complex (RISC).[1] miRISC and siRISC may contain distinct components, but both require an argonaute (Ago) protein for activity. miRNAs in animal systems generally act by binding with imperfect complementarity to sites in the 3′UTR of target mRNAs, leading to repression of translation and mRNA degradation, whereas siRNAs bind to exactly complementary sites and cleave the mRNA.[1] These functions are interchangeable, however; miRNAs presented with an exactly complementary target will mediate site-specific RNA cleavage, whereas siRNAs will function as miRNAs if the target is miRNA-like.

This chapter will consider the emerging importance of miRNAs in viral infections of mammalian systems. We first provide a broad overview of the metazoan miRNA pathway and an introduction to the role of miRNAs in human disease, before focusing on the interactions between viruses and the miRNA pathway.

## A. Discovery of MicroRNAs

The first suggestion that noncoding RNAs might be important in the regulation of gene expression came in 1993, as a result of research on the *Caenorhabditis elegans* gene *lin-4*. The *lin-4* gene is important in the determination of cell fate during development when it is expressed at a specific larval stage and functions by downregulating LIN-14 protein levels. It was shown that *lin-4* does not encode a protein, but instead produces a 22-nt RNA molecule, which has partial complementarity to seven sites in the 3′UTR of *lin-14* mRNA.[2]

A second gene that is important in the *C. elegans* heterochronic pathway, *let-7*, was also shown to encode a small, noncoding RNA that has partially complementary sites in the 3′UTR of downstream mRNA targets.[3] Evidence that regulation by small RNAs might be a much more widespread phenomenon came from the observation that the sequence of *let-7* is conserved across a wide range of animal species and that the expression of this RNA shows similar temporal regulation patterns in diverse organisms.[4] Cloning of RNA from size-fractionated libraries from *C. elegans*, *Drosophila*, and human cells then revealed that multiple different 21–23 nt RNA molecules are present in all these organisms.[5–7] Thousands of these molecules, now known as miRNAs, have since been isolated from a wide range of eukaryotic organisms, and many show conservation between species. The most recent release of the miRNA database (miRBase 16) contains 1048 different human miRNAs.[8] In the context of this review, it is important to note that miRNAs are also encoded by a number of viruses. This is discussed in detail in Section IV.

## B. miRNA Biogenesis

miRNAs are encoded within nuclear transcripts that may be several hundred nucleotides in length, known as pri-miRNAs. Pri-miRNAs are usually transcribed by RNA polymerase II, and may contain a single miRNA, or a cluster of several different miRNAs, often with related functions.[9] Pri-miRNAs may be unique transcripts, or may be derived from the intron of a pre-mRNA. miRNA sequences are located in double-stranded hairpin regions of RNA, which are recognized by the RNAse III enzyme Drosha in a nuclear processing step (Fig. 1). Drosha acts in association with a protein known as DGCR8 in mammals to cleave pri-miRNAs and generate pre-miRNAs, which are precursor hairpin molecules of approximately 70 nt.[9] Pre-miRNA export to the cytoplasm occurs via Exportin 5. The RNAse III enzyme Dicer then cleaves the pre-miRNA to generate a 21–23-nt duplex with 2 nt 3′overhangs (Fig. 1). Some examples of miRNAs generated by noncanonical pathways exist: a few miRNAs are located in "mirtrons," in which the pre-miRNA is excised directly

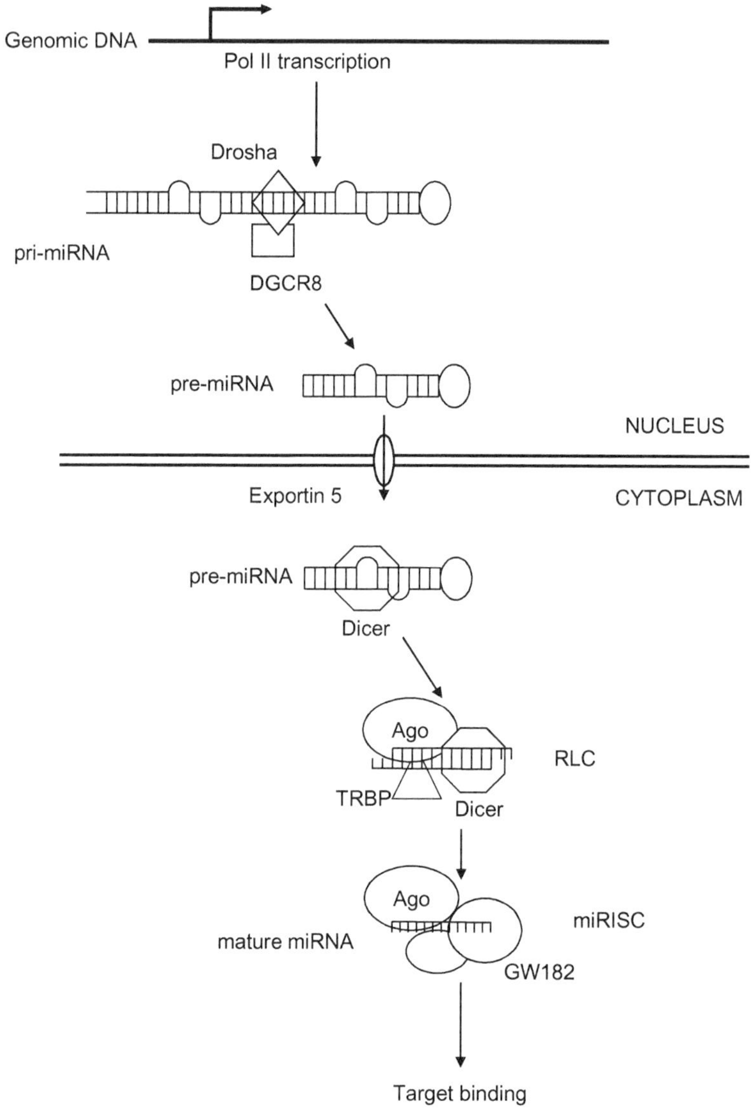


FIG. 1. The canonical pathway of miRNA biogenesis. Pri-miRNA transcripts are synthesized from cellular or viral genomic DNA by RNA polymerase II. The RNAse III enzyme Drosha, aided by DGCR8, cleaves the pri-miRNA to yield an approximately 70 nt pre-miRNA hairpin. The pre-miRNA is transported to the cytoplasm by Exportin 5, where it serves as a template for further processing by the RNAse III enzyme Dicer. This results in the production of an miRNA/miRNA* duplex. The mature, single-stranded, 21–23 nt miRNA molecule is loaded into the functional miRISC by the RISC-loading complex (RLC). There are exceptions to this biogenesis pathway, such as the MHV68 miRNAs (Section IV.E).

from a short intron by the splicing machinery, independent of Drosha processing,[10,11] and the mammalian miR-451 is generated from its pre-miRNA by Ago2 cleavage, and so is independent of Dicer.[12,13]

After Dicer processing occurs, the miRNA duplex associates with a RISC-loading complex (RLC), which comprises Dicer, Ago, and TAR RNA binding protein (TRBP) in mammals (Fig. 1).[1,9] The mature miRNA "guide" strand is then selected according to stability at the ends of the duplex, and is incorporated into the functional miRISC complex (Fig. 1). The "passenger" strand of the duplex is removed, probably by helicase activity or by Ago2-mediated cleavage. Some miRNA passenger strands, known as miR*, are present in significant quantities and mediate biological functions.

## C. miRNA Target Sites

miRNAs can function like siRNAs and cleave an exactly complementary target RNA. This is the major mechanism of miRNA activity in plants, but in animal systems there are very few examples of endogenous complementary targets for miRNAs. Instead, most miRNAs bind to mRNA targets with stretches of pairing interrupted by loops and bulges, and sites have evolved to avoid the exact complementarity to nt 10–11 of the miRNA that is required for target cleavage.[1,14] This has made it difficult to identify targets by computational methods, although some of the rules that govern effective miRNA targeting have now been determined.[14] An important indicator is exact Watson–Crick base-pairing to nt 2–7 of the miRNA, known as the "seed." This is enhanced by complementarity to nt 8, and by an adenine residue opposite position 1 of the miRNA (which is usually a uridine).[15,16] Although seed pairing is generally a good indicator of a functional miRNA target site, some sites do not obey this rule, and other features such as complementarity to the 3′ region of the miRNA and surrounding secondary structure also influence whether a site is used.[14,16] Conservation of predicted target sites between species is included in most algorithms and enhances the efficiency of prediction, but is limited to miRNAs and target mRNAs that are conserved, and will result in some sites being missed. It is always necessary to confirm predicted target sites experimentally, by determining the effects of the miRNA on the endogenous target protein and/or by using reporter constructs containing the 3′UTR under investigation. Multiple target sites for the same or different miRNAs within a 3′UTR frequently occur, and individual miRNAs target many different mRNAs.

The vast majority of miRNA target sites identified to date are found in the 3′UTR of target mRNAs, and there is still uncertainty as to whether, and to what extent, miRNAs function by binding to sites in the open reading frame (ORF) or 5′UTR of mRNAs. Synthetic miR-30 reporter sites that were introduced into either the ORF or 3′UTR of a reporter plasmid showed effective repression by miR-30 in the 3′UTR but not the ORF.[17] Introduction of rare codons upstream of the miR-30 sites in the ORF restored regulation, suggesting that miRNA sites in ORFs of efficiently translated mRNAs are nonfunctional because translating ribosomes displace the miRISC from the mRNA. However, a recent bioinformatic analysis found a similar number of conserved miRNA sites in *Drosophila* 3′UTRs and ORFs, and a reduced but significant number in 5′UTRs. miRNA-dependent repression was confirmed for some of these sites, although the strength of this repression was lower in the 5′UTR and ORF sites than in 3′UTRs.[18] The predicted number of ORF sites in human mRNAs was lower than in *Drosophila*. The functional importance of miRNA sites in the coding regions of *Nanog*, *Oct4*, and *Sox2* genes has been demonstrated in mouse embryonic stem cells.[19]

## D. miRNA Activity: Translation Inhibition

The consequence of an miRNA binding to a 3′UTR target site in animal systems is to repress the synthesis of the corresponding protein. There has been much controversy about the mechanism by which miRNAs inhibit translation and the extent to which translation inhibition and mRNA degradation contribute to the effects of miRNAs on protein levels.

Translation occurs in three stages: initiation, elongation, and termination.[20] Initiation is generally thought to be the rate-limiting step. Eukaryotic mRNAs are synthesized with a 5′ 7-methylguanosine cap and a 3′ poly(A) tail, both of which are usually required for translation initiation. In cap-dependent translation initiation, the eIF4F complex of proteins binds to the cap and recruits the 40S ribosomal subunit. Poly(A) binding protein (PABP) interacts with eIF4F and binds to the poly(A) tail, mediating mRNA circularization and stimulating translation.[20] The 40S subunit traverses the 5′UTR by a scanning mechanism until an AUG codon in the correct context is encountered, at which point 60S ribosome subunit joining takes place. Elongation then takes place as the 80S ribosome moves along the coding region, synthesizing the encoded polypeptide. Termination occurs when a stop codon is encountered, and the ribosome dissociates from the mRNA. An alternative method of translation initiation is internal ribosome entry, in which the 40S subunit is recruited directly to a structured region of the 5′UTR, known as an internal ribosome entry site (IRES). IRES-dependent translation initiation occurs in a number of viral RNAs, including the picornaviruses and hepatitis C virus (HCV), as well as in some cellular mRNAs. Different IRESs vary both in their mechanism of ribosome recruitment and in their requirement for initiation factors and other proteins.[20]

In different studies, miRNAs have been observed to inhibit translation either at the level of initiation, such that ribosomes dissociate from the mRNA, or at a postinitiation stage, with targeted mRNAs remaining in polysomal fractions. A number of potential mechanisms have been proposed for both these modes of inhibition. Most current evidence from cell culture and *in vitro* systems suggests that translation initiation is inhibited, but there is still much debate on this issue and on the mechanism involved. There is also conflicting evidence indicating either that miRNA repression can only occur on capped mRNAs, or that both cap- and IRES-driven transcripts are susceptible to inhibition.[1,21]

## E. miRNA Activity: mRNA Degradation

It was initially thought that miRNAs inhibit translation with no effects on mRNA levels, but in most recent studies some degree of mRNA destabilization has also been observed; the extent of this destabilization varies in different studies and between different miRNA:target pairs, so it is possible that it occurs to varying extents under different circumstances. miRNA binding results in deadenylation of target mRNAs, which is followed by decapping and degradation in association with cytoplasmic processing bodies (P bodies).[1,21] P bodies are sites of accumulation of components of the decapping and deadenylation machineries, where translationally repressed mRNAs are stored and/or degraded. Components of the miRISC, miRNAs, and miRNA-targeted mRNAs all localize to P bodies, but the integrity of microscopically visible P bodies is not essential for miRNA repression, suggesting that mRNA degradation might occur as a consequence of translational repression following miRNA binding.[22] This is supported by kinetic analysis in an *in vitro* system, in which a primary effect of miRNA binding on translation was succeeded by changes in mRNA stability.[23]

A reduction in the level of miRNA-targeted mRNAs is widespread in cells, and total mRNA microarrays have proved to be an effective method of identifying miRNA targets,[24] although it is possible that targets that are regulated purely at the level of translation are missed. A recent study used deep sequencing to analyze total and ribosome-associated mRNA, and demonstrated that for three mammalian miRNAs almost all target regulation occurs at the level of mRNA degradation, with minimal additional effects on translation.[25]

## F. The miRISC

miRNA function occurs in association with the miRISC. The core components of miRISC that are known to be required for miRNA-dependent repression are the Ago proteins and GW182.[1,21,26] In mammalian systems, there are four Ago proteins which share considerable sequence homology. Only Ago2 contains the catalytic site that directs RNA cleavage when an exactly complementary site is bound. It appears that all four proteins can function interchangeably to mediate miRNA-dependent repression of gene expression. Mammals also possess three different GW182 proteins, designated TNRC6A–C, which also appear to function interchangeably. miRNA-dependent repression requires interaction between Ago and GW182, and can be recapitulated by direct tethering of either Ago or GW182 proteins to 3′UTR sites. Recent evidence suggests that GW182 is the mediator of miRNA activity and that Ago proteins function to recruit GW182.[21] GW182 directs deadenylation of target mRNAs by association with PABP, in a process dependent on the CAF1 deadenylase.[23]

The exact identity of the miRISC is still not certain, and it is possible that it varies according to the cell state and the identity of the miRNA:mRNA pair. Several accessory factors have been identified that modulate the activity of miRNAs in specific circumstances; for example, the protein FXR1 is recruited to miRISC in quiescent cells and switches miRNA activity from repression to activation of translation.[27] Other RNA binding proteins that are recruited to 3′UTRs targeted by miRNAs can enhance or inhibit the activity of the miRNA.[26]

## II. Regulation of miRNAs and Their Association with Disease

miRNAs are a central feature of the control of gene expression in eukaryotes, and are estimated to regulate around 50% of mammalian mRNAs.[26] It is therefore unsurprising that miRNAs have been implicated in many different disease processes in humans. There is particularly strong evidence for a role for miRNAs in cancer,[28] but numerous other examples exist of the association between the expression of specific miRNAs and particular disease processes, such as cardiovascular disease. There is extensive interaction between the miRNA pathway and infection by various viruses, which is the subject of this chapter.

### A. Regulation of miRNA Expression

Many miRNAs in complex organisms show highly specific, tightly controlled patterns of expression that allow them to regulate specific biological processes. Some miRNAs are expressed ubiquitously, but many are switched on and off at particular times during development, and are present only in certain tissue or cell types.[29] Expression of specific miRNAs at specific time points is crucial in driving correct differentiation. Altered regulation of miRNA expression has a role in many disease processes.

Much regulation of miRNA expression takes place at the level of pri-miRNA transcription, and is similar to the regulation of mRNA transcription. Pri-miRNAs that encode a cluster of miRNAs allow coordinated expression of these molecules. Some transcription factors that regulate the expression of specific pri-miRNAs have been identified, and computational prediction has been applied to detect pri-miRNA promoters.[9,26] Regulation of pri- and pre-miRNA processing also occurs. The levels of Drosha and Dicer, and their respective cofactors DGCR8 and TRBP, are all subject to regulation in mammals. Dicer mRNA is targeted by *let-7*, forming a negative feedback loop in miRNA biogenesis. Several Drosha and Dicer cofactors that regulate the generation of specific groups of miRNAs have been isolated. These include the Drosha cofactors p68 and p72, which stimulate the processing of around one-third of pri-miRNAs by an unknown mechanism. LIN-28 specifically represses the expression of *let-7* family members in undifferentiated embryonic stem cells. LIN-28 interacts with the terminal loop of pri-*let-7* and inhibits Drosha processing, while in the cytoplasm this interaction results in 3′-terminal poly-uridylation of pre-*let-7*, which blocks Dicer processing and leads to pre-*let-7* degradation.[30,31] The biogenesis of some miRNAs is also regulated by RNA editing, which can affect Drosha and Dicer cleavage.[9,26,32]

Although miRNA stability is generally high, a few examples have emerged of regulated miRNA turnover in animal systems, and this is likely to be an important regulatory mechanism for specific miRNAs in certain circumstances. miR-122, a liver-specific miRNA, is subject to the addition of a single 3′ adenine by the enzyme GLD-2, resulting in enhanced miRNA stability.[33] Many neuronal miRNAs show rapid turnover, which is regulated in response to neuronal activity. This accelerated decay in retinal neurons is important for retinal function.[34] The enzymes and regulatory factors that govern miRNA turnover in mammals have not been identified, but the 5′-3′ exonuclease Xrn-2 mediates miRNA decay in *C. elegans*.[35] Association of miRISC with a target mRNA protects the miRNA from Xrn-2 degradation.

## B. miRNAs and Cancer

The importance of miRNA expression in the control of development and differentiation makes it unsurprising that changes in specific miRNAs are frequently associated with cancer. This first became evident when a chromosomal region that is frequently lost in chronic lymphocytic leukemia (CLL), 13q14, was found to contain two miRNAs which act as tumor suppressors.[36] These miRNAs, namely, miR-15a and miR-16-1, are lost in 70% of CLLs, and their loss occurs at an early stage in tumorigenesis.[36]

Other miRNAs are also frequently located at fragile sites or in cancer-associated genomic regions,[37] and it is now apparent that many different miRNAs function as either oncogenes or tumor suppressors in specific tissues (for review, see Ref. 28). The miRNA targets that regulate tumorigenesis are gradually being elucidated. Important examples of miRNAs that function as tumor suppressors include *let-7*, which represses tumor growth in the lung.[38] LIN-28, which represses *let-7* biogenesis, is an oncoprotein, and *let-7* targets include *lin-28* and the oncogenes *ras* and *HMGA2*.[39–41] miR-155 and the miR-17-92 cluster are amplified in a number of B-cell lymphomas. They have subsequently been shown to function as oncogenes in a range of different tumor types.[28] Several other oncogenic miRNAs have been identified, in particular miR-21, which is overexpressed in most solid tumors. Targets of these miRNAs that are involved in their oncogenesis are beginning to be elucidated.[28] miRNA expression profiles are proving to be useful and accurate in determining the stage, and predicting the outcome of therapy, in specific cancers.

## C. miRNAs and Viral Infection

Viruses rely on host cell functions for many aspects of their life cycle. As the miRNA pathway is an important part of the host's regulatory systems, there is much potential for interplay between miRNAs and viruses.[42] There are several ways in which viruses could conceivably interact with the host miRNA machinery: (1) Viral translation, or other aspects of the viral life cycle, may be regulated by host miRNAs that bind directly to sites in the viral RNA or mRNA. (2) The virus may alter the expression of host miRNAs, thereby changing the regulation of host or viral RNA targets. (3) The virus may express its own miRNAs, which then regulate host or viral RNA targets. All these mechanisms have been observed during infection with specific animal viruses, and will be discussed in detail in Sections IV–VI. Some viruses induce transformation of infected cells, and there are various examples of associations between viral infection, cancer, and miRNA expression, which will also be discussed.

## D. miRNA-Based Therapeutics

The recent interest in miRNA biology has been accompanied by a rapidly expanding selection of new technologies directed at analyzing and regulating the activity of miRNAs. Given the importance of specific miRNAs in viral infection, cancer, and other diseases, much effort has been directed toward building new diagnostic or therapeutic platforms based on these molecules.

miRNAs can be effectively sequestered and inhibited by the introduction of chemically modified antisense oligonucleotides. This was initially demonstrated using 2′-*O*-methylated oligonucleotides, which can be introduced into cultured cells by liposome-mediated transfection and are bound by the miRNA in complex with the miRISC, preventing interaction with its endogenous targets.[43,44] This technology has proven to be very useful in the identification and characterization of targets of specific miRNAs in cultured cells and animal models. It also has broader implications as a potential therapeutic. Different modifications to the oligomer backbone or the bases can influence its binding affinity and stability, and conjugation to molecules such as cholesterol can also influence delivery.[45,46] For example, locked nucleic acid (LNA) oligonucleotides with a phosphorothioate backbone are delivered to the liver following intravenous injection in mice and primates and effectively target the liver-specific miR-122.[47–49] Inhibition of miRNAs in other organs is likely to be more difficult, and some therapeutic applications may require the targeted inhibition of a ubiquitous miRNA within a particular organ. There are many barriers to be crossed before miRNA-based therapeutics become a practical approach, but the possibility of directly and specifically targeting these small, highly-conserved molecules using antisense technology makes them a highly attractive target.

There are also potential therapeutic applications based on the overexpression of specific miRNAs. In cell culture, miRNAs can be ectopically expressed by transfection of the pri-miRNA or pre-miRNA under the control of appropriate promoters. It is also possible to use lipofection to directly introduce the miRNA/miRNA* product of Dicer cleavage directly into the cytoplasm as a synthetic RNA duplex. These strategies have been useful in analyzing the activity of specific miRNAs, but require modification before they can be used in whole animals. The principles involved are the same as in the delivery of siRNAs for therapeutic purposes, and both synthetic miRNA duplexes and pre-miRNA delivery using viral vectors show promise.[46]

## III. Introduction to Viral Infection

Viruses are obligate intracellular parasites that require host cell functions for many different aspects of their life cycles. Different viruses have evolved numerous mechanisms to influence and subvert the host cell machinery. Before considering the means by which viruses can interact with the cellular miRNA pathway, it is useful to consider a generalized scheme of infection for both DNA and RNA viruses.

### A. Viral Entry

All viruses must first enter a permissive host cell in order for a full cycle of replication to be achieved. Entry into a cell requires attachment of the virion to the cell via the interaction of viral proteins and molecules on the cellular surface.[50,51] A wide range of cell-surface molecules are recognized by viruses and may determine the tropism of infection.[50] Upon binding to cell-surface receptors, viruses must cross the plasma membrane to gain access to the cytoplasm. Numerous pathways and mechanisms exist to achieve this, including direct membrane fusion between the virus and host,[52,53] endocytosis, and clathrin-dependent and independent routes.[50,51] Once within the cytoplasm, the fates of DNA and RNA virus genomes differ.

### B. DNA Viruses

For DNA viruses, the genome, encased within the nucleocapsid, must cross the cytoplasm. Molecular crowding in the cytoplasm restricts the maximum size of freely diffusing entities to around 500 kDa,[54,55] so large viruses use the host cell cytoskeleton and motor proteins to access the nucleus.[51,56,57] Once at the nucleus, large DNA viruses such as adenoviruses and herpesviruses interact with the nuclear pore complex (NPC) and uncoat the genome for release into the nucleoplasm.[51] The nucleocapsids of smaller DNA viruses such as polyomaviruses can enter the nucleus through the NPC and then may uncoat the genome within the nucleoplasm.[51] Once the viral genome is in the nucleus, the transcription of viral genes begins, leading to the expression of viral proteins, and culminating in replication of the viral genome, production of progeny particles, and eventual release of progeny virions from the cell. DNA viruses commonly use host RNA polymerase II to transcribe viral mRNAs, from which viral proteins are synthesized using the host machinery. Both host and viral DNA polymerases and accessory factors are involved in replication of DNA virus genomes.[58,59] There are exceptions to the nuclear replication of DNA viruses, such as poxviruses, which replicate entirely in the cytoplasm.

## C. RNA Viruses

RNA virus infections utilize different strategies after entry into the cell. Most RNA viruses, with the exception of influenza virus and retroviruses, replicate entirely in the cytoplasm. The genomes of positive (+) and negative (−) sense ssRNA viruses and those of dsRNA viruses undergo different initial fates.

In all cases, the genome and associated proteins are thought to be transported through the cell to the site of viral replication.[56] The genome of a + ssRNA virus acts directly as a template for translation of the viral proteins using the host protein biosynthesis machinery. + ssRNA viruses may recruit the cellular translation machinery by cap- or IRES-dependent mechanisms in order to synthesize viral proteins (see Section I.D).[60–62] The genome of − ssRNA viruses must first be replicated by virion-derived, RNA-dependent RNA polymerases (RdRp) to convert the antisense genomic RNA to positive-sense mRNAs, from which viral protein synthesis can occur.[63] The genomic RNAs of dsRNA viruses are not released into the cytoplasm of the infected cell but are instead transcribed by viral RdRps within the capsid, with the transcripts subsequently released into the cytoplasm for translation.[64]

In all RNA viral infections, formation of replication complexes and compartments follows the onset of viral protein synthesis, and replication of the viral genome begins with the formation of the complementary strand to act as a template for the synthesis of the genomic RNA.[65] Progeny virions are assembled from replicated genomes, viral proteins, and host membranes (where appropriate) and are ultimately released from the cell.

### D. Retroviruses

Unlike other RNA viruses, retroviruses need to access the nucleus for the completion of their infectious cycle.[66] Retroviruses are characterized by the reverse transcription of the viral RNA genome using a virally encoded reverse transcriptase to form a DNA provirus, which is integrated into the host genome by the activity of the viral integrase. The integration of the provirus allows long-term and latent infections to be established, and transcription from the integrated provirus subsequently generates mature progeny virions.[66]

### E. Virus–Host Interactions

Owing to the numerous stages of viral life cycles, viruses are able to influence the cell at many different levels, and through diverse pathways, in order to favor their own replication. Some viruses, for example, adenoviruses and polyomavirus, encode proteins capable of stimulating the cell cycle to create conditions favorable for replication.[67,68] The alphaherpesviruses employ viral endoribonucleases to degrade host mRNA in order to clear the translational machinery of host mRNA and to allow viral mRNAs access for translation.[69] Viral RNAs can compete effectively with host mRNAs for the translation machinery by various means, such as IRESs, which mediate translation in viruses such as HCV using a reduced set of eukaryotic translation initiation factors.[62] Many viruses encode proteins to modulate the immune response or interfere with apoptotic signaling.[70,71] The discovery of miRNAs has led to the identification of another cellular pathway that can be influenced by viral infection and that is utilized by many viruses to modulate their own life cycles or the host environment.

## IV. Viral miRNAs

Expression of miRNAs provides an attractive strategy for a virus to manipulate its host environment. The small size of miRNAs means that they do not need much coding capacity, and they are nonimmunogenic. miRNAs also have the potential to evolve to target numerous host and viral RNAs to allow sensitive modulation of the infectious cycle. A number of viruses, predominantly of the large dsDNA herpesvirus family, have been found to encode their own miRNAs. These tend to be processed in the same manner as cellular miRNAs from pri-miRNAs transcribed in the nucleus. The targets of viral miRNAs may be viral or cellular mRNAs. The identity of these targets is beginning to be elucidated, revealing that viruses can use miRNAs to modulate the cellular environment in order to support aspects of the viral life cycle such as latency.

## A. Discovery of Viral miRNAs

The first example of virally encoded miRNAs came from a cloning experiment in human B cells latently infected with the herpesvirus Epstein–Barr virus (EBV).[72] These miRNAs are encoded in clusters in two regions of the EBV genome, and miRNAs deriving from each region show different expression patterns according to the stage of latent or lytic infection. The presence of miRNAs in EBV suggested that such virally encoded miRNAs might be present in other herpesviruses or, more widely, in large DNA viruses.[72] Indeed, small RNA cloning from cells infected with Kaposi's sarcoma herpesvirus (KSHV), mouse gammaherpesvirus 68 (MHV68), and human cytomegalovirus (HCMV) revealed that these herpesviruses each express several miRNAs.[73] Many other viral miRNAs have subsequently been identified, predominantly in large dsDNA viruses, and 235 viral miRNA sequences are listed in the most recent release of miRBase (release 16).[8]

Viral miRNAs have been found in herpesviruses, polyomaviruses, and ascoviruses, and potentially in adenovirus.[74] In both herpesviruses and polyomaviruses, the encoded miRNAs use the cellular miRNA pathway as a means to regulate viral gene expression, and some herpesvirus miRNAs have also been shown to have cellular targets. Not all DNA viruses appear to encode miRNAs, with no miRNAs yet reported in papillomaviruses.[75] Although miRNA expression appears to be a common feature of most herpesvirus infections, no miRNAs were detected in cells latently infected with the herpesvirus Varicella Zoster virus (VZV); the presence of VZV miRNAs in lytic infection remains a possibility.[76] miRNAs also appear to be absent from poxviruses, which are DNA viruses that replicate in the cytoplasm.[74]

There is little evidence for miRNAs produced by RNA viruses. miRNAs derived from the HIV-1 TAR RNA have been reported, although their existence is controversial.[73,77,78] Deep sequencing of cells infected by a number of RNA viruses did not reveal any viral miRNAs.[79] Although it remains possible that some RNA viruses do encode miRNAs, there are several reasons why this strategy is unlikely to be used by cytoplasmic RNA viruses. Pri-miRNAs that are expressed in the nucleus are available to enter the canonical miRNA biogenesis machinery via Drosha, whereas the genomes of cytoplasmic viruses are physically separated from Drosha and its cofactors. The processing of miRNAs from RNA viral genomes in the cytoplasm would result in the cleavage of the genome during excision of the miRNA, rendering it useless as a template for translation or replication.[42] Current information about viral miRNA targets suggests that they tend to be involved in processes such as viral latency that are a feature of long-term infections established by many DNA viruses, whereas RNA viral infections are usually far more rapid and would have less need for the delicate manipulation of the host and viral machinery afforded by viral miRNAs.[74]

## B. Herpesvirus miRNAs and Viral Latency

miRNAs are expressed by almost all herpesviruses examined, including representatives of the alpha-, beta-, and gamma-*herpesvirinae.*[74] Herpesvirus infection involves both lytic and latent cycles, and several herpesvirus miRNAs seem to be involved in the regulation of viral latency. Lytic replication proceeds via the successive expression of three classes of genes: immediate early (IE), early, and late. IE gene products prepare the cell for infection and stimulate the expression of early genes, the products of which are required for DNA replication. The late genes encode the structural proteins of the virus.

All herpesviruses are able to establish nonreplicative infections in particular cell populations. Alphaherpesviruses generally establish latent infections in sensory ganglia; betaherpesviruses can establish latent infections in a broad range of tissues; gammaherpesviruses establish latent infections in lymphoid tissues.[80] Latent herpesvirus infections are characterized by the minimal expression of viral gene products, and latent infections can reactivate to cause renewed lytic (symptomatic or asymptomatic) infection, although the triggers of reactivation are currently poorly defined.[81] Latency is commonly divided into three stages: establishment, maintenance, and reactivation.[82] The events leading to the establishment of a latent infection remain incompletely understood, but current opinion suggests that in may be associated with the failure to express the first wave of viral master regulators, the IE genes. This failure would result in loss of subsequent expression of viral early and late genes.[81]

## C. Alphaherpesvirus miRNAs and Their Targets

Most analysis of miRNAs expressed by this family of viruses has been directed toward the prototypical alphaherpesvirus herpes simplex virus 1 (HSV-1). Cloning of small RNAs from cells undergoing a productive HSV-1 infection led to the identification of miR-H1, which is a late gene product.[83] HSV-1 forms latent infections in the sensory ganglia, enervating the site of initial infection, usually the trigeminal ganglia. This infection is characterized by not only the almost complete absence of lytic gene expression but the expression of a particular viral transcript, the latency associated transcript or LAT.[81] LAT maps to the long repeat region at the terminal end of the unique long genome segment, and is transcribed antisense to the IE gene *ICP0* (Fig. 2).[81] The LAT transcript exists in three isoforms generated by alternative splicing and showing different expression patterns (Fig. 2).[81] To date, there are no reports of a peptide being translated from LAT.[81,84] However, recent work has shown that LAT can be processed to give rise to six functional miRNAs in HSV-1 latently infected cells (Fig. 2; reviewed in Refs. 85,86,76). These are termed miR-H2, miR-H3, miR-H4, miR-H5, miR-H7, and miRH8.

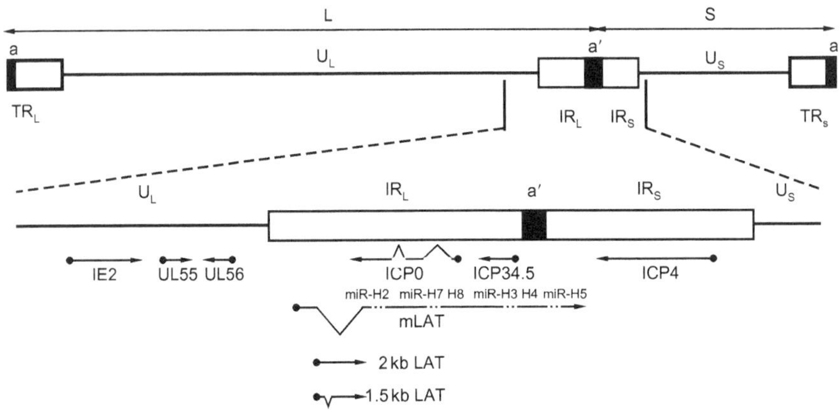


FIG. 2. Location of LAT miRNAs in the HSV-1 genome (not shown to scale). The long (L) and short (S) genome regions are indicated with the unique ($U_L$ and $U_S$) sequences shown as solid lines, and major repeat elements ($TR_L$, $IR_L$, $IR_S$, and $TR_S$) as open boxes. The terminal "a" sequences and internal "a′" sequence are marked by filled boxes. The approximate positions of viral transcripts in the enlarged region around $IR_L$ and $IR_S$ are shown below for both lytic genes and for the latency associated transcripts (LATs). mLAT is transcribed antisense to the transactivator ICP0 and is spliced to release the 2-kb LAT, from which a further intron is spliced to yield the 1.5-kb LAT. Within mLAT, the approximate positions of the HSV miRNAs miR-H2, H3, H4, H5, H7, and H8 are indicated as a broken line.

The LAT-derived miR-H2 is in a genomic location antisense to the viral transactivator *ICP0* and causes a reduction in ICP0 protein levels. Despite perfect complementarity to its target, miR-H2 does not affect the levels of *ICP0* mRNA, so appears to function at the level of translation.[86] miR-H7 and miR-H8 also map to positions antisense to *ICP-* but are unlikely to be functional against the *ICP0* transcript, as the miRNA complementary sites are located in the first intron of *ICP-* and so should not be accessible to miRNAs acting in the cytoplasm. There is some evidence for alternative splicing in *ICP0*, so it remains possible that particular splice variants might be targeted by these miRNAs.[42]

miR-H3 and miR-H4 are expressed antisense to the viral *ICP34.5* transcript.[86] To date, there is no direct demonstration of an effect of miR-H3 or H4 on ICP34.5, but prior mutational studies support this proposition.[42,86] Targeting of *ICP34.5* by these miRNAs may reduce the neurovirulence of HSV-1 infection, as ICP34.5 is known to regulate this process. Similarly to ICP0, another IE master regulator of HSV-1 gene expression, ICP4, is the target of an HSV-1 miRNA, miR-H6. miR-H6 has seed match rather than exact complementarity to *ICP4* mRNA, and mediates a reduction in levels of ICP4 protein, but not mRNA.[86]

ICP0 is an E3 ubiquitin ligase, which, although not essential for virus replication, is important for the establishment of lytic replication at low multiplicity of infection (MOI). In addition to stimulating the expression of viral early genes, ICP0 appears to mediate its function by remodeling repressive bodies in the nucleoplasm, known as ND10 or PML bodies.[81] ICP4 functions primarily as a transactivator for the expression of early genes. By downregulating expression of both ICP0 and ICP4, it appears likely that HSV-1 miR-H2 and miR-H6 would provide a potent block to entry into lytic replication and maintain the virus in an established latent state.

The closely related virus herpes simplex virus 2 (HSV-2) also expresses miRNAs from its LAT. There is functional conservation of miRNA targeting between the two viruses, as HSV-2 miRNAs downregulate expression of ICP0 and ICP34.5, suggesting similar mechanisms for establishing and maintaining latency.[87,88] Despite their conserved location and viral targets, there is only limited sequence homology between miRNAs expressed by the two viruses, so they are unlikely to have common cellular targets.[88]

## D. Betaherpesvirus miRNAs and Their Targets

The human betaherpesvirus HCMV expresses a number of miRNAs that are located throughout the viral genome, in contrast to alpha- and gammaherpesvirus miRNAs which are clustered.[89] HCMV miRNAs were identified in productively infected cells. The HCMV miRNA miR-UL112-1 represses the translation of IE1, an IE gene product that transactivates expression of early genes during lytic infection, by binding to a single 3′UTR site.[90,91] IE1 is not essential except at low MOI,[92] and is expressed highly at early times after infection, with levels declining as expression of miR-UL112-1 increases at late times.[89] Premature expression of miR-UL112-1 was found to adversely affect viral replication.[93] It has been suggested that this represents a possible role for miR-UL112-1 in contributing to the establishment or maintenance of latency, although it is not yet known whether this miRNA is expressed in latently infected cells.[91] It has been suggested that miRNA-mediated control of the expression of IE transactivators, such as ICP0 and IE1, is a common mechanism used by different classes of herpesvirus to regulate entry into lytic or latent infection.[91]

A cellular target for miR-UL112-1 has been identified in the form of MICB, a cell-surface protein recognized by natural killer (NK) cells and induced by stresses such as viral infection. miRUL112-1 binds to the 3′UTR of *MICB* transcripts and leads to decreased levels of MICB protein, presumably to evade an NK response against infected cells.[94] This HCMV miRNA can therefore modulate viral infection directly via a viral target, and indirectly via cellular targets, and it is possible that other viral miRNAs will also have multiple functions.

A RISC immunoprecipitation strategy was used to identify targets of another HCMV miRNA, miR-US25-1. Interestingly, the targets tended to be located in the 5′UTR of cellular mRNAs, and miR-US25-1 could mediate repression of reporter gene expression when some of these 5′UTRs were placed upstream of the luciferase coding region.[95] As there are only a very few examples of miRNAs targeting 5′UTRs and the consequences of this targeting are varied, this raises interesting questions regarding the mechanism of repression via these 5′UTR sites, and whether such targeting is a specific feature of this miRNA. The targets include several mRNAs encoding proteins involved in the cell cycle, one of which, cyclin E2, was shown to be repressed by miR-US25-1 during viral infection.[95] It was proposed that this miRNA might regulate the cell cycle in order to influence the differentiation of latently infected progenitor cells.

## E. Gammaherpesvirus miRNAs and Their Targets

A number of targets have been identified for miRNAs expressed by the gammaherpesviruses EBV and KSHV, both of which establish persistent infections in humans that can lead to malignancy.

EBV miRNAs were initially found in two clusters: one in the Bam HI fragment H rightward open reading frame 1 (*BHRF1*) gene and one in the Bam HI-A-region rightward transcript (*BART*) gene.[72] Both viral and cellular targets of EBV miRNAs have been identified. miR-BART2 is encoded antisense to the *BALF5* gene and acts to decrease the levels of *BALF5* DNA polymerase.[96] This has been suggested as a potential means of stabilizing latency in EBV by suppressing viral replication and thus stalling the lytic cycle of the virus.[42] Several BART mRNAs have target sites in the 3′UTR of the EBV *LMP1* mRNA, and expression of the BART cluster led to a reduction in late membrane protein (LMP1) protein levels.[97] LMP1 contributes to transformation and modulates the NF-κB signaling pathway, which was also regulated by BART miRNA expression.[97] miR-BART5 downregulates the expression of the cellular protein PUMA,[98] which is involved in apoptosis downstream of p53 and can activate a p53-independent pathway.[99] This implies that this miRNA interferes with the apoptotic response to viral infection and allows the infected cells to survive and continue producing progeny virions. miR-BHRF1-3 targets CXCL-11,[100] a T-cell attractant,[101] the downregulation of which may assist in the evasion of the host immune response.

The broader effects of EBV miRNA expression on the host cell were studied by infecting human B cells with EBV mutants in which miRNAs of either the *BART* or the *BHRF* cluster, or both, were inactivated.[102] Interestingly, the rate of lytic reactivation was unaffected by these mutations, implying that EBV miRNAs are not important in the maintenance of viral latency, in contrast to those of other herpesviruses. miRNAs did have a role in modulation of the host environment, however, as the *BHRF* miRNAs were found to inhibit apoptosis and promote proliferation early in infection.[102]

A number of EBV miRNAs show a high level of sequence conservation to miRNAs of the closely related rhesus lymphocryptovirus (rLCV), particularly in the seed regions.[103] This is unusual among viral miRNAs identified to date, and may indicate that targeting of cellular mRNAs is an important feature of miRNAs expressed by these viruses and has exerted evolutionary pressure to conserve seed sequences.

A number of targets for miRNAs encoded by KSHV have been identified and provide an insight into the infectious pathway. The cellular mRNA *BCLAF1* is regulated by miR-K5, K9, and K10 binding to sites in the 3′UTR.[104] Inhibition of any of these miRNAs resulted in an increase in expression of BCLAF1. This led to poor recovery of progeny virus following activation from latency, suggesting that BCLAF1 might be involved in KSHV latency. Another KSHV miRNA target that is likely to have a role in regulation of viral latency is the major lytic switch protein RTA, which is repressed by miR-K9*.[105] KSHV miR-K1, K3-3p, K6-3p, and K11 regulate the expression of THBS1.[106] THBS1 is a multifunctional protein and can act as a chemoattractant, recruiting monocytes and T cells to sites of infection,[107] suggesting that these miRNAs may have a role in aiding immune evasion by KSHV. miR-K1 targets the cyclin-dependent kinase inhibitor p21 in latently infected B cells, and inhibits p21-mediated cell cycle arrest.[108] Multiple KSHV miRNAs repress expression of the transcriptional repressor MAF, and thus regulate differentiation of infected cells.[109]

The gammaherpesvirus murine herpesvirus 68 (MHV68) encodes nine miRNAs that are all expressed by an unusual mechanism.[73,110] The MHV68 pri-miRNAs are transcribed by RNA polymerase III from internal promoters within a tRNA-like structure 5′ of the pre-miRNA. The resulting pri-miRNAs have a 5′ tRNA moiety, and are processed in the nucleus by tRNAse Z instead of Drosha to yield canonical pre-miRNAs.[110]

## F. Common Targets of Herpesvirus miRNAs

Interestingly, EBV and KSHV both regulate targets of the cellular miRNA miR-155, but use different mechanisms to do so. miR-155 is considered an oncomir as its overexpression is associated with cancer.[111] EBV upregulates miR-155 expression (see Section V.C), while KSHV miR-K12-11 has a seed sequence identical to that of miR-155,[112,113] and the more distantly related alphaherpesvirus Marek's disease virus-1 (MDV-1) encodes an miR-155 ortholog.[114,115] KSHV miR-K12-11 targeted almost all the mRNAs regulated by miR-155, but there were slight differences between the target sets, which could be attributed to pairing outside the seed.[112] The ability of all three of these viruses to mimic or enhance miR-155 activity points to an important function for the targets of miR-155 in the context of infection. These herpesviruses all have oncogenic potential and the role of miR-155 in B cell transformation suggests that regulation of miR-155 targets may be important in promoting tumorigenesis.

A second common regulatory mechanism used by miRNAs expressed by different herpesviruses is the targeting of MICB, presumably as a means of evading an NK cell response. miRNAs encoded by HCMV, KSHV, and EBV are all able to downregulate expression of this protein. This is an intriguing example of functional conservation, as the miRNAs encoded by these viruses have different sequences and therefore target different sites in the 3′UTR of *MICB* mRNA.[116]

## G. Polyomavirus miRNAs

Polyomaviruses establish persistent infections in a range of host species, including humans, and can transform infected cells.[117] They have small dsDNA genomes that express two early genes, the small and large T-antigens, which initiate DNA replication and are involved in oncogenesis, and three late genes that encode the viral structural proteins. The early and late mRNAs are both transcribed from the same region of the genome, in opposite directions.[117] SV40 is a well-studied monkey polyomavirus that has been found to encode miRNAs capable of regulating viral infection.[118] A single functional SV40 pre-miRNA is encoded 3′ of the SV40 late mRNA poly(A) cleavage site, and expressed with late kinetics. The pre-miRNA is antisense to the SV40 early mRNA, mapping to a position corresponding to the 3′ end of the large T-antigen. miRNAs from both sides of the pre-miRNA hairpin are incorporated into RISC, and both are functional in cleavage of the large T-antigen mRNA.[118] This leads to a reduction in the level of both large T-antigen mRNA and protein, although disruption of the miRNA function does not affect the production of infectious virus. Instead, SV40 miRNA activity appears to be involved in evasion of the immune response. The large T-antigen is a target for the cytotoxic T-lymphocyte (CTL) response, so the miRNA-mediated reduction in large T-antigen levels might help infected cells to escape immune surveillance. This hypothesis was confirmed under conditions mimicking primary infection, in which cells infected with wild-type SV40 experienced lower susceptibility to CTL-mediated lysis than cells infected with a mutant virus that lacked the miRNA.[118]

Similar miRNAs have been identified in several other polyomaviruses, including BK virus, JC virus, and Merkel cell polyomavirus, which cause disease in humans.[119–121] The genomic location, expression pattern, and large T-antigen targeting of these polyomaviral miRNAs are conserved, although the miRNA sequence is not.

## H. Adenovirus miRNAs

Adenovirus is a dsDNA virus that expresses two structured noncoding RNA molecules, VAI and VAII. Both these RNAs are processed at low efficiency by Dicer to yield miRNAs which associate with the RISC.[122,123] The functions of these miRNAs remain unclear, raising the possibility that they might exist only as a consequence of low-level Dicer processing of highly expressed structured RNAs. The VAI-derived mivaRI-138 was recently shown to repress the expression of a number of genes, most of which do not have seed matches for the miRNA and are therefore unlikely to be direct targets.[124] The one direct target identified was TIA-1, an RNA metabolism factor with a proapoptotic function; however, the authors did not find any evidence of repression of apoptosis during adenoviral infection mediated by this miRNA.[124] Further research will be necessary to determine the importance of adenoviral miRNA expression during viral infection.

## I. HIV miRNAs

The evidence for miRNA expression by HIV-1 remains controversial. The HIV-1 TAR RNA has a hairpin structure and can be processed by Dicer *in vitro* to generate an miRNA.[125] miRNAs derived from both the 5′ and 3′ sides of the TAR hairpin were detected in HIV-1-infected CD4+ T cells.[77] On the other hand, two separate studies in which small RNAs were cloned from HIV-1 infected cells failed to reveal any virally derived miRNA sequences.[73,78] Cells infected with a different retrovirus, namely, the human T-cell leukemia virus type 1 (HTLV-1), were also tested, and no viral miRNAs were detected.[78]

## J. Common Themes of Viral miRNAs

The herpesviral miRNA targets identified to date show some common biological functions. Most viral and some cellular targets appear to be important in viral latency, and most evidence would suggest that viral miRNAs have a major role in regulating latency, although EBV miRNAs may be an exception to this. Cellular targets of herpesviral miRNAs appear to be involved in cell growth, differentiation, apoptosis, and immune evasion, all of which are processes that would be desirable for the virus to regulate. Certain common targets have emerged across the herpesvirus miRNAs. HCMV, EBV, and KSHV all use miRNAs to repress expression of MICB, which is involved in attracting NK cells, suggesting a common mechanism of immune evasion. Modulation of the immune response also appears to be the major function of the polyomavirus miRNAs. Three oncogenic herpesviruses use different methods to downregulate targets of the oncogenic miR-155, suggesting that they have evolved to use an existing oncogenic pathway.

Viral miRNAs appear generally not to be conserved in sequence identity, but frequently show conservation of position and function. This is clearly apparent in the polyomaviruses, where conserved miRNA location results in conserved targeting of the large T-antigen. Related herpesviruses also tend to have miRNAs in similar genomic locations. Although there is generally little sequence conservation between herpesvirus miRNAs, there are a few exceptions. These include some of the EBV and rLCV miRNAs as well as the HSV-1 and HSV-2 versions of miR-H2, which share a seed sequence. Conservation of miRNA seeds suggests that these miRNAs are likely to have important cellular mRNA targets that have exerted evolutionary pressure on these viruses. The ability of different miRNAs encoded by different viruses to target MICB by binding to different sites indicates that manipulation of the host cell environment may be conserved even in the absence of any sequence conservation.

## V. Regulation of Cellular miRNAs by Viral Infection

Viruses are known to regulate their cellular environment by altering the expression of host mRNAs, and it is now becoming apparent that miRNAs can also be regulated. There is increasing evidence that cellular miRNAs binding to cellular targets may regulate viral life cycles, and viral regulation of cellular miRNAs may result in changes in expression of cellular targets that affect the course of infection. This raises the possibility of therapeutic antiviral intervention directed against specific host miRNAs. In this section, we will consider the effects of viral infection on cellular miRNAs and examples of cellular miRNAs that indirectly regulate the life cycle of specific viruses.

### A. Regulation of Host miRNAs During CMV Infection

In addition to expressing their own miRNAs, some herpesviruses have been shown to modulate expression of specific host miRNAs. HCMV infection did not affect the expression of most miRNAs as measured by microarray, but did result in changes in the level of specific miRNA species.[126] Some miRNAs with identical seed sequences were coordinately regulated, despite different chromosomal locations; this suggests that common targets are likely to be regulated by modulation of miRNA expression. miR-100 and miR-101, both of which are downregulated on HCMV infection, were analyzed in more detail. Both miRNAs repress the expression of proteins in the mTOR pathway, and both negatively regulate HCMV replication when overexpressed.[126]

Murine cytomegalovirus (MCMV), which itself encodes a number of miRNAs, was recently shown to mediate changes in host miRNA expression.[127] Microarray analysis indicated that miR-27a levels are strongly downregulated during lytic MCMV infection in NIH-3T3 cells. Two other miRNAs located within the same cluster, miR-23a and miR-24, did not show any changes in expression, suggesting that miR-27a is controlled at a posttranscriptional level.[127] This was confirmed by the observation that pri-miR-27a levels do not change significantly during infection. miR-27b, which is expressed from a different genomic locus but only differs from miR-27a by one nucleotide, also showed decreased levels on MCMV infection, implying that sequence elements within the mature miRNA drive the regulation. Levels of the passenger miR-27a* and miR-27b* strands were unchanged, leading to the hypothesis that miR-27 is regulated by MCMV at the level of mature miRNA stability.[127]

It is not yet clear whether miR-27 is downregulated directly by the virus or as part of a host response to viral infection. miR-27 levels were reduced following MCMV infection of several different cell types, but it has not been seen to change in any other viral infections.[127] Overexpression of miR-27 in MCMV-infected cells leads to a reduction in viral titre, although the miR-27 targets involved in this antiviral effect have not yet been identified.[127] This suggests that the virus regulates the cellular miRNA pathway to evade a host antiviral response. No changes in miR-27 levels were observed during HCMV infection, and miR-100 and miR-101 were unaltered during MCMV infection.[126,127] This can be explained by the evolutionary distance of the two viruses, and emphasizes the specificity of viral effects on the miRNA pathway.

## B. Regulation of Host miRNAs by *Herpesvirus Saimiri* Noncoding RNAs

*Herpesvirus saimiri*, a gammaherpesvirus, expresses seven U-rich noncoding RNAs (HSURs) in transformed T cells.[128–130] These are similar in structure to small nuclear RNAs (snRNAs) and their function was unknown until an unexpected link to the miRNA pathway was recently revealed. The most highly-conserved HSURs, namely, HSUR1 and 2, both interact with host miRNAs.[131] miR-142-3p binds to the 5′ end of both HSUR1 and 2, while miR-27a binds to the central stem-loop of HSUR1, and miR-16 to the equivalent location in HSUR2. miR-27a binding to HSUR1 leads to the degradation of the miRNA by an unknown mechanism, whereas miR-142-3p and miR-16 levels are unaffected by the HSURs.[131] The specific regulation of miR-27a stability is interesting, as it parallels the miR-27 degradation induced by MCMV infection (Section V.A). It will be important to follow up these observations to determine what governs the regulation of miR-27 stability and what the biological consequences are for herpesvirus infection.

## C. Effects of EBV Infection on Host miRNA Expression

EBV is another example of a herpesvirus that, in addition to encoding its own miRNAs (Section IV.E), regulates the levels of specific host miRNAs. EBV infects naïve B cells and drives their proliferation and differentiation into memory B cells. Persistent latent EBV infection can result in various cancers. EBV gene expression follows a specific pattern during the three stages of latency, and LMP1, expressed during type III latency, shows the strongest association with tumorigenesis.[132] A subtractive hybridization method applied to human B cells indicated that several miRNAs were upregulated on EBV infection; miR-155, miR-146a, and miR-21 showed the most significant increase in levels.[133] Microarray analysis confirmed these results and showed that these miRNAs were induced in type III latency, whereas there were no significant differences in miRNA profile between type I latency and uninfected cells.[134] Interestingly, both studies also found that EBV infection induced an increase in the level of miR-27a and the other miRNAs encoded in the same cluster, that is, miR-23a and miR-24.[133,134] This is in contrast to the reduction in miR-27a levels observed in MCMV and *Herpesvirus saimiri* infections. The biological role of this miRNA is not yet clear in any of these viral infections.

The induction in miR-146a levels is driven by LMP1, which regulates pri-miR-146a transcription and requires two NF-κB sites in the pri-miR-146a promoter.[133,135] miR-146a overexpression suppresses the interferon response, and may thus contribute to EBV immune evasion and to tumorigenesis.[135] Similar to miR-146a, miR-155 is overexpressed in type III latency. miR-155 induction occurs at the level of transcription and is dependent on EBV latency gene products, but neither LMP1 nor LMP2A was sufficient to mediate this induction, so a complex regulatory mechanism is likely to be involved.[136] An AP-1 site in the pri-miR-155 promoter is critical for the induction, and an NF-κ B site also contributes.[136] miR-155 regulates the expression of several transcription factors, and KSHV encodes an miR-155 mimic (Section IV.F), suggesting that the effects of miR-155 on gene expression may have broad importance for oncogenic herpesvirus infection. The induction in miR-21 levels takes place partly at the level of pri-miR-21 transcription, but the effect on mature miR-21 is of greater magnitude, suggesting that other aspects of miR-21 biogenesis may also be affected.[134] miR-21 has oncogenic properties and is associated with several malignancies.[137]

## D. Other Herpesviruses

MDV is an oncogenic alphaherpesvirus that infects poultry. In addition to expressing virally encoded miRNAs, an MDV-transformed lymphoblastoid cell line showed increased expression of cellular miR-221 and miR-222.[137] These two miRNAs are associated with human tumors and repress $p27^{Kip1}$, a cell cycle regulatory protein, leading to increased growth and proliferation. This targeting was shown to be conserved in chicken, and $p27^{Kip1}$ levels are reduced in MDV-transformed cell lines. These results suggest that regulation of cellular miRNAs may contribute to the oncogenic effects of MDV infection, although miR-221 and miR-222 were only regulated in one of the cell lines tested, so other mechanisms must also be involved.

## E. Human Papillomavirus

Human papillomaviruses (HPVs) are small DNA viruses that establish persistent infections in basal epithelial cells, and high-risk strains of HPV show a strong association with cervical cancer.[138] During HPV infection, the viral genome enters the nucleus and exists as an episome, with low-level transcription of the viral early genes. Viral DNA replication and transcription of the late genes occur on differentiation of the host cells to suprabasal epithelia. The cell cycle exit that would normally occur on differentiation is prevented in a process involving the E6 and E7 oncoproteins, which induce the degradation of the tumor suppressors p53 and pRB, respectively.[139,140] As DNA viruses that establish long-term infections, the papillomaviruses were good candidates for viral miRNA production, but cloning approaches did not identify any HPV miRNAs.[75,141] However, HPV infection does regulate the expression of specific cellular miRNAs.

Total miRNA cloning and microarray analysis in HPV-infected cervical carcinoma cell lines and tissue samples showed changes in expression of several miRNAs. In particular, miR-143 and miR-145, which are expressed from the same pri-miRNA, were downregulated in cervical carcinoma.[141] miR-34a levels were also reduced in cervical cancer tissues and cell lines, as well as in cultured keratinocytes immortalized with high-risk HPV16 or 18.[142] p53 stimulates pri-miR-34a transcription, and destabilization of p53 by E6 was responsible for the modulation of miR-34a expression by HPV.[142] Ectopic expression of miR-34a, miR-143, or miR-145 repressed proliferation in HeLa cells, which are derived from a cervical carcinoma.[141,142] miR-218 was also specifically reduced in HPV-positive cell lines and cervical carcinoma, and targets the mRNA encoding the epithelial cell-specific protein LAMB3, suggesting a possible role in control of differentiation and oncogenesis.[143]

Two groups have recently shown that HPV infection inhibits the induction of miR-203 levels that normally occurs during keratinocyte differentiation. There is debate regarding the viral proteins that mediate this effect: Melar-New *et al.* attributed the response to the viral E7 protein, acting via inhibition of the mitogen-activated protein kinase (MAPK)/protein kinase C (PKC) pathway, whereas McKenna *et al.* observed a role for E6 acting via destabilization of p53.[144,145] The details of transcription factors that bind to and regulate pri-miR-203 have not yet been determined, and these studies emphasize that the regulatory mechanisms governing miRNA expression may be complex. miR-203 inhibits cellular proliferation, an activity that is likely to be related to its downregulation of the transcription factor p63.[144] The HPV life cycle is linked to the differentiation status of the epithelial cells the virus infects, and it seems likely that regulation of miR-203 contributes to this by increasing the proliferative capacity of infected cells. This hypothesis was supported by overexpression of miR-203, which led to a reduction in HPV genomes when cells were differentiated or cultured for several passages.[144]

## F. Adenovirus

The adenovirus VAI RNA is present at high levels in infected cells. Ectopic expression of VAI at levels consistent with adenoviral infection led to a reduction in levels of miRNAs and siRNAs expressed from transfected DNA precursors. This was due in part to competition between VAI and pre-miRNAs or shRNAs for Exportin 5, and in part to inhibition of Dicer function.[146] As adenoviral VA RNAs are partially processed to miRNAs (Section IV.H), it is not clear what the functional consequences of the engagement between VA RNAs and the miRNA pathway are.

## G. HIV

HIV-1 infection provides an example of an antiviral effect of cellular miRNAs that the virus can partially overcome. Evidence of a role for the miRNA pathway in HIV-1 replication was found when siRNA-mediated depletion of the enzymes Drosha and Dicer, both of which are important for miRNA biogenesis, led to an increase in HIV-1 replication kinetics.[147] HIV-1 infection resulted in increased expression of several cellular miRNAs, including miR-122, miR-370, miR-373*, and miR-297, although the consequences of this effect remain unclear. HIV-1 also suppresses expression of the miR-17-92 cluster of miRNAs via an unknown mechanism. Knockdown of miR-17-92 led to an increase in HIV-1 replication, suggesting that decreased levels of these miRNAs might contribute to the phenotype of Drosha and Dicer depletion. Two members of the miR-17-92 cluster, miR-17-5p and miR-20a, bind to target sites in the 3′UTR of the histone acetyltransferase *PCAF* mRNA and inhibit the expression of the protein.[147] The effects of Drosha and Dicer knockdown on HIV-1 replication were attenuated when PCAF was also depleted. PCAF is a cofactor for the HIV-1 Tat protein,[72] and Tat-mediated transactivation of an integrated HIV-1 LTR construct increased when Drosha was knocked down. These results indicate that miR-17-5p and miR-20a regulate HIV-1 infection via PCAF/Tat, although Drosha and Dicer may also affect HIV-1 replication by additional mechanisms.[147,148]

# VI. Regulation of Viruses by Cellular miRNAs

The possibility that miRNAs might bind directly to viral RNA targets has been the subject of many investigations. Both proviral and antiviral roles of specific miRNAs binding to specific viruses have been revealed, although it is not yet clear whether many of the antiviral responses occur in physiologically relevant conditions.[74] Indirect modulation of viruses by the miRNAs of an infected cell is also likely, as the cellular pathways that control the survival of infected cells or the virus life cycle may be regulated by miRNAs. Some examples of indirect regulation of viruses by miRNAs are discussed in Section V. This section is concerned with direct regulation by host miRNAs binding to viral RNA.

## A. Positive Regulation of HCV by miR-122

The potential for direct and positive regulation of a virus by a host miRNA has been demonstrated by the role of the liver-specific miR-122 in HCV replication.[149]

HCV is a positive-sense RNA virus with a 9.6-kb genome (Fig. 3), which establishes persistent infections in the human liver. Around 3% of the global population is infected with HCV, and chronic infection may eventually lead to liver cirrhosis and hepatocellular carcinoma.[150] miR-122 is highly expressed in the liver, and was shown to positively regulate HCV replication in cultured

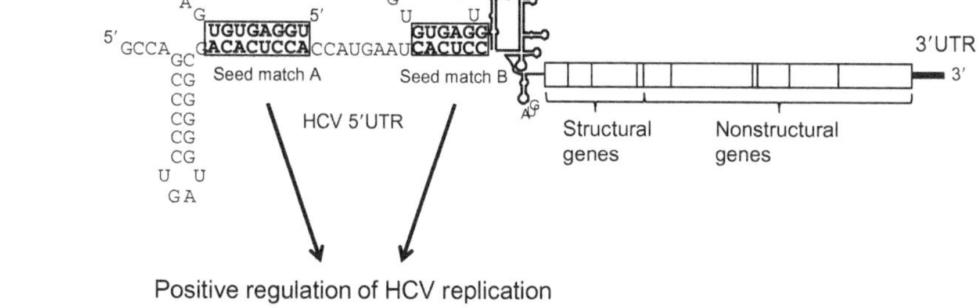


FIG. 3. Regulation of HCV by miR-122. The structure of the 9.6-kb positive-sense RNA genome of HCV is shown. A single open reading frame encoding the viral structural and nonstructural proteins is flanked by structured 5′ and 3′UTRs. Translation is driven by an IRES in the 5′UTR. The sequence of the 45-nt upstream of the IRES, containing the two miR-122 binding sites, is shown. In this model, two miR-122 molecules are shown interacting simultaneously with the two miR-122 seed matches, but it is possible that the two seed matches are not occupied at the same time, and that miR-122 makes additional contacts with HCV RNA outside the seed match. The miR-122 interaction promotes HCV replication by an unknown mechanism.

Huh7 human liver cells.[149] It was found that this regulation was governed by a direct interaction between miR-122 and two adjacent sites in the 5′UTR of the viral RNA (Fig. 3).[149,151] RNAi-mediated knockdown of components of the miRNA machinery resulted in reduced HCV replication, which is likely to be due to a reduction in miR-122 levels and/or activity.[152]

While a mechanism that fully explains the effects of miR-122 on HCV replication remains unknown, a few studies hint at the details behind this process. The two miR-122 binding sites are located upstream of the HCV IRES which drives translation of the viral genome (Fig. 3). Initial observations did not show any effects of miR-122 on HCV protein synthesis,[149] but it was subsequently shown that there is a moderate level of translation stimulation on miR-122 binding.[153] This translational regulation was confirmed in a second study, but this indicated that the translation defect in the absence of miR-122 binding was not sufficient to account for the absence of detectable HCV replication under these conditions.[154] This implies that miR-122 must also positively regulate another stage of the HCV life cycle. The identity of this stage remains unclear, as there were no detectable changes in HCV RNA synthesis in the absence of miR-122.[155] A study of HCV RNA synthesis in isolated replication complexes established that miR-122 does not exert its positive effect at the elongation phase of HCV RNA synthesis in this cell-free system.[156]

This is currently the only example of an miRNA that directly binds to a viral genome and exerts a positive effect on the viral life cycle. It will be interesting to see whether other RNA viruses, especially flaviviruses, which are related to HCV, engage in similar interactions with host miRNAs and/or the host miRNA processing machinery.

## B. Potential for HCV Therapeutics Based on Inhibition of miR-122

Chemically modified oligonucleotides complementary to target miRNA sequences have been used successfully to silence miRNAs *in vivo* (Section II. C). Different studies used this strategy to sequester miR-122 in mice and African green monkeys, and effective inhibition was demonstrated when endogenous targets of miR-122 were derepressed.[47,48,157,158]

Specific therapeutic silencing of miR-122 has thus far emerged as a potentially effective anti-HCV medication. Recent trials concerning the efficacy of miR-122 silencing in chronically infected chimpanzees delivered promising results for potential HCV therapeutics.[49] Using anti-miR-122 LNA oligonucleotides, the investigators were able to show effective suppression of miR-122 in chimpanzee liver tissues and a significant reduction in HCV viremia.[49] A lack of rebound over the course of a 12-week treatment program and the appearance of no resistant viral strains, coupled with an absence of liver toxicity, lend this approach great promise.[49] This study demonstrates the feasibility of therapeutic miRNA inhibition, but this research remains at an early stage. The interplay between miR-122 and HCV may be more complex in human patients, as liver biopsies from HCV-infected individuals showed no positive correlation between miR-122 expression and HCV RNA.[159]

## C. Negative Regulation of Viral Infection by Direct Binding of Cellular miRNAs

The possibility that miRNAs interact with viral RNAs and thereby limit replication is still somewhat controversial.[160] Experimentally introduced siRNAs are known to inhibit virus infection, suggesting at least in principle that inhibition of viruses by miRNAs is possible under normal cellular conditions.[161] However, given the very short regions of complementarity that are required for miRNA targeting, and the rapid evolution of viral sequences, it is unlikely that a virus would retain a target site for an miRNA that had detrimental effects on its life cycle.

Targeting of viral RNA by a tissue-specific miRNA could be an effective determinant of viral tissue tropism by restricting viral infection to cells in which the miRNA is not expressed.[162] The regulation of viral tropism by tissue-specific miRNAs has been established in principle using engineered picornaviruses. These viruses contain muscle-specific miRNA target sequences in their 3′UTRs and these abolish infection in cells that contain this miRNA.[163]

Endogenous cellular miR-32 was shown to limit the replication of the retrovirus primate foamy virus (PFV) in HeLa and BHK21 cells through an interaction with a poorly conserved region in the 3′ region of the PFV genome.[164] Whether such an interaction would occur in the context of a natural infection and thereby limit PFV replication remains to be shown. Recent observations have shown that Dicer-deficient mice are hypersensitive to infection by the − ssRNA vesicular stomatitis virus (VSV).[165] Furthermore, miR- 24 and miR-93 were shown to downregulate viral mRNAs.[165] Although this demonstrates that cellular miRNAs may directly interact with VSV, there is currently no evidence that these cellular miRNAs restrict VSV replication in its natural hosts.[165]

Several miRNAs have target sites in the 3′UTR of HIV-1 mRNAs, and can function to repress expression of these HIV-1 genes in CD4+ T cells, contributing to viral latency.[166] HIV-1 latency is important in resistance to highly active antiretroviral therapy (HAART), but it is questionable whether these interactions would be relevant in the absence of HAART.[74] An abundant miRNA in T cells, miR-29a, was also shown to interact with the HIV-1 3′UTR and to inhibit viral replication.[167] This interaction resulted in the recruitment of HIV-1 RNA to P bodies, which could be relevant in releasing the viral RNA from the translational machinery during the viral life cycle.[74] There is emerging evidence that P bodies may be important in infection with various RNA viruses.[168]

Increased expression of a number of cellular miRNAs was observed upon the treatment of hepatocytes with interferon beta (IFNβ).[169] Several of these miRNAs possess target sequences in the HCV genome, and were shown to inhibit viral replication. miR-122 levels were downregulated by IFNβ treatment, suggesting that modulation of cellular miRNA expression might contribute to the antiviral effect of IFNβ.[169] However, most of these changes were not observed in a subsequent study of human liver cells treated with either IFNα or IFNβ, and some of the miRNAs reported to be upregulated by IFN treatment were found to be expressed at very low levels in these cells, and are therefore unlikely to be physiologically relevant.[159]

## VII. Conclusions

It is becoming increasingly apparent that viruses of diverse origin can utilize the host cell miRNA pathway in numerous different ways in order to regulate viral infection and the host response (Fig. 4). Several important questions remain to be answered. It is still unclear whether the production of viral miRNAs is an important mechanism outside the herpesviruses and polyomaviruses. Moreover, although some of the targets of herpesvirus miRNAs have now been identified, this research is still at an early stage and it is likely that many more targets are still to be revealed. The exact interplay between herpesvirus miRNA expression and processes such as viral latency and immune evasion remains uncertain. Current evidence suggests that different herpesviruses can use miRNAs to regulate some common mechanisms, but the extent to which miRNAs from different viruses are used for the same function is not yet clear.

The regulation of viral life cycles by cellular miRNAs will be another important area for future research. Regulation of cellular miRNA expression by viruses appears to be a complex and widely used mechanism that may affect the host environment and the viral life cycle by changes in the expression of host miRNA targets. The mechanisms by which specific viral infections modulate the expression of specific cellular miRNAs are largely unknown and may be due to direct regulation by viral proteins or to indirect regulation caused by cellular responses to viral infection, such as the induction of apoptosis.

Several examples of miRNAs that bind directly to viral RNA and negatively regulate the virus have been uncovered, but in many instances these experiments were conducted in cells or organisms other than the natural host, and the physiological relevance of these interactions remains uncertain. It is likely that evolutionary pressure would prevent a virus retaining target sites for host

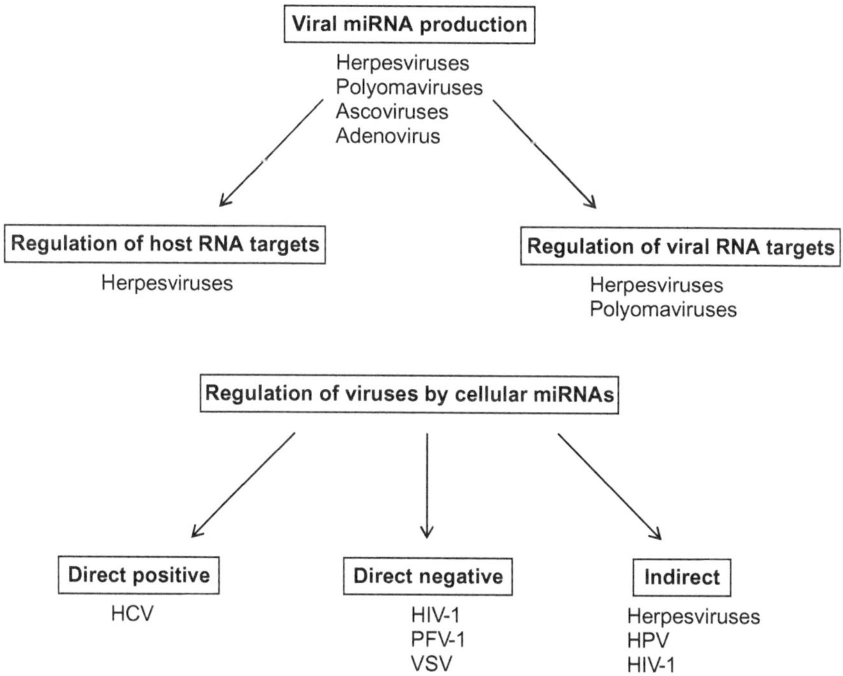


FIG. 4. Overview of viruses and the miRNA pathway. Different mechanisms by which viruses can interact with the miRNA pathway are shown in boxes, with examples of viruses that have been shown to use each mechanism below.

miRNAs that negatively regulate its infection, but it is possible that such a site would be conserved if miRNA-mediated repression of the virus at certain points in the life cycle was beneficial.

Finally, the regulation of HCV by miR-122 is currently the only example of a direct positive role for an miRNA in a viral life cycle. It would be very interesting to establish whether other viruses use similar mechanisms.

## REFERENCES

1. Carthew RW, Sontheimer EJ. Origins and mechanisms of miRNAs and siRNAs. *Cell* 2009;**136**:642–55.
2. Lee RC, Feinbaum RL, Ambros V. The *C. elegans* heterochronic gene lin-4 encodes small RNAs with antisense complementarity to lin-14. *Cell* 1993;**75**:843–54.
3. Reinhart BJ, Slack FJ, Basson M, Pasquinelli AE, Bettinger JC, Rougvie AE, et al. The 21-nucleotide let-7 RNA regulates developmental timing in *Caenorhabditis elegans*. *Nature* 2000;**403**:901–6.

4. Pasquinelli AE, Reinhart BJ, Slack F, Martindale MQ, Kuroda MI, Maller B, et al. Conservation of the sequence and temporal expression of let-7 heterochronic regulatory RNA. *Nature* 2000;**408**:86–9.
5. Lau N, Lim L, Weinstein E, Bartel D. An abundant class of tiny RNAs with probable regulatory roles in *Caenorhabditis elegans*. *Science* 2001;**294**:858–62.
6. Lee RC, Ambros V. An extensive class of small RNAs in *Caenorhabditis elegans*. *Science* 2001;**294**:862–4.
7. Lagos-Quintana M, Rauhut R, Lendeckel W, Tuschl T. Identification of novel genes coding for small expressed RNAs. *Science* 2001;**294**:853–8.
8. Griffiths-Jones S, Saini HK, van Dongen S, Enright AJ. miRBase: tools for microRNA genomics. *Nucleic Acids Res* 2008;**36**:D154–8.
9. Kim VN, Han J, Siomi MC. Biogenesis of small RNAs in animals. *Nat Rev Mol Cell Biol* 2009;**10**:126–39.
10. Ruby J, Jan C, Bartel D. Intronic microRNA precursors that bypass Drosha processing. *Nature* 2007;**448**:83–6.
11. Okamura K, Hagen J, Duan H, Tyler D, Lai E. The mirtron pathway generates microRNA-class regulatory RNAs in Drosophila. *Cell* 2007;**130**:89–100.
12. Cheloufi S, Dos Santos C, Chong M, Hannon G. A dicer-independent miRNA biogenesis pathway that requires Ago catalysis. *Nature* 2010;**465**:584–9.
13. Cifuentes D, Xue H, Taylor DW, Patnode H, Mishima Y, Cheloufi S, et al. A novel miRNA processing pathway independent of Dicer requires Argonaute2 catalytic activity. *Science* 2010;**328**:1694–8.
14. Bartel DP. MicroRNAs: target recognition and regulatory functions. *Cell* 2009;**136**:215–33.
15. Lewis BP, Shih IH, Jones-Rhoades MW, Bartel DP, Burge CB. Prediction of mammalian microRNA targets. *Cell* 2003;**115**:787–98.
16. Brennecke J, Stark A, Russell RB, Cohen SM. Principles of MicroRNA-target recognition. *PLoS Biol* 2005;**3**:404–18.
17. Gu S, Jin L, Zhang F, Sarnow P, Kay M. Biological basis for restriction of microRNA targets to the 3′ untranslated region in mammalian mRNAs. *Nat Struct Mol Biol* 2009;**16**:144–50.
18. Schnall-Levin M, Zhao Y, Perrimon N, Berger B. Conserved microRNA targeting in Drosophila is as widespread in coding regions as in 3′UTRs. *Proc Natl Acad Sci USA* 2010;**107**:15751–6.
19. Tay Y, Zhang J, Thomson A, Lim B, Rigoutsos I. MicroRNAs to Nanog, Oct4 and Sox2 coding regions modulate embryonic stem cell differentiation. *Nature* 2008;**455**:1124–8.
20. Jackson R, Hellen C, Pestova T. The mechanism of eukaryotic translation initiation and principles of its regulation. *Nat Rev Mol Cell Biol* 2010;**11**:113–27.
21. Fabian M, Sonenberg N, Filipowicz W. Regulation of mRNA translation and stability by microRNAs. *Annu Rev Biochem* 2010;**79**:351–79.
22. Eulalio A, Behm-Ansmant I, Schweizer D, Izaurralde E. P-body formation is a consequence, not the cause, of RNA-mediated gene silencing. *Mol Cell Biol* 2007;**27**:3970–81.
23. Fabian MR, Mathonnet G, Sundermeier T, Mathys H, Zipprich JT, Svitkin YV, et al. Mammalian miRNA RISC recruits CAF1 and PABP to affect PABP-dependent deadenylation. *Mol Cell* 2009;**35**:868–80.
24. Lim LP, Lau NC, Garrett-Engele P, Grimson A, Schelter JM, Castle J, et al. Microarray analysis shows that some microRNAs downregulate large numbers of target mRNAs. *Nature* 2005;**433**:769–73.
25. Guo H, Ingolia N, Weissman J, Bartel D. Mammalian microRNAs predominantly act to decrease target mRNA levels. *Nature* 2010;**466**:835–40.
26. Krol J, Loedige I, Filipowicz W. The widespread regulation of microRNA biogenesis, function and decay. *Nat Rev Genet* 2010;**11**:597–610.

27. Vasudevan S, Tong Y, Steitz J. Switching from repression to activation: microRNAs can up-regulate translation. *Science* 2007;**318**:1931–4.
28. Croce C. Causes and consequences of microRNA dysregulation in cancer. *Nat Rev Genet* 2009;**10**:704–14.
29. Landgraf P, Rusu M, Sheridan R, Sewer A, Iovino N, Aravin A, et al. A mammalian microRNA expression atlas based on small RNA library sequencing. *Cell* 2007;**129**:1401–14.
30. Viswanathan SR, Daley GQ, Gregory RI. Selective blockade of MicroRNA processing by Lin28. *Science* 2008;**320**:97–100.
31. Heo I, Joo C, Cho J, Ha M, Han J, Kim VN. Lin28 mediates the terminal uridylation of let-7 precursor MicroRNA. *Mol Cell* 2008;**32**:276–84.
32. Slezak-Prochazka I, Durmus S, Kroesen BJ, van den Berg A. MicroRNAs, macrocontrol: regulation of miRNA processing. *RNA* 2010;**16**:1087–95.
33. Katoh T, Sakaguchi Y, Miyauchi K, Suzuki T, Kashiwabara SI, Baba T, et al. Selective stabilization of mammalian microRNAs by 3′ adenylation mediated by the cytoplasmic poly(A) polymerase GLD-2. *Genes Dev* 2009;**23**:433–8.
34. Krol J, Busskamp V, Markiewicz I, Stadler MB, Ribi S, Richter J, et al. Characterizing light-regulated retinal microRNAs reveals rapid turnover as a common property of neuronal microRNAs. *Cell* 2010;**141**:618–31.
35. Chatterjee S, Grosshans H. Active turnover modulates mature microRNA activity in *Caenorhabditis elegans*. *Nature* 2009;**461**:U546–9.
36. Calin GA, Dumitru CD, Shimizu M, Richi R, Zupo S, Noch E, et al. Frequent deletions and down-regulation of micro-RNA genes miR15 and miR16 at 13q14 in chronic lymphocytic leukemia. *Proc Natl Acad Sci USA* 2002;**99**:15524–9.
37. Calin GA, Sevignani C, Dumitru CD, Hyslop T, Noch E, Yendamuri S, et al. Human microRNA genes are frequently located at fragile sites and genomic regions involved in cancers. *Proc Natl Acad Sci USA* 2004;**101**:2999–3004.
38. Esquela-Kerscher A, Trang P, Wiggins JF, Patrawala L, Cheng A, Ford L, et al. The let-7 microRNA reduces tumor growth in mouse models of lung cancer. *Cell Cycle* 2008;**7**:759–64.
39. Johnson SM, Grosshans H, Shingara J, Byrom M, Jarvis R, Cheng A, et al. RAS is regulated by the let-7 microRNA family. *Cell* 2005;**120**:635–47.
40. Lee Y, Dutta A. The tumor suppressor microRNA let-7 represses the HMGA2 oncogene. *Genes Dev* 2007;**21**:1025–30.
41. Rybak A, Fuchs H, Smirnova L, Brandt C, Pohl EE, Nitsch R, et al. A feedback loop comprising lin-28 and let-7 controls pre-let-7 maturation during neural stem-cell commitment. *Nat Cell Biol* 2008;**10**:987–93.
42. Umbach JL, Cullen BR. The role of RNAi and microRNAs in animal virus replication and antiviral immunity. *Genes Dev* 2009;**23**:1151–64.
43. Hutvágner G, Simard M, Mello C, Zamore P. Sequence-specific inhibition of small RNA function. *PLoS Biol* 2004;**2**:E98.
44. Meister G, Landthaler M, Dorsett Y, Tuschl T. Sequence-specific inhibition of microRNA- and siRNA-induced RNA silencing. *RNA* 2004;**10**:544–50.
45. Esau C. Inhibition of microRNA with antisense oligonucleotides. *Methods* 2008;**44**:55–60.
46. Seto A. The road toward microRNA therapeutics. *Int J Biochem Cell Biol* 2010;**42**:1298–305.
47. Elmen J, Lindow M, Silahtaroglu A, Bak M, Christensen M, Lind-Thomsen A, et al. LNA-mediated microRNA silencing in non-human primates. *Nature* 2008;**452**:896–9.
48. Elmen J, Lindow M, Schutz S, Lawrence M, Petri A, Obad S, et al. Antagonism of microRNA-122 in mice by systemically administered LNA-antimiR leads to up-regulation of a large set of predicted target mRNAs in the liver. *Nucleic Acids Res* 2008;**36**:1153–62.

49. Lanford RE, Hildebrandt-Eriksen ES, Petri A, Persson R, Lindow M, Munk ME, et al. Therapeutic silencing of MicroRNA-122 in primates with chronic hepatitis C virus infection. *Science* 2010;**327**:198–201.
50. Marsh M, Helenius A. Virus entry: open sesame. *Cell* 2006;**124**:729–40.
51. Puntener D, Greber UF. DNA-tumor virus entry-from plasma membrane to the nucleus. *Semin Cell Dev Biol* 2009;**20**:631–42.
52. Maurer UE, Sodeik B, Grunewald K. Native 3D intermediates of membrane fusion in herpes simplex virus 1 entry. *Proc Natl Acad Sci USA* 2008;**105**:10559–64.
53. Sodeik B, Ebersold MW, Helenius A. Microtubule-mediated transport of incoming herpes simplex virus 1 capsids to the nucleus. *J Cell Biol* 1997;**136**:1007–21.
54. Sodeik B. Mechanisms of viral transport in the cytoplasm. *Trends Microbiol* 2000;**8**:465–72.
55. Sodeik B. Unchain my heart, baby let me go—the entry and intracellular transport of HIV. *J Cell Biol* 2002;**159**:393–5.
56. Greber UF, Way M. A superhighway to virus infection. *Cell* 2006;**124**:741–54.
57. Lyman MG, Enquist LW. Herpesvirus interactions with the host cytoskeleton. *J Virol* 2009;**83**:2058–66.
58. Boehmer PE, Lehman IR. Herpes simplex virus DNA replication. *Annu Rev Biochem* 1997;**66**:347–84.
59. Wobbe CR, Dean F, Weissbach L, Hurwitz J. Invitro replication of duplex circular DNA containing the simian-virus 40 DNA origin site. *Proc Natl Acad Sci USA* 1985;**82**:5710–4.
60. Hobman T, Chantler J. Rubella Virus. In: Knipe DM, Howley PM, editors. *Fields virology*. Philadelphia: Lippincott Williams & Wilkins; 2007.
61. Lai MMC, Perlman S, Anderson LJ. Knipe DM, Howley PM, editors. *Fields Virology*. Lippincott Williams & Wilkins; 2007.
62. Balvay L, Rifo RS, Ricci EP, Decimo D, Ohlmann T. Structural and functional diversity of viral IRESes. *Biochim Biophys Acta* 2009;**1789**:542–57.
63. Engelhardt OG, Fodor E. Functional association between viral and cellular transcription during influenza virus infection. *Rev Med Virol* 2006;**16**:329–45.
64. Reinisch KM. The dsRNA viridae and their catalytic capsids. *Nat Struct Biol* 2002;**9**:714–6.
65. Miller S, Krijnse-Locker J. Modification of intracellular membrane structures for virus replication. *Nat Rev Microbiol* 2008;**6**:363–74.
66. Cullen BR. Journey to the center of the cell. *Cell* 2001;**105**:697–700.
67. Blackford AN, Grand RJA. Adenovirus E1B 55-kilodalton protein: multiple roles in viral infection and cell transformation. *J Virol* 2009;**83**:4000–12.
68. Khalili K, Sariyer IK, Safak M. Small tumor antigen of polyomaviruses: role in viral life cycle and cell transformation. *J Cell Physiol* 2008;**215**:309–19.
69. Smiley JR. Herpes simplex virus virion host shutoff protein: immune evasion mediated by a viral RNase? *J Virol* 2004;**78**:1063–8.
70. Galluzzi L, Brenner C, Morselli E, Touat Z, Kroemer G. Viral control of mitochondrial apoptosis. *PLoS Pathog* 2008;**4**:e1000018.
71. Hansen TH, Bouvier M. MHC class I antigen presentation: learning from viral evasion strategies. *Nat Rev Immunol* 2009;**9**:503–13.
72. Pfeffer S, Zavolan M, Grasser FA, Chien M, Russo JJ, Ju J, et al. Identification of virus-encoded microRNAs. *Science* 2004;**304**:734–6.
73. Pfeffer S, Sewer A, Lagos-Quintana M, Sheridan R, Sander C, Grasser FA, et al. Identification of microRNAs of the herpesvirus family. *Nat Methods* 2005;**2**:269–76.
74. Skalsky RL, Cullen BR. Viruses, microRNAs, and host interactions. *Annu Rev Microbiol* 2010;**17**:17.

75. Cai X, Li G, Laimins L, Cullen B. Human papillomavirus genotype 31 does not express detectable microRNA levels during latent or productive virus replication. *J Virol* 2006;**80**:10890–3.
76. Umbach J, Nagel M, Cohrs R, Gilden D, Cullen B. Analysis of human alphaherpesvirus microRNA expression in latently infected human trigeminal ganglia. *J Virol* 2009;**83**:10677–83.
77. Ouellet DL, Plante I, Landry P, Barat C, Janelle ME, Flamand L, et al. Identification of functional microRNAs released through asymmetrical processing of HIV-1 TAR element. *Nucleic Acids Res* 2008;**36**:2353–65.
78. Lin J, Cullen B. Analysis of the interaction of primate retroviruses with the human RNA interference machinery. *J Virol* 2007;**81**:12218–26.
79. Parameswaran P, Sklan E, Wilkins C, Burgon T, Samuel MA, Lu R, et al. Six RNA viruses and forty-one hosts: viral small RNAs and modulation of small RNA repertoires in vertebrate and invertebrate systems. *PLoS Pathog* 2010;**6**:e1000764.
80. Pellett PE, Roizman B. The Herpesviridae family: A Brief Introduction. In: Knipe DM, Howley PM, editors. *Fields virology*. Philadelphia: Lippincott Williams & Wilkins; 2007.
81. Efstathiou S, Preston CM. Towards an understanding of the molecular basis of herpes simplex virus latency. *Virus Res* 2005;**111**:108–19.
82. Roizman B, Knipe DM, Whitley RJ. Herpes Simplex Viruses. In: Knipe DM, Howley PM, editors. *Fields virology*. Philadelphia: Lippincott Williams & Wilkins; 2007.
83. Cui C, Griffiths A, Li G, Silva LM, Kramer MF, Gaasterland T, et al. Prediction and identification of herpes simplex virus 1-encoded microRNAs. *J Virol* 2006;**80**:5499–508.
84. Jones C. Herpes simplex virus type 1 and bovine herpesvirus 1 latency. *Clin Microbiol Rev* 2003;**16**:79–95.
85. Grey F, Hook L, Nelson J. The functions of herpesvirus-encoded microRNAs. *Med Microbiol Immunol* 2008;**197**:261–7.
86. Umbach JL, Kramer MF, Jurak I, Karnowski HW, Coen DM, Cullen BR. MicroRNAs expressed by herpes simplex virus 1 during latent infection regulate viral mRNAs. *Nature* 2008;**454**:780–3.
87. Tang S, Bertke AS, Patel A, Wang K, Cohen JI, Krause PR. An acutely and latently expressed herpes simplex virus 2 viral microRNA inhibits expression of ICP34.5, a viral neurovirulence factor. *Proc Natl Acad Sci USA* 2008;**105**:10931–6.
88. Tang S, Patel A, Krause P. Novel less-abundant viral microRNAs encoded by herpes simplex virus 2 latency-associated transcript and their roles in regulating ICP34.5 and ICP0 mRNAs. *J Virol* 2009;**83**:1433–42.
89. Grey F, Antoniewicz A, Allen E, Saugstad J, McShea A, Carrington JC, et al. Identification and characterization of human cytomegalovirus-encoded microRNAs. *J Virol* 2005;**79**:12095–9.
90. Grey F, Meyers H, White E, Spector D, Nelson J. A human cytomegalovirus-encoded microRNA regulates expression of multiple viral genes involved in replication. *PLoS Pathog* 2007;**3**:e163.
91. Murphy E, Vanicek J, Robins H, Shenk T, Levine AJ. Suppression of immediate-early viral gene expression by herpesvirus-coded microRNAs: implications for latency. *Proc Natl Acad Sci USA* 2008;**105**:5453–8.
92. Greaves RF, Mocarski ES. Defective growth correlates with reduced accumulation of a viral DNA replication protein after low-multiplicity infection by a human cytomegalovirus ie1 mutant. *J Virol* 1998;**72**:366–79.
93. Grey F, Meyers H, White EA, Spector DH, Nelson J. A human cytomegalovirus-encoded microRNA regulates expression of multiple viral genes involved in replication. *PLoS Pathog* 2007;**3**:1593–602.

94. Stern-Ginossar N, Elefant N, Zimmermann A, Wolf DG, Saleh N, Biton M, et al. Host immune system gene targeting by a viral miRNA. *Science* 2007;**317**:376–81.
95. Grey F, Tirabassi R, Meyers J, Wu G, McWeeney S, Hook L, et al. A viral microRNA down-regulates multiple cell cycle genes through mRNA 5′UTRs. *PLoS Pathog* 2010;**6**:e1000967.
96. Barth S, Pfuhl T, Mamiani A, Ehses C, Roemer K, Kremmer E, et al. Epstein-Barr virus-encoded microRNA miR-BART2 down-regulates the viral DNA polymerase BALF5. *Nucleic Acids Res* 2008;**36**:666–75.
97. Lo AK, To KF, Lo KW, Lung RW, Hui JW, Liao G, et al. Modulation of LMP1 protein expression by EBV-encoded microRNAs. *Proc Natl Acad Sci USA* 2007;**104**:16164–9.
98. Choy EY, Siu KL, Kok KH, Lung RW, Tsang CM, To KF, et al. An Epstein-Barr virus-encoded microRNA targets PUMA to promote host cell survival. *J Exp Med* 2008;**205**:2551–60.
99. Han J, Flemington C, Houghton AB, Gu Z, Zambetti GP, Lutz RJ, et al. Expression of bbc3, a pro-apoptotic BH3-only gene, is regulated by diverse cell death and survival signals. *Proc Natl Acad Sci USA* 2001;**98**:11318–23.
100. Xia T, O'Hara A, Araujo I, Barreto J, Carvalho E, Sapucaia JB, et al. EBV MicroRNAs in primary lymphomas and targeting of CXCL-11 by ebv-mir-BHRF1-3. *Cancer Res* 2008;**68**:1436–42.
101. Cole KE, Strick CA, Paradis TJ, Ogborne KT, Loetscher M, Gladue RP, et al. Interferon-inducible T cell alpha chemoattractant (I-TAC): a novel non-ELR CXC chemokine with potent activity on activated T cells through selective high affinity binding to CXCR3. *J Exp Med* 1998;**187**:2009–21.
102. Seto E, Moosmann A, Gromminger S, Walz N, Grundhoff A, Hammerschmidt W. Micro RNAs of Epstein-Barr virus promote cell cycle progression and prevent apoptosis of primary human B cells. *PLoS Pathog* 2010;**6**. doi:10.1371/journal.ppat.1001063.
103. Cai X, Schafer A, Lu S, Bilello JP, Desrosiers RC, Edwards R, et al. Epstein-Barr virus microRNAs are evolutionarily conserved and differentially expressed. *PLoS Pathog* 2006;**2**:e23.
104. Ziegelbauer JM, Sullivan CS, Ganem D. Tandem array-based expression screens identify host mRNA targets of virus-encoded microRNAs. *Nat Genet* 2009;**41**:130–4.
105. Bellare P, Ganem D. Regulation of KSHV lytic switch protein expression by a virus-encoded microRNA: an evolutionary adaptation that fine-tunes lytic reactivation. *Cell Host Microbe* 2009;**6**:570–5.
106. Samols MA, Skalsky RL, Maldonado AM, Riva A, Lopez MC, Baker HV, et al. Identification of cellular genes targeted by KSHV-encoded microRNAs. *PLoS Pathog* 2007;**3**:611–8.
107. Narizhneva NV, Razorenova OV, Podrez EA, Chen J, Chandrasekharan UM, DiCorleto PE, et al. Thrombospondin-1 up-regulates expression of cell adhesion molecules and promotes monocyte binding to endothelium. *FASEB J* 2005;**19**:1158–60.
108. Gottwein E, Cullen B. A human herpesvirus microRNA inhibits p21 expression and attenuates p21-mediated cell cycle arrest. *J Virol* 2010;**84**:5229–37.
109. Hansen A, Henderson S, Lagos D, Nikitenko L, Coulter E, Roberts S, et al. KSHV-encoded miRNAs target MAF to induce endothelial cell reprogramming. *Genes Dev* 2010;**24**:195–205.
110. Bogerd H, Karnowski HW, Cai X, Shin J, Pohlers M, Cullen BR. A mammalian herpesvirus uses noncanonical expression and processing mechanisms to generate viral MicroRNAs. *Mol Cell* 2010;**37**:135–42.
111. Kluiver J, Poppema S, de Jong D, Blokzijl T, Harms G, Jacobs S, et al. BIC and miR-155 are highly expressed in Hodgkin, primary mediastinal and diffuse large B cell lymphomas. *J Pathol* 2005;**207**:243–9.
112. Gottwein E, Mukherjee N, Sachse C, Frenzel C, Majoros WH, Chi JT, et al. A viral microRNA functions as an orthologue of cellular miR-155. *Nature* 2007;**450**:1096–9.

113. Skalsky RL, Samols MA, Plaisance KB, Boss IW, Riva A, Lopez MC, et al. Kaposi's sarcoma-associated herpesvirus encodes an ortholog of miR-155. *J Virol* 2007;**81**:12836–45.
114. Morgan R, Anderson A, Bernberg E, Kamboj S, Huang E, Lagasse G, et al. Sequence conservation and differential expression of Marek's disease virus MicroRNAs. *J Virol* 2008;**82**:12213–20.
115. Zhao Y, Yao Y, Xu H, Lambeth L, Smith LP, Kgosana L, et al. A functional MicroRNA-155 ortholog encoded by the oncogenic Marek's disease virus. *J Virol* 2009;**83**:489–92.
116. Nachmani D, Stern-Ginossar N, Sarid R, Mandelboim O. Diverse herpesvirus microRNAs target the stress-induced immune ligand MICB to escape recognition by natural killer cells. *Cell Host Microbe* 2009;**5**:376–85.
117. Johnson E. Structural evaluation of new human polyomaviruses provides clues to pathobiology. *Trends Microbiol* 2010;**18**:215–23.
118. Sullivan C, Grundhoff A, Tevethia S, Pipas J, Ganem D. SV40-encoded microRNAs regulate viral gene expression and reduce susceptibility to cytotoxic T cells. *Nature* 2005;**435**:682–6.
119. Seo G, Fink L, O'Hara B, Atwood W, Sullivan C. Evolutionarily conserved function of a viral microRNA. *J Virol* 2008;**82**:9823–8.
120. Seo G, Chen C, Sullivan C. Merkel cell polyomavirus encodes a microRNA with the ability to autoregulate viral gene expression. *Virology* 2009;**383**:183–7.
121. Sullivan CS, Sung CK, Pack CD, Grundhoff A, Lukacher AE, Benjamin TL, et al. Murine polyomavirus encodes a microRNA that cleaves early RNA transcripts but is not essential for experimental infection. *Virology* 2009;**387**:157–67.
122. Aparicio O, Razquin N, Zaratiegui M, Narvaiza I, Fortes P. Adenovirus virus-associated RNA is processed to functional interfering RNAs involved in virus production. *J Virol* 2006;**80**:1376–84.
123. Xu N, Segerman B, Zhou X, Akusjärvi G. Adenovirus virus-associated RNAII-derived small RNAs are efficiently incorporated into the RNA-induced silencing complex and associate with polyribosomes. *J Virol* 2007;**81**:10540–9.
124. Aparicio O, Carnero E, Abad X, Razquin N, Guruceaga E, Segura V, et al. Adenovirus VA RNA-derived miRNAs target cellular genes involved in cell growth, gene expression and DNA repair. *Nucleic Acids Res* 2010;**38**:750–63.
125. Klase Z, Kale P, Winograd R, Gupta MV, Heydarian M, Berro R, et al. HIV-1 TAR element is processed by Dicer to yield a viral micro-RNA involved in chromatin remodeling of the viral LTR. *BMC Mol Biol* 2007;**8**:63.
126. Wang FZ, Weber F, Croce C, Liu CG, Liao X, Pellett PE. Human cytomegalovirus infection alters the expression of cellular microRNA species that affect its replication. *J Virol* 2008;**82**:9065–74.
127. Buck AH, Perot J, Chisholm MA, Kumar DS, Tuddenham L, Cognat V, et al. Post-transcriptional regulation of miR-27 in murine cytomegalovirus infection. *RNA* 2010;**16**:307–15.
128. Lee S, Murthy S, Trimble J, Desrosiers R, Steitz J. Four novel U RNAs are encoded by a herpesvirus. *Cell* 1988;**54**:599–607.
129. Wassarman D, Lee S, Steitz J. Nucleotide sequence of HSUR 5 RNA from herpesvirus saimiri. *Nucleic Acids Res* 1989;**17**:1258.
130. Albrecht J, Fleckenstein B. Nucleotide sequence of HSUR 6 and HSUR 7, two small RNAs of herpesvirus saimiri. *Nucleic Acids Res* 1992;**20**:1810.
131. Cazalla D, Yario T, Steitz J, Steitz J. Down-regulation of host microRNA by a herpesvirus saimiri noncoding RNA. *Science* 2010;**328**:1563–6.
132. Morris M, Dawson C, Young L. Role of the Epstein-Barr virus-encoded latent membrane protein-1, LMP1, in the pathogenesis of nasopharyngeal carcinoma. *Future Oncol* 2009;**5**:811–25.

133. Mrázek J, Kreutmayer S, Grässer F, Polacek N, Hüttenhofer A. Subtractive hybridization identifies novel differentially expressed ncRNA species in EBV-infected human B cells. *Nucleic Acids Res* 2007;**35**:e73.
134. Cameron JE, Fewell C, Yin Q, McBride J, Wang X, Lin Z, et al. Epstein-Barr virus growth/latency III program alters cellular microRNA expression. *Virology* 2008;**382**:257–66.
135. Cameron JE, Yin Q, Fewell C, Lacey M, McBride J, Wang X, et al. Epstein-Barr virus latent membrane protein 1 induces cellular MicroRNA miR-146a, a modulator of lymphocyte signaling pathways. *J Virol* 2008;**82**:1946–58.
136. Yin Q, McBride J, Fewell C, Lacey M, Wang X, Lin Z, et al. MicroRNA-155 is an Epstein-Barr virus-induced gene that modulates Epstein-Barr virus-regulated gene expression pathways. *J Virol* 2008;**82**:5295–306.
137. Lambeth L, Yao Y, Smith L, Zhao Y, Nair V. MicroRNAs 221 and 222 target p27Kip1 in Marek's disease virus-transformed tumour cell line MSB-1. *J Gen Virol* 2009;**90**:1164–71.
138. Hebner CM, Laimins LA. Human papillomaviruses: basic mechanisms of pathogenesis and oncogenicity. *Rev Med Virol* 2006;**16**:83–97.
139. Scheffner M, Werness B, Huibregtse J, Levine A, Howley P. The E6 oncoprotein encoded by human papillomavirus types 16 and 18 promotes the degradation of p53. *Cell* 1990;**63**:1129–36.
140. Gonzalez S, Stremlau M, He X, Basile J, Münger K. Degradation of the retinoblastoma tumor suppressor by the human papillomavirus type 16 E7 oncoprotein is important for functional inactivation and is separable from proteasomal degradation of E7. *J Virol* 2001;**75**:7583–91.
141. Wang X, Tang S, Le SY, Lu R, Rader JS, Meyers C, et al. Aberrant expression of oncogenic and tumor-suppressive microRNAs in cervical cancer is required for cancer cell growth. *PLoS ONE* 2008;**3**:e2557.
142. Wang X, Wang HK, McCoy JP, Banerjee NS, Rader JS, Broker TR, et al. Oncogenic HPV infection interrupts the expression of tumor-suppressive miR-34a through viral oncoprotein E6. *RNA* 2009;**15**:637–47.
143. Martinez I, Gardiner AS, Board KF, Monzon FA, Edwards RP, Khan SA. Human papillomavirus type 16 reduces the expression of microRNA-218 in cervical carcinoma cells. *Oncogene* 2008;**27**:2575–82.
144. Melar-New M, Laimins L. Human papillomaviruses modulate expression of microRNA 203 upon epithelial differentiation to control levels of p63 proteins. *J Virol* 2010;**84**:5212–21.
145. McKenna D, McDade S, Patel D, McCance D. MicroRNA 203 expression in keratinocytes is dependent on regulation of p53 levels by E6. *J Virol* 2010;**84**:10644–52.
146. Lu S, Cullen B. Adenovirus VA1 noncoding RNA can inhibit small interfering RNA and MicroRNA biogenesis. *J Virol* 2004;**78**:12868–76.
147. Triboulet R, Mari V, Lin YL, Chable-Bessia C, Bennasser Y, Lebrigand K, et al. Suppression of MicroRNA-silencing pathway by HIV-1 during virus replication. *Science* 2007;**315**:1579–82.
148. Kiernan RE, Vanhulle C, Schiltz L, Adam E, Xiao H, Maudoux F, et al. HIV-1 tat transcriptional activity is regulated by acetylation. *EMBO J* 1999;**18**:6106–18.
149. Jopling CL, Yi MK, Lancaster AM, Lemon SM, Sarnow P. Modulation of hepatitis C virus RNA abundance by liver-specific microRNA. *Science* 2005;**309**:1577–81.
150. Tang HL, Grise H. Cellular and molecular biology of HCV infection and hepatitis. *Clin Sci* 2009;**117**:49–65.
151. Jopling CL, Schuetz S, Sarnow P. Position-dependent function for a tandem microRNA miR-122-binding site located in the hepatitis C virus RNA genome. *Cell Host Microbe* 2008;**4**:77–85.
152. Randall G, Panis M, Cooper JD, Tellinghuisen TL, Sukhodolets KE, Pfeffer S, et al. Cellular cofactors affecting hepatitis C virus infection and replication. *Proc Natl Acad Sci USA* 2007;**104**:12884–9.

153. Henke JI, Goergen D, Zheng J, Song Y, Schuttler CG, Fehr C, et al. microRNA-122 stimulates translation of hepatitis C virus RNA. *EMBO J* 2008;**27**:3300–10.
154. Jangra RK, Yi M, Lemon SM. Regulation of hepatitis C virus translation and infectious virus production by the microRNA miR-122. *J Virol* 2010;**84**:6615–25.
155. Norman KL, Sarnow P. Modulation of hepatitis C virus RNA abundance and the isoprenoid biosynthesis pathway by microRNA miR-122 involves distinct mechanisms. *J Virol* 2010;**84**:666–70.
156. Villanueva RA, Jangra RK, Yi M, Pyles R, Bourne N, Lemon SM. miR-122 does not modulate the elongation phase of hepatitis C virus RNA synthesis in isolated replicate complexes. *Antiviral Res* 2010;**88**:119–23.
157. Esau C, Davis S, Murray SF, Yu XX, Pandey SK, Pear M, et al. miR-122 regulation of lipid metabolism revealed by *in vivo* antisense targeting. *Cell Metab* 2006;**3**:87–98.
158. Krutzfeldt J, Rajewsky N, Braich R, Rajeev KG, Tuschl T, Manoharan M, et al. Silencing of microRNAs *in vivo* with 'antagomirs'. *Nature* 2005;**438**:685–9.
159. Sarasin-Filipowicz M, Krol J, Markiewicz I, Heim MH, Filipowicz W. Decreased levels of microRNA miR-122 in individuals with hepatitis C responding poorly to interferon therapy. *Nat Med* 2009;**15**:31–3.
160. Cullen BR. Viruses and microRNAs. *Nat Genet* 2006;**38**:S25–30.
161. Gitlin L, Andino R. Nucleic acid-based immune system: the antiviral potential of mammalian RNA silencing. *J Virol* 2003;**77**:7159–65.
162. Brown BD, Venneri MA, Zingale A, Sergi LS, Naldini L. Endogenous microRNA regulation suppresses transgene expression in hematopoietic lineages and enables stable gene transfer. *Nat Med* 2006;**12**:585–91.
163. Kelly EJ, Hadac EM, Greiner S, Russell SJ. Engineering microRNA responsiveness to decrease virus pathogenicity. *Nat Med* 2008;**14**:1278–83.
164. Lecellier CH, Dunoyer P, Arar K, Lehmann-Che J, Eyquem S, Himber C, et al. A cellular microRNA mediates antiviral defense in human cells. *Science* 2005;**308**:557–60.
165. Otsuka M, Jing Q, Georgel P, New L, Chen J, Mols J, et al. Hypersusceptibility to vesicular stomatitis virus infection in Dicer1-deficient mice is due to impaired miR24 and miR93 expression. *Immunity* 2007;**27**:123–34.
166. Huang J, Wang F, Argyris E, Chen K, Liang Z, Tian H, et al. Cellular microRNAs contribute to HIV-1 latency in resting primary CD4+ T lymphocytes. *Nat Med* 2007;**13**:1241–7.
167. Nathans R, Chu CY, Serquina AK, Lu CC, Cao H, Rana TM. Cellular microRNA and P bodies modulate host-HIV-1 interactions. *Mol Cell* 2009;**34**:696–709.
168. Beckham CJ, Parker R. P bodies, stress granules, and viral life cycles. *Cell Host Microbe* 2008;**3**:206–12.
169. Pedersen IM, Cheng G, Wieland S, Volinia S, Croce CM, Chisari FV. Interferon modulation of cellular microRNAs as an antiviral mechanism. *Nature* 2007;**449**:919–22.

# Toward a Durable Treatment of HIV-1 Infection Using RNA Interference

JULIA J.M. EEKELS AND
BEN BERKHOUT

*Laboratory of Experimental Virology, Department of Medical Microbiology, Center for Infection and Immunity Amsterdam, Academic Medical Center, University of Amsterdam, Amsterdam, The Netherlands*



RNA interference (RNAi) is a cellular mechanism that mediates sequence-specific gene silencing at the posttranscriptional level. RNAi can be used as an antiviral approach against human pathogens. An attractive target for RNAi therapeutics is the human immunodeficiency virus type 1 (HIV-1), and the first clinical trial using a lentiviral gene therapy was initiated in early 2008. In this chapter, we focus on some basic principles of such an RNAi-based gene therapy against HIV-1. This includes the subjects of target site selection within the viral RNA genome, the phenomenon of viral escape, and therapeutic strategies to prevent viral escape. The latter antiescape strategies include diverse combinatorial RNAi approaches that are all directed against the HIV-1 RNA genome. As an alternative strategy, we also discuss the possibilities and restrictions of targeting cellular cofactors that are essential for virus replication, but less important for cell physiology.

Progress in Molecular Biology
and Translational Science, Vol. 102
DOI: 10.1016/B978-0-12-415795-8.00001-5

# I. RNAi: From Natural Pathway to Therapeutic Method

The development of RNAi-based therapies against a wide variety of diseases, including cancer as well as neurological, autoimmune, and infectious diseases, was triggered by the discovery of the RNAi mechanism and RNAi-mediated gene silencing in mammalian cells.[1–7] RNAi also holds promise for antiviral therapy against pathogenic viruses such as HIV-1. However, in order to properly evaluate RNAi therapeutic strategies—including the risks involved—it is important to describe in some detail the ins and outs of the cellular RNAi mechanism. We first describe the natural microRNA (miRNA) pathway.

It is estimated that human cells express more than 500 miRNAs (http://microrna.sanger.ac.uk). These miRNAs are important in the process of cell differentiation and development for regulating gene expression at the posttranscriptional level.[8–13] The natural miRNA pathway uses RNA polymerase II or III to produce a primary transcript or pri-miRNA that encodes one or multiple miRNAs, the latter in case of a polycistronic transcript.[14] The pri-miRNA is processed by the microprocessor complex into a pre-miRNA with 5′-monophospate and 3′-hydroxyl 2-nucleotide (nt) overhang.[15] This enzyme complex contains the RNase III-like enzyme Drosha and the dsRNA-binding protein DGCR8/Pasha.[16–20] Some miRNAs that are encoded within introns (mirtrons) are processed via a distinct route that uses the splicing machinery.[21,22] Pre-miRNA processing occurs in the nucleus and the product is subsequently exported by Exportin-5 (Exp-5) to the cytoplasm.[23–25] The cytoplasmic RNase III-like endonuclease Dicer cleaves the base-paired stem approximately 22-base pairs (bp) away from its base, generating a 2-nt overhang at the 3′-end.[26] Dicer is associated with the TAR RNA-binding protein (TRBP), which is required to recruit Argonaute 2 (Ago2).[27] The Ago2–RNA complex forms the minimal core of the RNA-induced silencing complex (RISC).[28,29] RISC unwinds the miRNA and keeps one RNA strand (guide strand) in the complex, while the other (passenger strand) is cleaved and subsequently degraded.[30] One of the two RNA strands gets preferentially incorporated into the complex to execute the subsequent mRNA silencing step.[31,32]

RNAi-mediated gene silencing is thought to be elicited by translational repression of the targeted mRNA in mammals.[9] However, a recent study has suggested that mRNA cleavage is the favorite mode of miRNA action.[33] An important determinant of RNAi action is the level of base-pair complementarity between the miRNA and the targeted mRNA, leading either to mRNA cleavage with a perfect complementarity or translational repression with a near-perfect complementarity.[34–38] The "seed" region of the miRNA (5′ terminal nucleotides 2–8) typically finds multiple target sequences in the 3′ untranslated region (3′UTR) of the mRNA. The silencing efficiency is determined by the overall arrangement of these 3′UTR targets, including the number and the intermotif distance.[39] Although most mammalian miRNAs base-pair with imperfect complementarity to the mRNA to cause translational repression, at least one human case of perfect complementarity and subsequent mRNA cleavage has been reported.[40] Degradation of the targeted mRNA is initiated by endonucleolytic cleavage in the 3′UTR opposite nucleotide position 10–11 of the miRNA.

In contrast to the action of natural miRNAs, man-made small interfering RNAs (siRNAs) that are designed to have full base-pair complementarity can direct mRNA cleavage with only a single target sequence, and this site can be positioned anywhere within the mRNA molecule. Such designer double-stranded (ds) RNA molecules can be produced or synthesized by several methods. Synthetic siRNAs can be transfected directly into the cytoplasm of cells, where the dsRNA will be picked up by RISC.[41] Short hairpin RNA (shRNA) transcripts[42,43] and man-made miRNA-mimics can be expressed intracellularly from a transgene construct.[44] The transcripts are expressed in the nucleus and should thus be processed and transported to the cytoplasm in order to meet RISC and induce RNAi action. The natural miRNA processing and transport pathway can be instructed with these man-made inhibitors for downregulation of a specific mRNA. This therapeutic possibility is relevant for diseases caused by overexpression of a specific mRNA. Alternatively, RNAi action can be specifically induced to target the RNA genome of invading microbes such as HIV-1 and other pathogenic viruses. Several of these types of inhibitors have been employed against HIV-1.[45–49]

Instruction of the cellular miRNA pathway with new siRNA specificity is associated with certain risks. A general problem is that the artificial siRNA molecules can compete with the endogenous siRNAs, and siRNA overexpression may lead to saturation of the miRNA pathway. This can have unwanted side effects because the miRNA pathway is important in the control of cellular gene expression. One could expect a disturbance in the cellular differentiation program, possibly cell death or even cancer.[50] Saturation of the miRNA pathway was reported to cause the death of mice when high doses of shRNAs were delivered by an adeno-associated virus (AAV) vector.[4,51–55] Thus, exogenous RNAi inducers should be expressed at a balanced level. Another potential problem that is harder to avoid is the targeting of other unintended mRNAs. Such off-target effects can occur because the miRNAs require only a seed sequence complementarity of 7–8 bp within the 3′UTR of a given mRNA.[56] Off-target effects are difficult to predict and should be screened for in appropriate experimental models. Note that off-targeting can be elicited not only by the siRNA guide strand but also by the passenger strand.[57–59] Another problem relates to the induction of an immune response by siRNAs and shRNAs.[60,61] This effect can be minimized and possibly avoided by optimal design of the si/shRNA molecule.[62]

## II. Antiviral RNAi Strategies: Toward a Gene Therapy for HIV-1

HIV-1 causes a chronic infection that ultimately leads to AIDS and death. Disease progression can be halted effectively with antiviral drugs, and, in particular, a combinatorial approach can avoid the evolution of drug-resistant HIV-1 variants. Problems associated with such drug regimens include serious toxicity during long-term follow-up. In the absence of any breakthrough at the anti-HIV vaccine front, one should think about the design of more durable therapeutic measures. The goal of an RNAi-based gene therapy approach against HIV-1 is to durably protect the cells of the immune system that are susceptible to HIV-1 infection. This includes the CD4+ T cells, monocytes, macrophages, and dendritic cells. Such "intracellular immunization" will prevent the depletion of these immune cells during chronic HIV-1 infection. Maintenance of the immune function should prevent opportunistic infections and disease progression toward AIDS.

HIV-1 causes a persistent infection, and no spontaneous viral clearance has been reported. Thus, a continuously active treatment regime is required. Repeated delivery of exogenous siRNAs as anti-HIV therapy has been described in a mouse model with a humanized immune system (HIS).[63] Effective virus inhibition was observed with a concomitant prevention of the loss of human CD4+ T cells. The systemic delivery of siRNAs was recently reported in a human phase I clinical trial via targeted nanoparticles to patients with solid cancers.[64] Tumor biopsies from patients with melanoma obtained after treatment show the presence of intracellular nanoparticles, and a reduction was found in both the targeted messenger RNA and the protein levels when compared to predosing tissue.

We, however, seriously doubt whether such an siRNA approach would be suitable in the setting of an HIV-infected patient, where the prevention of viral escape requires the continuous presence of an effective dose of multiple siRNAs in all human cell types that can be infected, which are in fact located in many different tissues and body compartments. Instead, we would like to advocate the gene therapy concept with continuous expression of anti-HIV molecules after a single transduction of HIV-susceptible cells with an appropriate delivery vector. We think that the lentiviral vector system is particularly suited for this anti-HIV therapeutic action. The lentiviral vector is based on the genome of HIV-1 itself. The pathogenic genes were replaced by novel control and therapeutic sequences. The lentiviral vector can efficiently infect most target cell types and deposit the transgene payload in a stable manner, but it cannot replicate. A benefit of the lentiviral vector compared to other viral delivery systems is that dividing and nondividing cell types can be transduced efficiently. Furthermore, the lentiviral vector stably integrates in the host cell genome at a random position, thus yielding permanent transduction.[65,66]

Some specific problems can be encountered when using the HIV-based lentiviral vector system to target the HIV-1 RNA genome with RNAi reagents. This includes self-targeting of the vector RNA in the producer cell by the antiviral shRNAs. One can of course avoid the use of shRNAs which target HIV-1 sequences that are present in the vector genome. These problems and solutions have previously been discussed in detail by Ter Brake *et al.*[67] The introduction of miRNA cassettes also raises specific problems that can be countered by appropriate vector design.[68] Alternative viral vector systems are available for delivery of a therapeutic RNAi transgene and these systems have been extensively discussed by others.[69,70]

We have depicted a possible gene therapy procedure for HIV-infected individuals in Fig. 1. Hematopoietic stem cells seed the different lineages of immune cells in the blood and organs and are therefore interesting target cells for an *ex vivo* gene therapy, followed by autologous transplantation of these CD34+ cells back into the patient. The hematopoietic stem cells equipped with the lentiviral vector will durably supply all derived immune cells with the antiviral arsenal. In the presence of HIV-1, one expects the preferential survival of these shRNA-expressing immune cells over untreated cells because the latter are likely to become infected and will subsequently be removed by the immune system. This survival benefit should result in a gradual increase in the percentage of protected cells. It remains unclear what percentage of cells can be transduced, but the treatment should result in partial reconstitution of the immune system. When reconstitution reaches a certain level, one should be able to block HIV-1 infection from progressing toward AIDS. Ideally, a single gene therapy treatment should achieve a durable effect because the transduced stem cells will continue to generate HIV-protected immune cells in the diverse lineages.

Hematopoietic stem cells transduced with a retroviral vector encoding an anti-HIV-1 ribozyme have already been evaluated in clinical trials.[71,72] These trials demonstrate the feasibility and safety of the proposed stem cell approach, although little therapeutic effect was scored for the ribozyme antiviral. A recent study demonstrated the safety of the lentiviral vector in combination with *ex vivo* targeting of CD34+ cells.[73] Another option is the treatment of the mature CD4+ T-cell population, which represents the major target cell population for HIV-1. In this scenario, the gene therapy should be applied repetitively because T cells have only a limited life span.[74]

Potent and sequence-specific HIV-1 inhibition has been reported with RNAi-inducing reagents in cell culture infections, but it soon became apparent that HIV-1 is prone to viral escape when a single shRNA inhibitor is applied.[75–81] The ease of HIV-1 escape mimics what happens in patients treated with a single antiretroviral drug, but we know that a combinatorial drug regimen can prevent viral escape and therapy failure. Thus, the therapeutic RNAi vector to

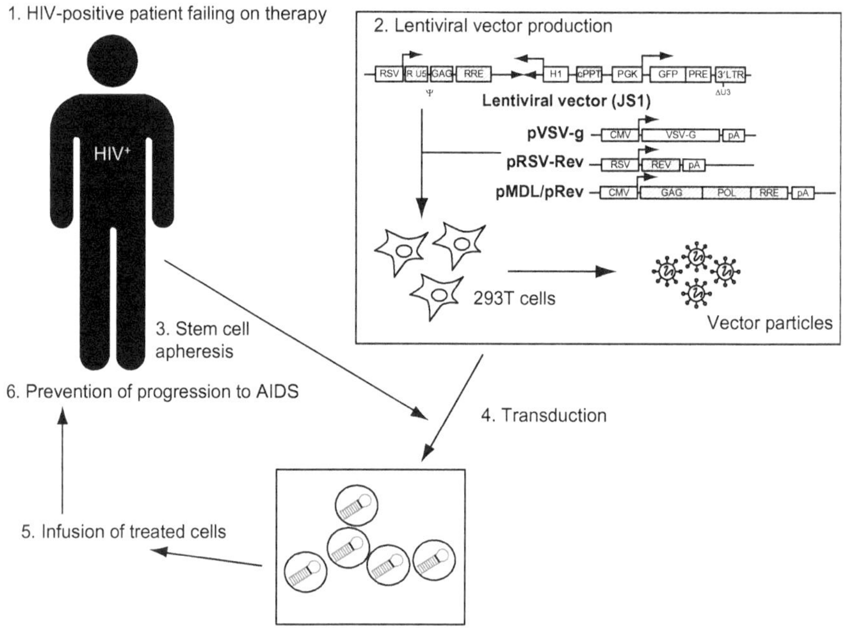


FIG. 1. RNAi gene therapy for HIV-1. The HIV-1 infected patient who fails on regular antiretroviral therapy[1] could be offered the RNAi-based gene therapy with a lentiviral vector. The lentiviral vector is produced in 293T cells[2] transfected with the lentiviral vector (e.g., JS1) and a standard set of packaging plasmids (pRSV-Rev, pVSV-g, and pSYNGP). The lentiviral vector will produce viral genomes and the packaging plasmids will produce the proteins required to assemble new viral particles. pVSV-g produces the vesicular stomatitis virus glycoprotein that is used for virus pseudotyping. The patient will undergo an apheresis for the collection of hematopoietic stem cells after pretreatment with granulocyte colony stimulatory factor (GCSF) which mobilizes these cells from the bone marrow into the periphery.[3] The hematopoietic stem cells will be purified and transduced with the therapeutic lentiviral construct.[4] This "intracellular immunization" with the antiviral shRNA will protect these cells against HIV-1. Transduced cells will be infused back into the patient[5] and the HIV-resistant immune cells will hopefully prevent disease progression toward AIDS.[6]

be developed for clinical testing should tackle the virus with multiple shRNA inhibitors at the same time. Such a combinatorial RNAi attack can target the virus at multiple genome positions,[82] but one can also add an attack against host-encoded cofactors.[83,84] One could also combine RNAi molecules with other RNA effector molecules such as decoys and ribozymes.[85,86] Different RNA-based inhibitors can also be combined in a single transcript such as the conjugate of an antiviral aptamer that binds the HIV-1 Envelope protein and an antiviral siRNA.[87] The aptamer not only blocks the Envelope protein on virion particles, but it also selectively ferries the siRNA to HIV-infected cells that express the Envelope protein on their surface. This conjugate demonstrated good antiviral activity in the preclinical model of the humanized mouse, although uncertainty remains about the efficiency of the intracellular delivery of the siRNA.[88] Another elegant solution to avoid viral escape is the use of the second-generation shRNAs that specifically target viral escape variants.[89] The relatively high number of viral escape routes available to HIV-1 may limit the feasibility of this approach.[90] In fact, we recently demonstrated the power of the second-generation concept by effectively blocking favorite viral escape routes, but little therapeutic benefit was achieved because HIV-1 selected alternative escape routes.[91]

## III. Where to Target the HIV-1 RNA Genome?

Several criteria can be formulated to identify the optimal target sites on the 9-kb HIV-1 RNA genome for RNAi attack. One could propose to select target sequences present in the multiply spliced HIV-1 mRNAs that are synthesized early upon infection and that encode the early viral proteins Tat, Rev, and Nef. The idea is that such an early block in viral gene expression will severely impact the subsequent expression of unspliced and singly spliced mRNAs that encode the structural HIV-1 proteins. Alternatively, one could target HIV-1 genomic regions that are represented in all the spliced viral mRNAs, which is the case for small sequence stretches in the untranslated 5′-leader and 3′-trailer domains.[92] Target RNA structure can effectively block an RNAi attack.[81,93] Thus, targeting of "open" RNA domains is beneficial, and this selection could be helped by the recent description of the RNA secondary structure of the complete HIV-1 RNA genome.[94] Another important selection criterion concerns the variability of target sequences. One should select targets that are highly conserved among virus isolates, simply because one wants to inhibit as many virus strains as possible. Targeting of highly conserved genome regions may also restrict the evolution of viral escape mutants because well-conserved sequences will likely exhibit an important function in HIV-1 biology, such that RNAi-induced sequence variation may be expected to have an impact on the viral replication capacity and fitness.[95]

An extensive shRNA screen against highly conserved sequences of the HIV-1 genome has been performed, yielding approximately 20 potent shRNAs.[79] Stable shRNA-expressing T-cell lines were generated that were subsequently infected with HIV-1, which yielded four durable shRNA inhibitors that restricted virus replication for more than 100 days.[96] Other groups have also screened large sets of anti-HIV shRNAs.[97] We and others have identified effective shRNAs and siRNAs targeting regulatory HIV-1 sequences, for example, in the long terminal repeat (LTR) and untranslated leader

RNA[79,98] and most viral genes, including gag,[79,99–101] pol,[99,102,103] vif,[98] tat,[79,103–105] rev,[79,104,105] vpu,[99] env,[101] and nef.[98] Follow-up analyses should include prolonged culturing of stably transduced T cells to score the impact on cell viability. To address safety in more detail, the off-target effects of the antiviral shRNAs on human mRNAs can be evaluated.[57] Prolonged culturing in the presence of HIV-1 should be done to test the likelihood of viral escape, which is in fact a possibility of which the likelihood is difficult to predict. Care should be taken not to misinterpret the results of such escape studies. For instance, we previously reasoned that the appearance of point mutations in the viral target sequence forms proof of viral escape, and in fact it demonstrates the exquisite sequence specificity of RNAi action. However, sometimes viral break-through is observed without the acquisition of escape mutations, which may be an indication of suboptimal virus inhibition.[90,96] Thus, detailed phenotypic and genotypic analyses are required to satisfactorily address this issue. Guidelines for the proper testing and selection of potent and safe shRNA inhibitors against HIV-1 have been formulated.[96]

## IV. Combinatorial RNAi Approaches

Combinatorial drug/RNAi approaches are essential to restrict HIV-1 evolution and to prevent viral escape, which will lead to therapeutic failure. A variety of strategies have been described for multiplexing of shRNA cassettes in a single therapeutic vector. As repeat sequences should be avoided in the lentiviral vector to prevent recombination-mediated deletions, the multiple shRNA cassettes generally use separate polymerase III promoters or a combination of polymerase II and III promoters.[82] Multiplexed siRNAs can also be expressed from a single transcript. We and others developed extended shRNAs that are processed into two or maximally three functional siRNAs.[46,106] Another strategy uses truly long hairpin RNAs (lhRNAs) that should encode numerous siRNAs.[48,107,108] A disadvantage of the lhRNA approach is that it is not known whether the produced siRNAs will be active inhibitors,[89] and it was recently demonstrated that a very low level of siRNAs is produced from such constructs.[46] Polycistronic miRNA transcripts have also been developed.[109] Various groups have reported toxicity of shRNAs,[51,52,54,55] which can perhaps be solved by inserting the siRNA sequence into a natural miRNA backbone.[52]

Conditional expression of the siRNA molecules will increase the safety of a therapeutic vector. For instance, one would like to avoid shRNA expression in transduced hematopoietic stem cells that still have to undergo hematopoiesis, a process that will be particularly sensitive to changes in the RNAi machinery. Tissue-specific miRNA expression has been described for several organs, including the liver.[110] Another option is the design of constructs that are induced by HIV-1 infection.[80] Selective expression in HIV-1 susceptible cells would be an elegant way to restrict putative saturation and off-target effects. Yet another option is the use of inducible gene expression systems such as the doxycycline-controlled Tet system.[111,112] While shRNAs are generally expressed from polymerase III promoters, miRNAs are expressed from polymerase II promoters. These polymerase II systems are better equipped for tissue-specific or drug-regulated expression.

Comprehensive reviews on combinatorial RNA approaches are available.[84,113] Other types of inhibitory RNA molecules can be added to the RNAi-inducing antiviral regimen, and we already mentioned the anti-HIV aptamer-siRNA conjugate transcript.[87,88] The currently ongoing phase I clinical trial at the City of Hope uses a lentiviral vector that encodes a TAR-decoy, CCR5-ribozyme, and an shRNA targeting the HIV-1 genome in the tat–rev region.[85,86] The TAR-decoy is a small nucleolar RNA molecule that absorbs the viral Tat protein, which will prevent the Tat–TAR interaction that is essential for enhanced viral promoter activity.[114] The ribozyme cleaves the CCR5-encoding mRNA to cause reduced expression of this important HIV-1 receptor on the cell surface.[115] Alternative antiviral RNA molecules include antisense transcripts,[116,117] decoys,[118] ribozymes,[115] and aptamers.[119] A new addition to this arsenal is an antisense molecule that can elicit transcriptional gene silencing of the viral LTR promoter.[120] Another promising approach is the novel RNAu method that is based on the expression of a modified U1 small nuclear RNA that blocks polyadenylation of the targeted mRNA, which is subsequently degraded.[121]

## V. Targeting Cellular Cofactors of HIV-1 Replication

An advantage of targeting host cell cofactors that are important for HIV-1 replication is the reduced change of viral escape. Silencing of several cofactors resulted in HIV-1 inhibition: nuclear factor kappa B,[103] CD4,[100,122] CXCR4,[122–125] DDX3,[126] LEDGF/p75,[127] CCR5,[122,124,128] and stable expression of shRNAs against several cofactors could inhibit HIV-1 replication *in vitro* for up to 2 months.[129] CCR5 is a critical receptor for HIV-1 entry and a promising and well-studied target. Individuals with the delta-32 mutation in the CCR5 gene are not susceptible to HIV-1 infection. Perhaps strikingly, the gene deletion does not cause any health problems; only an increased risk for infection with the West Nile virus was reported.[130] A potent shRNA targeting the mRNA for this host cell factor has been developed.[122,128] The potential of downregulation of CCR5 is supported by the cure of an HIV-1 infected patient who had leukemia in addition to AIDS. This patient was treated in Berlin and received a bone marrow transplantation of a matching donor who was homozygous for the 32-bp deletion in the CCR5 gene. Surprisingly, HIV-1 has not been detected in the patient's plasma for 600 days posttransfusion.[131]

CCR5-tropic viruses are generally responsible for HIV-1 transmission, but the virus can also use the alternative CXCR4 receptor. Downregulation of the CCR5 receptor will potentially set the stage for selection of CXCR4-tropic HIV-1 variants, but this evolutionary route was not observed in the Berlin patient. The same virus escape route was discussed when CCR5-blocking drugs were developed, but no such escape has been reported yet in patients treated with the CCR5 antagonist maraviroc.[132]

Many cellular targets will obviously not be proper candidates for gene therapy because they are essential for the cell and the host. For example, CXCR4 is required for homing of hematopoietic stem cells to the bone marrow and subsequent T-cell differentiation.[133] Although HIV-1 is a well-studied virus, many details of the viral replication cycle as well as the cellular cofactors involved remain elusive. Recently, three high-throughput RNAi gene knockdown screens were published identifying many new candidate cofactors.[134–136] Although each of the three screens reported hundreds of new candidates, the overlap between them was surprisingly small, in fact yielding only three proteins: MED7, MED8, and RELA (Fig. 2). A number of reasons for this enormous variation in experimental results have been discussed (reviewed in Ref. 137). Different experimental setups were used in the three studies, including different cell lines (293T vs. HeLa cells), lab-adapted viral strains

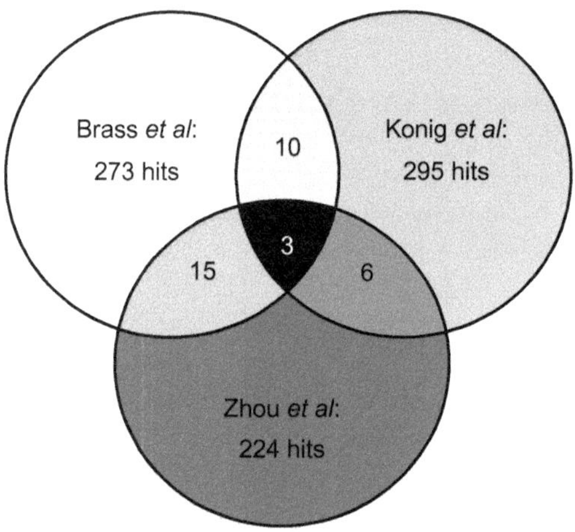


FIG. 2. Overlap of HIV-1 cofactors identified in genomewide RNAi screens. Venn-diagram of the host cofactors for HIV-1 replication identified in three RNAi screens.[134,135] Thirty-four genes were identified in at least two screens, and only three hits (MED7, MED8, and RELA) were scored in all three screens.

versus viral vectors, and various incubation times after siRNA transfection. It was also shown in a meta-analysis of these RNAi screens that experimental variation between replicates and the stringency of filtering thresholds contributed to the many discrepancies between the screens.[138]

Thus, important candidate cofactors may have been missed in the screens, as illustrated by the fact that several well-known cofactors were not found in any of the screens, such as LEDGF/p75. A reason for this could be that, as siRNAs target a single mRNA, those proteins whose function is redundant with other cellular proteins will not be found. Cofactors that need a complete knockdown before an effect on HIV-1 replication can be measured will also be missed, which is the case for the Integrase cofactor LEDGF/p75. Furthermore, many siRNAs can have toxic effects, for instance by off-target effects, and are thus excluded from further analysis, while the targeted cofactor can in fact be important for viral replication.

Although the overlap between individual cofactors identified in these screen was remarkably small, the overlap of the cellular pathways involved seems more significant. Comparing the three screens might not identify the individual cofactors that are important for HIV-1 replication, but it does give clues about the pathways that are important for viral replication. A gene ontology analysis of the cofactors reported in at least two siRNA screens highlighted the cellular processes of nuclear pore transport, GTP binding, and protein complex assembly as being important for HIV-1 replication.[138] Interfering with these pathways could be of therapeutic use.

The attack on cellular cofactors may have a more general advantage for the attack on other human pathogenic viruses that may use the same cellular pathways and/or cellular cofactors. Similar genomewide screens have been performed for other human viruses, including influenza virus A, hepatitis virus C (HCV), dengue virus, and West Nile virus.[139–143] The disadvantages discussed above for the HIV-1 RNAi screens also apply to these screens. For instance, the overlap between the two influenza screens is remarkably restricted, which could be explained by the use of *Drosophila* versus a human cell line. When comparing all screens performed for the different viruses, proteins involved in cytoskeleton complexes invariably present themselves as candidate cofactors with a broad antiviral impact.[140] Ten genes that were identified in at least one of the HIV-1 screens were also picked up in the HCV screen by Li *et al.*, and could thus be of double therapeutic value, which is important as an estimated 25–30% of HIV-infected individuals are coinfected with HCV.[141]

Thus, genomewide RNAi screens provide a powerful tool to identify novel druggable cellular cofactors, but the antiviral potential needs to be confirmed in more physiologically experimental systems. For instance, HIV-1 follow-up experiments should use T-cell lines and primary T cells in combination with primary virus isolates.

## VI. Preclinical Test Systems and Safety Concerns

When potent antiviral shRNAs have been identified in cell culture experiments, one can move to relevant preclinical models to critically assess the safety and efficacy of the proposed therapy. A simple and efficient *in vitro* test system to measure the impact of shRNA expression on cell viability is to perform a coculture of the GFP+ transduced cells and nontransduced cells (J.J.M. Eekels and B. Berkhout, unpublished results). A reduction in the percentage of GFP+ cells over time forms an indication of delayed cell growth and RNAi toxicity. Outgrowth of the transduced and thereby protected cells should occur in the presence of HIV-1, which can also be screened by using simple FACS analysis of a mixed cell culture.

The SIV/macaque model[144] has been used extensively for vaccination studies, but it can also be considered for testing of an anti-HIV-1 RNAi gene therapy. One should realize that this model has several limitations. First, anti-HIV shRNAs cannot easily be tested against SIV because of sequence dissimilarity, and the same likely holds for the genes encoding important cofactors in man versus macaque. Thus, one should either convert the anti-HIV shRNAs into anti-SIV shRNAs, which may affect their inhibitory power, or HIV-1 target sequences should be incorporated into the SIV test genome. Second, transduction of the HIV-based lentiviral vector is restricted by TRIM5$\alpha$ in macaque cells.[145] Third, macaque experiments are rather expensive, and the number of animals that can be used is restricted. A minimally modified simian-tropic (st) HIV-1 strain has recently been developed which produces an acute viremia and persistent infection in pig-tailed macaques.[146] In contrast to most infected humans, stHIV-1 infection is controlled in the macaque model after several months.

Most of these limitations do not apply to HIS mouse models.[147,148] All major human myeloid and lymphoid cellular compartments develop and mature from input human stem cells in the most recent HIS mouse.[149–151] This model provides access to *in vivo* and *ex vivo* experimentation on human T cells.[152] HIS mice can be infected by intravenous injection of the virus but also via rectal and vaginal transmission routes. Infection results in viremia and the depletion of human CD4+ cells as seen in the disease course of infected patients.[153–158] We used this model to test the safety and efficacy of a lentiviral-based gene therapy of human hematopoietic stem cells.[4] These and other animal models, including their potential and limitations, have recently been reviewed.[159]

An important lesson to be learned from various siRNA tests concerns the inclusion of appropriate control experiments. Several studies on the inhibition of infections and inflammation used a control siRNA that targets GFP. Results were in favor of a therapeutic effect, but it turned out that the GFP control is a rather special siRNA with very low immunogenicity compared to most other shRNAs. The therapeutic siRNAs trigger the TLR7/8 interferon pathway, but the GFP siRNA control does not.[160] In other words, the observed therapeutic effect was not elicited by downregulation of the targeted mRNA. Another lesson comes from a study on an siRNA therapeutic designed for the treatment of age-related macular degeneration in the eye. The siRNA exhibited a therapeutic effect, but this was not likely elicited by the RNAi mechanism since the charged siRNA molecule cannot easily penetrate cells. Instead, the clinical effect was reported to occur through TLR3 signaling.[161] Both examples illustrate the importance of selecting the correct controls to ensure that one is looking at RNAi-specific effects. For HIV-1 therapies that target the viral genome, exclusive specificity can be demonstrated by the selection or construction of escape variants with a single point mutation in the target sequence. The sequence specificity of a particular RNAi effector molecule must be demonstrated *in vitro* and in the preclinical animal model before one can move forward to the clinical test phase.[4,82,90]

## VII. Safety Issues Raised in Clinical Trials

The first patient treated with a gene therapy in 1990 suffered from adenosine deaminase deficiency, a form of severe combined immunodeficiency (SCID).[162] A patient died in 1999 as a result of the administration of a gene therapy. This patient was treated for a genetic liver disease—ornithine transcarbamylase deficiency—and received an adenovirus treatment with the wild-type gene. He died 4 days later of a massive immune response, most likely triggered by the use of the viral vector.[163] Another trial started in 2000 with SCID patients who received a gamma-retroviral gene transfer with the wild-type interleukin 2 gene. Although this procedure improved the condition of all the patients—a true success—two patients developed a leukemia-like condition of clonal lymphocyte proliferation.[164] Both cases were caused by integration of the retroviral vector near the promoter of the LMO2 proto-oncogene, leading to enhanced expression of the LMO2 protein, which has a crucial role in hematopoietic development.[165] More patients in this and a similar trial subsequently developed leukemia-like conditions due to insertional oncogenesis. After nearly 10 years of follow-up, gene therapy was shown to have corrected the immunodeficiency associated with SCID. Gene therapy may be an option for patients who do not have a donor with a compatible HLA-donor for hematopoietic stem-cell transplantation, although this treatment is associated with a risk of acute leukemia.[166]

By now, more than 1300 clinical trials involving a gene therapy have been performed.[167] From these clinical trials, lessons can be learned for future improvement of gene therapies. For instance, retroviral vectors have been replaced by lentiviral vectors, which are much safer because all transcriptional enhancer motifs have been removed ("self-inactivating" design[168]) and because these vectors tend to integrate in genes, and not near the promoter region. In addition, experiments with a lentiviral vector and hematopoietic stem cells in tumor-prone mice did not, in contrast to the retroviral vector, show signs of insertional oncogenesis.[169] Other safety and regulatory issues concerning lentiviral vectors are addressed in a comprehensive review based on the expertise gained in the first lentiviral trial.[170] The power of the lentiviral vector system to safely transduce CD34+ hematopoietic precursor cells was recently demonstrated in a gene therapy trial for children with adrenoleukodystrophy.[73]

## VIII. Gene Therapy Trials for HIV-AIDS

An overview of ongoing gene therapy trials for HIV-1 has been published.[171] While positive *in vitro* results were obtained for these antiviral gene therapies, the clinical trials failed to demonstrate a therapeutic benefit. In studies where T cells or hematopoietic stem cells were treated with the original retroviral vectors, one of the bottlenecks was effective gene delivery to a clinically relevant number of cells.[72] This problem has disappeared with the use of lentiviral vectors which have much higher transduction efficiency on a variety of cell types. In addition, many of the previously used inhibitory RNA molecules seem suboptimal when compared to antiviral shRNAs. RNAi is therefore a promising candidate for development of a future anti-HIV-1 gene therapy.

The first clinical trial with a lentiviral vector was in fact directed against HIV-1 by expression of an extended antisense transcript against the viral RNA genome. Persistent *in vivo* expression of the therapeutic antisense molecule was documented by the VirXsys company.[117] In addition, vector integration sites in blood cells revealed a preference for gene-rich regions, which is typical for a lentivirus, and no signs of insertional oncogenesis were observed. Another anti-HIV gene therapy trial that uses a triple RNA payload (ribozyme, decoy, shRNA) was performed at the City of Hope by the team of John Rossi. Patients with AIDS undergoing autologous transplantation for lymphoma were treated with gene-modified CD34+ hematopoietic progenitor cells. Persistent gene marking and sustained shRNA and ribozyme expression in blood cells for up to 24 months was documented for the first time.[86] No unexpected infusion-related toxicities were reported. For safety reasons, the transplant consisted of a mixture of genetically modified and unmanipulated cells. Future optimization of the transplant procedure, in particular the preferential infusion of transduced cells, should provide the setting for delivery of therapeutic levels of HIV-resistant cells.

## IX. Conclusion

We reviewed the current status of the development of an RNAi-based gene therapy to control HIV-1 infection and AIDS disease progression. Overall, an RNAi-based gene therapy against HIV-1 seems to be a promising candidate for a durable antiviral treatment, especially for a minority patient group for which the treatment options are exhausted because of drug toxicity or viral resistance. The potent and sequence-specific inhibition of HIV-1 with RNAi forms the corner stone for such a therapy. The superior transduction of hematopoietic stem cells with lentiviral vectors provides the means to deliver the transgene. The availability of several lentiviral production facilities is another promising development in the field. We are currently testing a candidate clinical vector that encodes multiple antiviral shRNAs to evaluate its safety and efficacy. This vector yielded very potent antiviral effects in prolonged *in vitro* cell cultures. The safety and efficacy are currently being addressed in a humanized mouse model and we expect to initiate a clinical trial within 2 years.

### Acknowledgments

RNAi research in the Berkhout lab is sponsored by ZonMw (Translational gene therapy program) and the Dutch AIDS Fund (Grant no. 2006006). We also thank the Belgian Federal Government for financial support through the Inter-University Attraction Pole, Grant no. P6/41.

### References

1. McCaffrey AP, Nakai H, Pandey K, Huang Z, Salazar FH, Xu H, et al. Inhibition of hepatitis B virus in mice by RNA interference. *Nat Biotechnol* 2003;**21**:639–44.
2. Kapadia SB, Brideau-Andersen A, Chisari FV. Interference of hepatitis C virus RNA replication by short interfering RNAs. *Proc Natl Acad Sci USA* 2003;**100**:2014–8.
3. Banerjea A, Li MJ, Bauer G, Remling L, Lee NS, Rossi J, et al. Inhibition of HIV-1 by lentiviral vector-transduced siRNAs in T lymphocytes differentiated in SCID-hu mice and CD34+ progenitor cell-derived macrophages. *Mol Ther* 2003;**8**:62–71.
4. Ter Brake O, Legrand N, von Eije KJ, Centlivre M, Spits H, Weijer K, et al. Evaluation of safety and efficacy of RNAi against HIV-1 in the human immune system (Rag-2(−/−)(c)(−/−)) mouse model. *Gene Ther* 2009;**16**:148–53.
5. Ding H, Schwarz DS, Keene A, Affar EB, Fenton L, Xia X, et al. Selective silencing by RNAi of a dominant allele that causes amyotrophic lateral sclerosis. *Aging Cell* 2003;**2**:209–17.

6. Takeshita F, Ochiya T. Therapeutic potential of RNA interference against cancer. *Cancer Sci* 2006;**97**:689–96.
7. Davidson BL, Paulson HL. Molecular medicine for the brain: silencing of disease genes with RNA interference. *Lancet Neurol* 2004;**3**:145–9.
8. Carrington JC, Ambros V. Role of microRNAs in plant and animal development. *Science* 2003;**301**:336–8.
9. Ambros V. The functions of animal microRNAs. *Nature* 2004;**431**:350–5.
10. Ambros V. microRNAs: tiny regulators with great potential. *Cell* 2001;**107**:823–6.
11. Bartel DP. MicroRNAs: genomics, biogenesis, mechanism, and function. *Cell* 2004;**116**:281–97.
12. Baehrecke EH. miRNAs: micro managers of programmed cell death. *Curr Biol* 2003;**13**: R473–5.
13. McManus MT. Small RNAs and immunity. *Immunity* 2004;**21**:747–56.
14. Lee Y, Jeon K, Lee JT, Kim S, Kim VN. MicroRNA maturation: stepwise processing and subcellular localization. *EMBO J* 2002;**21**:4663–70.
15. Bernstein E, Caudy AA, Hammond SM, Hannon GJ. Role for a bidentate ribonuclease in the initiation step of RNA interference. *Nature* 2001;**409**:363–6.
16. Denli AM, Tops BB, Plasterk RH, Ketting RF, Hannon GJ. Processing of primary microRNAs by the microprocessor complex. *Nature* 2004;**432**:231–5.
17. Gregory RI, Yan KP, Amuthan G, Chendrimada T, Doratotaj B, Cooch N, et al. The microprocessor complex mediates the genesis of microRNAs. *Nature* 2004;**432**:235–40.
18. Han J, Lee Y, Yeom KH, Kim YK, Jin H, Kim VN. The Drosha–DGCR8 complex in primary microRNA processing. *Genes Dev* 2004;**18**:3016–27.
19. Landthaler M, Yalcin A, Tuschl T. The human DiGeorge syndrome critical region gene 8 and its *D. melanogaster* homolog are required for miRNA biogenesis. *Curr Biol* 2004;**14**:2162–7.
20. Lee Y, Ahn C, Han J, Choi H, Kim J, Yim J, et al. The nuclear RNase III Drosha initiates microRNA processing. *Nature* 2003;**425**:415–9.
21. Ruby JG, Jan CH, Bartel DP. Intronic microRNA precursors that bypass Drosha processing. *Nature* 2007;**448**:83–6.
22. Chan SP, Slack FJ. And now introducing mammalian mirtrons. *Dev Cell* 2007;**13**:605–7.
23. Bohnsack MT, Czaplinski K, Gorlich D. Exportin 5 is a RanGTP-dependent dsRNA-binding protein that mediates nuclear export of pre-miRNAs. *RNA* 2004;**10**:185–91.
24. Lund E, Guttinger S, Calado A, Dahlberg JE, Kutay U. Nuclear export of microRNA precursors. *Science* 2004;**303**:95–8.
25. Yi R, Qin Y, Macara IG, Cullen BR. Exportin-5 mediates the nuclear export of pre-microRNAs and short hairpin RNAs. *Genes Dev* 2003;**17**:3011–6.
26. Zhang H, Kolb FA, Jaskiewicz L, Westhof E, Filipowicz W. Single processing center models for human Dicer and bacterial RNase III. *Cell* 2004;**118**:57–68.
27. Chendrimada TP, Gregory RI, Kumaraswamy E, Norman J, Cooch N, Nishikura K, et al. TRBP recruits the Dicer complex to Ago2 for microRNA processing and gene silencing. *Nature* 2005;**436**:740–4.
28. Gregory RI, Chendrimada TP, Cooch N, Shiekhattar R. Human RISC couples microRNA biogenesis and posttranscriptional gene silencing. *Cell* 2005;**123**:631–40.
29. Maniataki E, Mourelatos Z. A human, ATP-independent, RISC assembly machine fueled by pre-miRNA. *Genes Dev* 2005;**19**:2979–90.
30. Hammond SM, Boettcher S, Caudy AA, Kobayashi R, Hannon GJ. Argonaute2, a link between genetic and biochemical analyses of RNAi. *Science* 2001;**293**:1146–50.
31. Tomari Y, Matranga C, Haley B, Martinez N, Zamore PD. A protein sensor for siRNA asymmetry. *Science* 2004;**306**:1377–80.
32. Tomari Y, Zamore PD. Perspective: machines for RNAi. *Genes Dev* 2005;**19**:517–29.

33. Guo HL, Ingolia NT, Weissman JS, Bartel DP. Mammalian microRNAs predominantly act to decrease target mRNA levels. *Nature* 2010;**466**:835–40.
34. Brennecke J, Hipfner DR, Stark A, Russell RB, Cohen SM. Bantam encodes a developmentally regulated microRNA that controls cell proliferation and regulates the proapoptotic gene hid in Drosophila. *Cell* 2003;**113**:25–36.
35. Doench JG, Sharp PA. Specificity of microRNA target selection in translational repression. *Genes Dev* 2004;**18**:504–11.
36. Lewis BP, Shih IH, Jones-Rhoades MW, Bartel DP, Burge CB. Prediction of mammalian microRNA targets. *Cell* 2003;**115**:787–98.
37. Kiriakidou M, Nelson PT, Kouranov A, Fitziev P, Bouyioukos C, Mourelatos Z, et al. A combined computational-experimental approach predicts human microRNA targets. *Genes Dev* 2004;**18**:1165–78.
38. Lai EC. Micro RNAs are complementary to 3′ UTR sequence motifs that mediate negative post-transcriptional regulation. *Nat Genet* 2002;**30**:363–4.
39. Saetrom P, Heale BS, Snove Jr. O, Aagaard L, Alluin J, Rossi JJ. Distance constraints between microRNA target sites dictate efficacy and cooperativity. *Nucleic Acids Res* 2007;**35**:2333–42.
40. Yekta S, Shih IH, Bartel DP. MicroRNA-directed cleavage of HOXB8 mRNA. *Science* 2004;**304**:594–6.
41. Elbashir SM, Harborth J, Lendeckel W, Yalcin A, Weber K, Tuschl T. Duplexes of 21-nucleotide RNAs mediate RNA interference in cultured mammalian cells. *Nature* 2001;**411**:494–8.
42. Brummelkamp TR, Bernards R, Agami R. A system for stable expression of short interfering RNAs in mammalian cells. *Science* 2002;**296**:550–3.
43. Paddison PJ, Caudy AA, Bernstein E, Hannon GJ, Conklin DS. Short hairpin RNAs (shRNAs) induce sequence-specific silencing in mammalian cells. *Genes Dev* 2002;**16**:948–58.
44. Zeng Y, Wagner EJ, Cullen BR. Both natural and designed micro RNAs can inhibit the expression of cognate mRNAs when expressed in human cells. *Mol Cell* 2002;**9**:1327–33.
45. Schopman NC, Liu YP, Konstantinova P, Ter Brake O, Berkhout B. Optimization of shRNA inhibitors by variation of the terminal loop sequence. *Antiviral Res* 2010;**86**:204–11.
46. Liu YP, von Eije KJ, Schopman NC, Westerink JT, Ter Brake O, Haasnoot J, et al. Combinatorial RNAi against HIV-1 using extended short hairpin RNAs. *Mol Ther* 2009;**17**:1712–23.
47. Liu YP, Haasnoot J, Ter Brake O, Berkhout B, Konstantinova P. Inhibition of HIV-1 by multiple siRNAs expressed from a single microRNA polycistron. *Nucleic Acids Res* 2008;**36**:2811–24.
48. Barichievy S, Saayman S, von Eije KJ, Morris KV, Arbuthnot P, Weinberg MS. The inhibitory efficacy of RNA POL III-expressed long hairpin RNAs targeted to untranslated regions of the HIV-1 5′ long terminal repeat. *Oligonucleotides* 2007;**17**:419–31.
49. Saayman SM, Arbuthnot P, Weinberg MS. Effective Pol III-expressed long hairpin RNAs targeted to multiple unique sites of HIV-1. *Methods Mol Biol* 2010;**629**:159–74.
50. O'Connell RM, Chaudhuri AA, Rao DS, Gibson WS, Balazs AB, Baltimore D. MicroRNAs enriched in hematopoietic stem cells differentially regulate long-term hematopoietic output. *Proc Natl Acad Sci USA* 2010;**107**:14235–40.
51. Grimm D, Streetz KL, Jopling CL, Storm TA, Pandey K, Davis CR, et al. Fatality in mice due to oversaturation of cellular microRNA/short hairpin RNA pathways. *Nature* 2006;**441**:537–41.
52. McBride JL, Boudreau RL, Harper SQ, Staber PD, Monteys AM, Martins I, et al. Artificial miRNAs mitigate shRNA-mediated toxicity in the brain: implications for the therapeutic development of RNAi. *Proc Natl Acad Sci USA* 2008;**105**:5868–73.
53. Boudreau RL, Monteys AM, Davidson BL. Minimizing variables among hairpin-based RNAi vectors reveals the potency of shRNAs. *RNA* 2008;**14**:1834–44.

54. Vickers TA, Lima WF, Nichols JG, Crooke ST. Reduced levels of Ago2 expression result in increased siRNA competition in mammalian cells. *Nucleic Acids Res* 2007;**35**:6598–610.
55. Castanotto D, Sakurai K, Lingeman R, Li H, Shively L, Aagaard L, et al. Combinatorial delivery of small interfering RNAs reduces RNAi efficacy by selective incorporation into RISC. *Nucleic Acids Res* 2007;**35**:5154–64.
56. Brennecke J, Stark A, Russell RB, Cohen SM. Principles of microRNA-target recognition. *PLoS Biol* 2005;**3**:e85.
57. Jackson AL, Bartz SR, Schelter J, Kobayashi SV, Burchard J, Mao M, et al. Expression profiling reveals off-target gene regulation by RNAi. *Nat Biotechnol* 2003;**21**:635–7.
58. Fedorov Y, Anderson EM, Birmingham A, Reynolds A, Karpilow J, Robinson K, et al. Off-target effects by siRNA can induce toxic phenotype. *RNA* 2006;**12**:1188–96.
59. Jackson AL, Burchard J, Schelter J, Chau BN, Cleary M, Lim L, et al. Widespread siRNA "off-target" transcript silencing mediated by seed region sequence complementarity. *RNA* 2006;**12**:1179–87.
60. Bridge AJ, Pebernard S, Ducraux A, Nicoulaz AL, Iggo R. Induction of an interferon response by RNAi vectors in mammalian cells. *Nat Genet* 2003;**34**:263–4.
61. Sledz CA, Holko M, de Veer MJ, Silverman RH, Williams BR. Activation of the interferon system by short-interfering RNAs. *Nat Cell Biol* 2003;**5**:834–9.
62. Marques JT, Williams BR. Activation of the mammalian immune system by siRNAs. *Nat Biotechnol* 2005;**23**:1399–405.
63. Kumar P, Ban HS, Kim SS, Wu H, Pearson T, Greiner DL, et al. T cell-specific siRNA delivery suppresses HIV-1 infection in humanized mice. *Cell* 2008;**134**:577–86.
64. Davis ME, Zuckerman JE, Choi CH, Seligson D, Tolcher A, Alabi CA, et al. Evidence of RNAi in humans from systemically administered siRNA via targeted nanoparticles. *Nature* 2010;**464**:1067–70.
65. Naldini L, Blomer U, Gallay P, Ory D, Mulligan R, Gage FH, et al. In vivo gene delivery and stable transduction of nondividing cells by a lentiviral vector. *Science* 1996;**272**:263–7.
66. Berkhout B. HIV-based lentiviral vectors for anti-HIV gene therapy. *Future Virol* 2010;**5**:367–9.
67. Ter Brake O, Berkhout B. Lentiviral vectors that carry anti-HIV shRNAs: problems and solutions. *J Gene Med* 2007;**9**:743–50.
68. Liu YP, Vink MA, Westerink JT, Ramirez de AE, Konstantinova P, Ter Brake O, et al. Titers of lentiviral vectors encoding shRNAs and miRNAs are reduced by different mechanisms that require distinct repair strategies. *RNA* 2010;**16**:1328–39.
69. Nguyen T, Menocal EM, Harborth J, Fruehauf JH. RNAi therapeutics: an update on delivery. *Curr Opin Mol Ther* 2008;**10**:158–67.
70. de Fougerolles AR. Delivery vehicles for small interfering RNA in vivo. *Hum Gene Ther* 2008;**19**:125–32.
71. Amado RG, Mitsuyasu RT, Rosenblatt JD, Ngok FK, Bakker A, Cole S, et al. Anti-human immunodeficiency virus hematopoietic progenitor cell-delivered ribozyme in a phase I study: myeloid and lymphoid reconstitution in human immunodeficiency virus type-1-infected patients. *Hum Gene Ther* 2004;**15**:251–62.
72. Mitsuyasu RT, Merigan TC, Carr A, Zack JA, Winters MA, Workman C, et al. Phase 2 gene therapy trial of an anti-HIV ribozyme in autologous CD34(+) cells. *Nat Med* 2009;**15**:285–92.
73. Cartier N, Hacein-Bey-Abina S, Bartholomae CC, Veres G, Schmidt M, Kutschera I, et al. Hematopoietic stem cell gene therapy with a lentiviral vector in X-linked adrenoleukodystrophy. *Science* 2009;**326**:818–23.
74. Dropulic B. Lentivirus in the clinic. *Mol Ther* 2001;**4**:511–2.
75. Boden D, Pusch O, Lee F, Tucker L, Ramratnam B. Human immunodeficiency virus type 1 escape from RNA interference. *J Virol* 2003;**77**:11531–5.

76. Das AT, Brummelkamp TR, Westerhout EM, Vink M, Madiredjo M, Bernards R, et al. Human immunodeficiency virus type 1 escapes from RNA interference-mediated inhibition. *J Virol* 2004;**78**:2601–5.
77. Nishitsuji H, Kohara M, Kannagi M, Masuda T. Effective suppression of human immunodeficiency virus type 1 through a combination of short- or long-hairpin RNAs targeting essential sequences for retroviral integration. *J Virol* 2006;**80**:7658–66.
78. Sabariegos R, Gimenez-Barcons M, Tapia N, Clotet B, Martinez MA. Sequence homology required by human immunodeficiency virus type 1 to escape from short interfering RNAs. *J Virol* 2006;**80**:571–7.
79. Ter Brake O, Konstantinova P, Ceylan M, Berkhout B. Silencing of HIV-1 with RNA interference: a multiple shRNA approach. *Mol Ther* 2006;**14**:883–92.
80. Unwalla HJ, Li HT, Bahner I, Li MJ, Kohn D, Rossi JJ. Novel Pol II fusion promoter directs human immunodeficiency virus type 1-inducible coexpression of a short hairpin RNA and protein. *J Virol* 2006;**80**:1863–73.
81. Westerhout EM, Ooms M, Vink M, Das AT, Berkhout B. HIV-1 can escape from RNA interference by evolving an alternative structure in its RNA genome. *Nucleic Acids Res* 2005;**33**:796–804.
82. Ter Brake O, 't Hooft K, Liu YP, Centlivre M, von Eije KJ, Berkhout B. Lentiviral vector design for multiple shRNA expression and durable HIV-1 Inhibition. *Mol Ther* 2008;**16**:557–64.
83. Liu YP, Berkhout B. Lentiviral delivery of RNAi effectors against HIV-1. *Curr Top Med Chem* 2009;**9**:1130–43.
84. Grimm D, Kay MA. Combinatorial RNAi: a winning strategy for the race against evolving targets? *Mol Ther* 2007;**15**:878–88.
85. Li MJ, Kim J, Li S, Zaia J, Yee JK, Anderson J, et al. Long-term inhibition of HIV-1 infection in primary hematopoietic cells by lentiviral vector delivery of a triple combination of anti-HIV shRNA, anti-CCR5 ribozyme, and a nucleolar-localizing TAR decoy. *Mol Ther* 2005;**12**:900–9.
86. DiGiusto DL, Krishnan A, Li L, Li H, Li S, Rao A, et al. RNA-based gene therapy for HIV with lentiviral vector-modified CD34(+) cells in patients undergoing transplantation for AIDS-related lymphoma. *Sci Transl Med* 2010;**2**:36ra43.
87. Zhou J, Swiderski P, Li H, Zhang J, Neff CP, Akkina R, et al. Selection, characterization and application of new RNA HIV gp 120 aptamers for facile delivery of Dicer substrate siRNAs into HIV infected cells. *Nucleic Acids Res* 2009;**37**:3094–109.
88. Neff CP, Zhou J, Remling L, Kuruvilla J, Zhang J, Li H, et al. An aptamer-siRNA chimera suppresses HIV-1 viral loads and protects from helper CD4(+) T cell decline in humanized mice. *Sci Transl Med* 2011;**3**:66ra6.
89. Brake Ot, Berkhout B. A novel approach for inhibition of HIV-1 by RNA interference: counteracting viral escape with a second generation of siRNAs. *J RNAi Gene Silencing* 2005;**1**:56–65.
90. von Eije KJ, Ter Brake O, Berkhout B. Human immunodeficiency virus type 1 escape is restricted when conserved genome sequences are targeted by RNA interference. *J Virol* 2008;**82**:2895–903.
91. Schopman NC, Ter Brake O, Berkhout B. Anticipating and blocking HIV-1 escape by second generation antiviral shRNAs. *Retrovirology* 2010;**7**:52.
92. Muesing MA, Smith DH, Cabradilla CD, Benton CV, Lasky LA, Capon DJ. Nucleic acid structure and expression of the human AIDS/lymphadenopathy retrovirus. *Nature* 1985;**313**:450–8.
93. Westerhout EM, Berkhout B. A systematic analysis of the effect of target RNA structure on RNA interference. *Nucleic Acids Res* 2007;**35**:4322–30.

94. Watts JM, Dang KK, Gorelick RJ, Leonard CW, Bess Jr. JW, Swanstrom R, et al. Architecture and secondary structure of an entire HIV-1 RNA genome. *Nature* 2009;**460**:711–6.
95. Ter Brake O, von Eije KJ, Berkhout B. Probing the sequence space available for HIV-1 evolution. *AIDS* 2008;**22**:1875–7.
96. von Eije KJ, Ter Brake BO, Berkhout B. Stringent testing identifies highly potent and escape-proof anti-HIV short hairpin RNAs. *J Gene Med* 2009;**11**:459–67.
97. McIntyre GJ, Groneman JL, Yu YH, Jaramillo A, Shen S, Applegate TL. 96 shRNAs designed for maximal coverage of HIV-1 variants. *Retrovirology* 2009;**6**:55.
98. Jacque JM, Triques K, Stevenson M. Modulation of HIV-1 replication by RNA interference. *Nature* 2002;**418**:435–8.
99. Chang LJ, Liu X, He J. Lentiviral siRNAs targeting multiple highly conserved RNA sequences of human immunodeficiency virus type 1. *Gene Ther* 2005;**12**:1133–44.
100. Novina CD, Murray MF, Dykxhoorn DM, Beresford PJ, Riess J, Lee SK, et al. siRNA-directed inhibition of HIV-1 infection. *Nat Med* 2002;**8**:681–6.
101. Park WS, Miyano-Kurosaki N, Hayafune M, Nakajima E, Matsuzaki T, Shimada F, et al. Prevention of HIV-1 infection in human peripheral blood mononuclear cells by specific RNA interference. *Nucleic Acids Res* 2002;**30**:4830–5.
102. Berkhout B, Brake Ot. Towards an RNAi-based gene therapy. *Bioforum Eur* 2008;**4**:35–7.
103. Surabhi RM, Gaynor RB. RNA interference directed against viral and cellular targets inhibits human immunodeficiency virus type 1 replication. *J Virol* 2002;**76**:12963–73.
104. Coburn GA, Cullen BR. Potent and specific inhibition of human immunodeficiency virus type 1 replication by RNA interference. *J Virol* 2002;**76**:9225–31.
105. Lee NS, Dohjima T, Bauer G, Li H, Li MJ, Ehsani A, et al. Expression of small interfering RNAs targeted against HIV-1 rev transcripts in human cells. *Nat Biotechnol* 2002;**20**:500–5.
106. Liu YP, Haasnoot J, Berkhout B. Design of extended short hairpin RNAs for HIV-1 inhibition. *Nucleic Acids Res* 2007;**35**:5683–93.
107. Sano M, Li H, Nakanishi M, Rossi JJ. Expression of long anti-HIV-1 hairpin RNAs for the generation of multiple siRNAs: advantages and limitations. *Mol Ther* 2008;**16**:170–7.
108. Konstantinova P, de Vries W, Haasnoot J, Ter Brake O, de Haan P, Berkhout B. Inhibition of human immunodeficiency virus type 1 by RNA interference using long-hairpin RNA. *Gene Ther* 2006;**13**:1403–13.
109. Aagaard LA, Zhang J, von Eije KJ, Li H, Saetrom P, Amarzguioui M, et al. Engineering and optimization of the miR-106b cluster for ectopic expression of multiplexed anti-HIV RNAs. *Gene Ther* 2008;**15**:1536–49.
110. Snyder LL, Esser JM, Pachuk CJ, Steel LF. Vector design for liver-specific expression of multiple interfering RNAs that target hepatitis B virus transcripts. *Antiviral Res* 2008;**80**:36–44.
111. Wiznerowicz M, Trono D. Conditional suppression of cellular genes: lentivirus vector-mediated drug-inducible RNA interference. *J Virol* 2003;**77**:8957–61.
112. Zhou X, Symons J, Hoppes R, Krueger C, Berens C, Hillen W, et al. Improved single-chain transactivators of the Tet-On gene expression system. *BMC Biotechnol* 2007;**7**:6.
113. Liu YP, Berkhout B. Combinatorial RNAi strategies against HIV-1 and other escape-prone viruses. *Int J BioSci Technol* 2008;**1**:1–10.
114. Lisziewicz J, Sun D, Smythe J, Lusso P, Lori F, Louie A, et al. Inhibition of human immunodeficiency virus type 1 replication by regulated expression of a polymeric Tat activation response RNA decoy as a strategy for gene therapy in AIDS. *Proc Natl Acad Sci USA* 1993;**90**:8000–4.
115. Sarver N, Cantin EM, Chang PS, Zaia JA, Ladne PA, Stephens DA, et al. Ribozymes as potential anti-HIV-1 therapeutic agents. *Science* 1990;**247**:1222–5.
116. Chatterjee S, Johnson PR, Wong Jr. KK. Dual-target inhibition of HIV-1 in vitro by means of an adeno-associated virus antisense vector. *Science* 1992;**258**:1485–8.

117. Levine BL, Humeau LM, Boyer J, MacGregor RR, Rebello T, Lu X, et al. Gene transfer in humans using a conditionally replicating lentiviral vector. *Proc Natl Acad Sci USA* 2006;**103**:17372–7.
118. Kohn DB, Bauer G, Rice CR, Rothschild JC, Carbonaro DA, Valdez P, et al. A clinical trial of retroviral-mediated transfer of a rev-responsive element decoy gene into CD34(+) cells from the bone marrow of human immunodeficiency virus-1-infected children. *Blood* 1999;**94**:368–71.
119. Symensma TL, Giver L, Zapp M, Takle GB, Ellington AD. RNA aptamers selected to bind human immunodeficiency virus type 1 Rev in vitro are Rev responsive in vivo. *J Virol* 1996;**70**:179–87.
120. Weinberg MS, Villeneuve LM, Ehsani A, Amarzguioui M, Aagaard L, Chen ZX, et al. The antisense strand of small interfering RNAs directs histone methylation and transcriptional gene silencing in human cells. *RNA* 2006;**12**:256–62.
121. Abad X, Vera M, Jung SP, Oswald E, Romero I, Amin V, et al. Requirements for gene silencing mediated by U1 snRNA binding to a target sequence. *Nucleic Acids Res* 2008;**36**:2338–52.
122. Anderson J, Banerjea A, Akkina R. Bispecific short hairpin siRNA constructs targeted to CD4, CXCR4, and CCR5 confer HIV-1 resistance. *Oligonucleotides* 2003;**13**:303–12.
123. Martinez J, Patkaniowska A, Urlaub H, Luhrmann R, Tuschl T. Single-stranded antisense siRNAs guide target RNA cleavage in RNAi. *Cell* 2002;**110**:563–74.
124. Anderson J, Akkina R. CXCR4 and CCR5 shRNA transgenic CD34+ cell derived macrophages are functionally normal and resist HIV-1 infection. *Retrovirology* 2005;**2**:53.
125. Anderson J, Banerjea A, Planelles V, Akkina R. Potent suppression of HIV type 1 infection by a short hairpin anti-CXCR4 siRNA. *AIDS Res Hum Retroviruses* 2003;**19**:699–706.
126. Ishaq M, Hu J, Wu X, Fu Q, Yang Y, Liu Q, et al. Knockdown of cellular RNA helicase DDX3 by short hairpin RNAs suppresses HIV-1 viral replication without inducing apoptosis. *Mol Biotechnol* 2008;**39**:231–8.
127. Vandekerckhove L, Christ F, Van Maele B, De Rijck J, Gijsbers R, Van den HC, et al. Transient and stable knockdown of the integrase cofactor LEDGF/p75 reveals its role in the replication cycle of human immunodeficiency virus. *J Virol* 2006;**80**:1886–96.
128. An DS, Donahue RE, Kamata M, Poon B, Metzger M, Mao SH, et al. Stable reduction of CCR5 by RNAi through hematopoietic stem cell transplant in non-human primates. *Proc Natl Acad Sci USA* 2007;**104**:13110–5.
129. Eekels JJ, Geerts D, Jeeninga RE, Berkhout B. Long-term inhibition of HIV-1 replication with RNA interference against cellular co-factors. *Antiviral Res* 2011;**89**:43–53.
130. Lim JK, Glass WG, McDermott DH, Murphy PM. CCR5: no longer a "good for nothing" gene–chemokine control of West Nile virus infection. *Trends Immunol* 2006;**27**:308–12.
131. Hutter G, Nowak D, Mossner M, Ganepola S, Mussig A, Allers K, et al. Long-term control of HIV by CCR5 Delta32/Delta32 stem-cell transplantation. *N Engl J Med* 2009;**360**:692–8.
132. Westby M, Lewis M, Whitcomb J, Youle M, Pozniak AL, James IT, et al. Emergence of CXCR4-using human immunodeficiency virus type 1 (HIV-1) variants in a minority of HIV-1-infected patients following treatment with the CCR5 antagonist maraviroc is from a pretreatment CXCR4-using virus reservoir. *J Virol* 2006;**80**:4909–20.
133. Lapidot T. Mechanism of human stem cell migration and repopulation of NOD/SCID and B2mnull NOD/SCID mice. The role of SDF-1/CXCR4 interactions. *Ann N Y Acad Sci* 2001;**938**:83–95.
134. Brass AL, Dykxhoorn DM, Benita Y, Yan N, Engelman A, Xavier RJ, et al. Identification of host proteins required for HIV infection through a functional genomic screen. *Science* 2008;**319**:921–6.
135. Konig R, Zhou Y, Elleder D, Diamond TL, Bonamy GM, Irelan JT, et al. Global analysis of host-pathogen interactions that regulate early-stage HIV-1 replication. *Cell* 2008;**135**:49–60.

136. Zhou HL, Xu M, Huang Q, Gates AT, Zhang XHD, Castle JC, et al. Genome-scale RNAi screen for host factors required for HIV replication. *Cell Host Microbe* 2008;**4**:495–504.
137. Goff SP. Knockdown screens to knockout HIV-1. *Cell* 2008;**135**:417–20.
138. Bushman FD, Malani N, Fernandes J, D'Orso I, Cagney G, Diamond TL, et al. Host cell factors in HIV replication: meta-analysis of genome-wide studies. *PLoS Pathog* 2009;**5**:e1000437.
139. Karlas A, Machuy N, Shin Y, Pleissner KP, Artarini A, Heuer D, et al. Genome-wide RNAi screen identifies human host factors crucial for influenza virus replication. *Nature* 2010;**463**:818–22.
140. Konig R, Stertz S, Zhou Y, Inoue A, Hoffmann HH, Bhattacharyya S, et al. Human host factors required for influenza virus replication. *Nature* 2010;**463**:813–7.
141. Li Q, Brass AL, Ng A, Hu Z, Xavier RJ, Liang TJ, et al. A genome-wide genetic screen for host factors required for hepatitis C virus propagation. *Proc Natl Acad Sci USA* 2009;**106**:16410–5.
142. Sessions OM, Barrows NJ, Souza-Neto JA, Robinson TJ, Hershey CL, Rodgers MA, et al. Discovery of insect and human dengue virus host factors. *Nature* 2009;**458**:1047–50.
143. Krishnan MN, Ng A, Sukumaran B, Gilfoy FD, Uchil PD, Sultana H, et al. RNA interference screen for human genes associated with West Nile virus infection. *Nature* 2008;**455**:242–5.
144. Lackner AA, Veazey RS. Current concepts in AIDS pathogenesis: insights from the SIV/macaque model. *Annu Rev Med* 2007;**58**:461–76.
145. Stremlau M, Owens CM, Perron MJ, Kiessling M, Autissier P, Sodroski J. The cytoplasmic body component TRIM5alpha restricts HIV-1 infection in old world monkeys. *Nature* 2004;**427**:848–53.
146. Hatziioannou T, Ambrose Z, Chung NP, Piatak Jr. M, Yuan F, Trubey CM, et al. A macaque model of HIV-1 infection. *Proc Natl Acad Sci USA* 2009;**106**:4425–9.
147. Traggiai E, Chicha L, Mazzucchelli L, Bronz L, Piffaretti JC, Lanzavecchia A, et al. Development of a human adaptive immune system in cord blood cell-transplanted mice. *Science* 2004;**304**:104–7.
148. Gimeno R, Weijer K, Voordouw A, Uittenbogaart CH, Legrand N, Alves NL, et al. Monitoring the effect of gene silencing by RNA interference in human CD34+ cells injected into newborn RAG2−/− gammac−/− mice: functional inactivation of p53 in developing T cells. *Blood* 2004;**104**:3886–93.
149. Legrand N, Weijer K, Spits H. Experimental models to study development and function of the human immune system in vivo. *J Immunol* 2006;**176**:2053–8.
150. Shultz LD, Ishikawa F, Greiner DL. Humanized mice in translational biomedical research. *Nat Rev Immunol* 2007;**7**:118–30.
151. Manz MG. Human-hemato-lymphoid-system mice: opportunities and challenges. *Immunity* 2007;**26**:537–41.
152. Legrand N, Cupedo T, van Lent AU, Ebeli MJ, Weijer K, Hanke T, et al. Transient accumulation of human mature thymocytes and regulatory T cells with CD28 superagonist in "human immune system" Rag2(−/−)gammac(−/−) mice. *Blood* 2006;**108**:238–45.
153. Baenziger S, Tussiwand R, Schlaepfer E, Mazzucchelli L, Heikenwalder M, Kurrer MO, et al. Disseminated and sustained HIV infection in CD34+ cord blood cell-transplanted Rag2−/− gamma c−/− mice. *Proc Natl Acad Sci USA* 2006;**103**:15951–6.
154. Zhang L, Kovalev GI, Su L. HIV-1 infection and pathogenesis in a novel humanized mouse model. *Blood* 2007;**109**:2978–81.
155. Watanabe S, Terashima K, Ohta S, Horibata S, Yajima M, Shiozawa Y, et al. Hematopoietic stem cell-engrafted NOD/SCID/IL2Rgamma null mice develop human lymphoid systems and induce long-lasting HIV-1 infection with specific humoral immune responses. *Blood* 2007;**109**:212–8.

156. Berges BK, Wheat WH, Palmer BE, Connick E, Akkina R. HIV-1 infection and CD4 T cell depletion in the humanized Rag2−/−gamma c−/− (RAG-hu) mouse model. *Retrovirology* 2006;**3**:76.
157. Berges BK, Akkina SR, Folkvord JM, Connick E, Akkina R. Mucosal transmission of R5 and X4 tropic HIV-1 via vaginal and rectal routes in humanized Rag2(−/−)gammac(−/−) (RAG-hu) mice. *Virology* 2008;**373**:342–51.
158. An DS, Poon B, Ho Tsong FR, Weijer K, Blom B, Spits H, et al. Use of a novel chimeric mouse model with a functionally active human immune system to study human immunodeficiency virus type 1 infection. *Clin Vaccine Immunol* 2007;**14**:391–6.
159. Goldstein H. Summary of presentations at the NIH/NIAID new humanized rodent models 2007 workshop. *AIDS Res Ther* 2008;**5**:3.
160. Robbins M, Judge A, Ambegia E, Choi C, Yaworski E, Palmer L, et al. Misinterpreting the therapeutic effects of small interfering RNA caused by immune stimulation. *Hum Gene Ther* 2008;**19**:991–9.
161. Kleinman ME, Yamada K, Takeda A, Chandrasekaran V, Nozaki M, Baffi JZ, et al. Sequence- and target-independent angiogenesis suppression by siRNA via TLR3. *Nature* 2008;**452**:591–7.
162. Culliton BJ. Gene therapy begins. *Science* 1990;**249**:1372.
163. Marshall E. Gene therapy death prompts review of adenovirus vector. *Science* 1999;**286**:2244–5.
164. Hacein-Bey-Abina S, Le Deist F, Carlier F, Bouneaud C, Hue C, de Villartay JP, et al. Sustained correction of X-linked severe combined immunodeficiency by ex vivo gene therapy. *N Engl J Med* 2002;**346**:1185–93.
165. Hacein-Bey-Abina S, von Kalle C, Schmidt M, McCormack MP, Wulffraat N, Leboulch P, et al. LMO2-associated clonal T cell proliferation in two patients after gene therapy for SCID-X1. *Science* 2003;**302**:415–9.
166. Hacein-Bey-Abina S, Hauer J, Lim A, Picard C, Wang GP, Berry CC, et al. Efficacy of gene therapy for X-linked severe combined immunodeficiency. *N Engl J Med* 2010;**363**:355–64.
167. Edelstein ML, Abedi MR, Wixon J. Gene therapy clinical trials worldwide to 2007—an update. *J Gene Med* 2007;**9**:833–42.
168. Bushman F, Lewinski M, Ciuffi A, Barr S, Leipzig J, Hannenhalli S, et al. Genome-wide analysis of retroviral DNA integration. *Nat Rev Microbiol* 2005;**3**:848–58.
169. Montini E, Cesana D, Schmidt M, Sanvito F, Ponzoni M, Bartholomae C, et al. Hematopoietic stem cell gene transfer in a tumor-prone mouse model uncovers low genotoxicity of lentiviral vector integration. *Nat Biotechnol* 2006;**24**:687–96.
170. Manilla P, Rebello T, Afable C, Lu X, Slepushkin V, Humeau LM, et al. Regulatory considerations for novel gene therapy products: a review of the process leading to the first clinical lentiviral vector. *Hum Gene Ther* 2005;**16**:17–25.
171. Rossi JJ, June CH, Kohn DB. Genetic therapies against HIV. *Nat Biotechnol* 2007;**25**:1444–54.

# When Cellular Networks Run Out of Control: Global Dysregulation of the RNAi Machinery in Human Pathology and Therapy

Stefan Mockenhaupt,[1] Nina Schürmann,[1] and Dirk Grimm

*University of Heidelberg, Cluster of Excellence CellNetworks, Department of Infectious Diseases, Virology, Heidelberg, Germany*



RNA interference (RNAi) is an evolutionarily conserved fundamental cellular mechanism of potent gene and genome regulation whose misfunction is associated with numerous major human pathologies, from metabolic disorders and viral infections to cancers. Over the past 5 years, compelling evidence has been accumulated that this association is provided by dysregulations of specific mi(cro)RNAs and the ensuing aberrant expression of their target genes. Moreover, a string of interesting reports has now added proof that human disorders are also frequently characterized by global alterations in the RNAi machinery, comprising irregular expression and function of the key

[1]These two authors contributed equally.

Progress in Molecular Biology and Translational Science, Vol. 102
DOI: 10.1016/B978-0-12-415795-8.00006-4

protein players Drosha, DGCR8, Exportin-5, Dicer, TRBP, and Argonaute. Here, we comprehensively review these emerging findings in the specific contexts of cancers and infections with viral pathogens and, in addition, describe related observations in preclinical gene/RNAi therapy studies. Finally, we also thoroughly discuss the relevance of these results for future basic RNAi research as well as for the looming clinical translation of RNAi-based technologies and therapeutic concepts.

## I. Introduction: Quis Custodiet Ipsos Custodes or Who Watches the Watchmen?

Over the last decade, our view of genome regulation in humans and many other species has been revolutionized with the discovery of various distinct classes of noncoding RNAs (ncRNAs) that exert control at all major levels, from chromatin structure and transcription into mRNAs to RNA stability and translation into proteins. A particularly intriguing subclass of these ncRNAs are microRNAs (miRNAs), which are small (usually about 20–23 nucleotides long), endogenously encoded, single-stranded RNA molecules that specifically regulate gene expression. What makes miRNAs so unique and interesting is the fact that the rather small number of only 1048 (currently known) human miRNAs (according to version 16.0 of miRBase, www.mirbase.org) is capable of controlling at least 30% of all human genes (i.e., over 8000 genes). In fact, this number is likely even larger in reality since over 60% of all human genes are predicted to contain miRNA-binding sites in their 3′ untranslated region (3′UTR). The effect that miRNAs have on these genes is typically inhibitory and involves a variety of fundamentally different molecular mechanisms, including mRNA destabilization, changes in mRNA localization, and repression of mRNA translation, which have recently been reviewed extensively elsewhere.[1–7]

Unsurprisingly, in view of the propensity of individual miRNAs to potently control multiple (in fact, hundreds in some cases) genes simultaneously, these tiny molecules are critically involved in a plethora of essential cellular and organismal processes and regulate, among others, development, differentiation, apoptosis, metabolism, and cell division. Equally obvious is that dysregulation of miRNAs can readily cause, accelerate, maintain, or worsen human disorders. Evidence for this role in disease is indeed already overwhelming, with the most comprehensive and persuasive data available for cancers and infectious diseases. In this respect, one of the first hallmark studies was reported in 2005 by Lu and colleagues who systematically profiled all known (at the time) mammalian miRNAs in over 300 tumor samples including numerous different human cancers.[8] Their seminal finding was a characteristic widespread miRNA downregulation as compared to normal tissues, a result that was subsequently validated in many further descriptive as well as experimental studies by numerous other groups, and that paved the way for the development of miRNA expression profiles as novel biomarkers for human cancers. Noteworthy, miRNAs are not always downregulated in tumors or other human disorders, but, instead, there has been a flurry of opposite findings or of mixed observations as well. For instance, infection of human T-cell lines with HIV-1 results in concurrent up- and downregulation of different subsets of miRNAs in the same cell (see Section IV for details), instantly implying that miRNA expression is likely a highly intricate process and involves global as well as individual levels of control, and potentially multiple mechanistic layers in each case.

In fact, a series of papers that soon followed the initial profiling studies reported that the observed differences in amounts of mature miRNAs did not correlate with the levels of original miRNA transcripts in the same cell, strongly suggesting that it is not necessarily primary miRNA expression but rather processing that is frequently dysregulated in human diseases. For better understanding, it should briefly be recapitulated that miRNAs are typically transcribed by RNA polymerase II, which yields a roughly 70-nucleotide long pri(mary)-miRNA transcript consisting of one or more stem-loop (i.e., hairpin) structures. This pri-miRNA is first trimmed into a pre(cursor)-miRNA and then the mature miRNA in a series of two subsequent reactions: one occurring in the nucleus and exerted by the RNase Drosha and its partner DGCR8, producing the pre-miRNA, which is then transported to the cytoplasm by the karyopherin Exportin-5 where it is further cropped by the enzyme Dicer (together with its partner TRBP) into the mature miRNA. The latter then usually associates with one of four (in humans) related human Argonaute (Ago) proteins (Ago-1 to -4) and becomes part of the RNA-induced silencing complex (RISC), which then binds a target mRNA and mediates its silencing. Key players in this final step are the family of TNRC6 proteins (paralogs of GW182, three members in humans (A to C)) that interact with the Ago proteins and interfere with the function of the translation machinery as well as recruit general mRNA decay factors, thus representing the actual effectors of silencing.[9–15] In fact, artificial tethering of Ago or GW182 proteins to a target mRNA via fusion with specific RNA-binding proteins suffices to suppress gene expression in the absence of conventional small RNAs,[16] a concept that not only proves the essential role of Ago/GW182 in RNAi but also has potential therapeutic relevance (see outlook in Section VI). For more detailed information on the general RNAi machinery and mechanisms of miRNA biogenesis and function, we direct the reader to the accompanying chapters in this issue as well as to a collection of recent outstanding reviews.[17–24] Moreover, the canonical RNAi pathway in humans is schematically depicted in Fig. 1 (left).

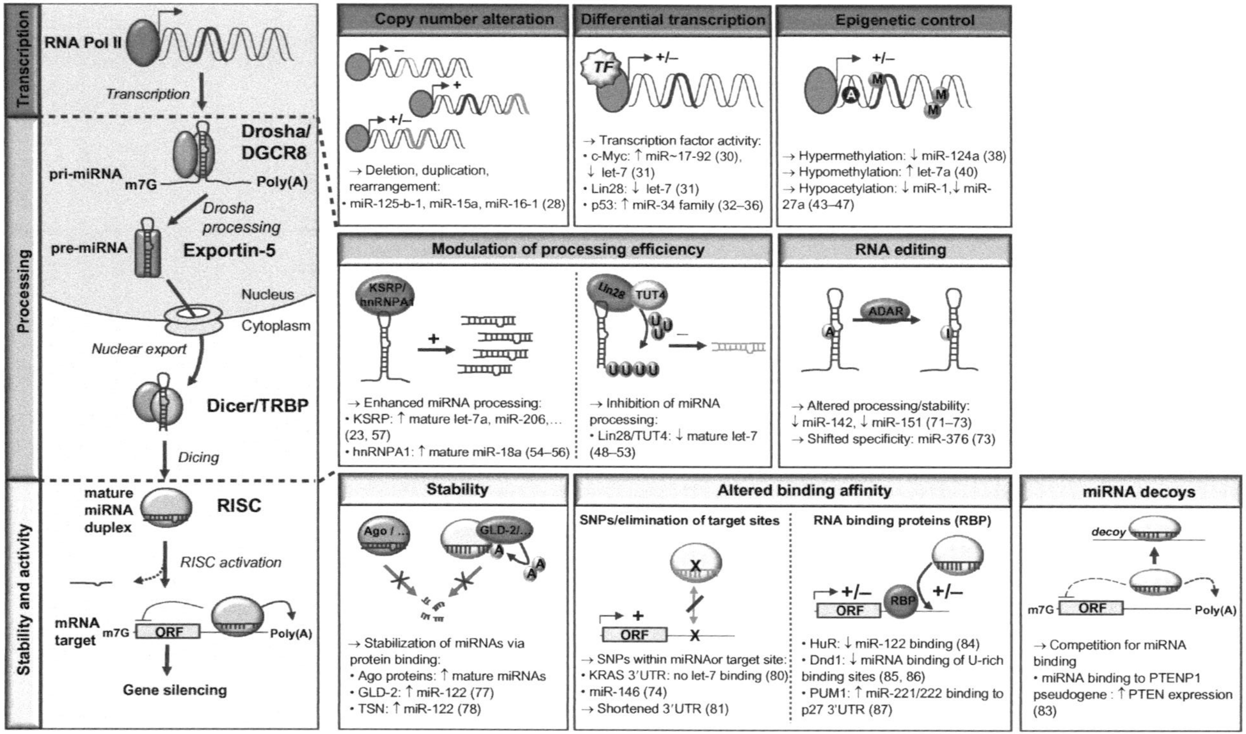


FIG. 1. Schematic overview over the RNAi pathway (left) and key steps (transcription, processing, as well as stability and activity) in miRNA biogenesis and function that are frequently dysregulated in human diseases (right), as discussed in more detail in the text. Also indicated at each step are selected examples for potential modulators as well as affected miRNAs, plus cognate references. ORF, open reading frame; m7G, 7-methylguanosine; TF, transcription factor. See text for all other abbreviations.

As mentioned, strong evidence has accumulated over the past few years that the reasons for the consistently observed disease-associated alterations in miRNA profiles are found in dysregulations somewhere along the processing cascade outlined above, from Drosha cleavage of the pri-miRNA to incorporation of the mature molecule into the RISC. Relevant proof was obtained via a combination of numerous descriptive as well as experimental studies, including a key report by Thomson *et al.* who performed correlation analyses between pri-miRNAs and their mature forms during early mouse development and available expression data from human tumors.[25] The fact that they actually found a striking absence of a clear correlation led them to suggest what we have already stated, namely, that miRNA dysregulations frequently do not derive from transcriptional aberrations but are rather due to posttranscriptional events. Further support for this important conclusion was provided in another salient study by Lee *et al.*, who comprehensively assessed the correlation of pre-miRNAs and mature forms in a large set of normal or tumor tissues, and who again found conspicuous discrepancies, in particular, in the cancer samples.[26] As a final notable example, an experimental study by Kumar *et al.* should be mentioned, in which they deliberately downregulated Drosha, DGCR8, and Dicer in murine or human cancer cells using short hairpin RNAs (shRNAs).[27] As expected, the ensuing defect in miRNA biogenesis resulted in a substantial decrease in mature miRNA levels, concurrent with a promotion of cellular transformation, proving that abrogation of global miRNA processing is sufficient to foster tumorigenesis in the absence of changes on the pri-miRNA expression level.

However, as always in nature, the situation is not black and white, and global downregulation of miRNA processing and widespread decreases in mature miRNA levels are by far not the only explanation for the involvement of miRNAs in human diseases. In fact, as already briefly indicated, infection of human cells with pathogens such as HIV-1 and others frequently results in upregulation of miRNAs as well, and increases rather than decreases of individual miRNAs have likewise been observed in many cancers and other diseases. Altogether, these puzzling and seemingly contradictory notions raise key questions about which mechanisms regulate miRNA expression and maturation, and how these are altered themselves in normal versus diseased cells: in other words, if miRNAs are the watchmen of cellular gene expression, then who watches these watchmen, and how?

To address these questions, and to reconcile the various reported findings, it is important to once again recapitulate that mature miRNA expression is the net sum of a cascade of events in the cell, from (1) pri-miRNA transcription, (2) Drosha trimming, (3) Exportin-5 transport, (4) Dicer cropping, and (5) RISC incorporation, to (6) target mRNA binding and regulation (see also Fig. 1, left). With this in mind, one can easily imagine manifold intricate possibilities of how a single miRNA, an entire group, or even the whole entity of miRNAs can become up- or downregulated in a cell, and one immediately realizes that global suppression of processing and reduction of all miRNAs, as highlighted by the early studies in cancer models, is indeed only one scenario. To date, just a few years after these initial findings, we are in fact well aware of a multitude of complex regulatory mechanisms that individually or globally perturb miRNA biogenesis and activity in pathological settings. Notably, we have also learned that some of these mechanisms not only occur during cancer or infections, but can also be inadvertently triggered by our attempts to actually cure these diseases, namely, when using gene therapy vectors to express extrinsic RNAi triggers, which may in some cases interfere with cellular miRNAs.

Considering the immense relevance of all these diverse observations for physiology, and in particular for human pathology and therapy, this review summarizes and critically discusses our current knowledge of the mechanisms and outcomes of dysregulated human RNAi machinery and the resulting loss of control of cellular networks. However, because the literature has grown tremendously over the past 5 years, and because the overall topic is very complex, we restrict ourselves to an update on those mechanisms that had in fact been proposed initially: that is, global dysregulations simultaneously affecting a majority of miRNAs and hence causing significant phenotypes. More specifically, we focus on quantitative adverse alterations in the key RNAi machinery comprising pathology-associated changes in the overall expression of central RNAi factors (Section III) as well as saturation-related events that occur during viral infections and gene therapies (Sections IV and V, respectively), and that can result in outcompetition of RNAi components. Still, before we describe and discuss the relevant studies, we briefly highlight some of the most important and intriguing alternative processes affecting individual miRNAs or single RNAi components in Section II. For any further detailed information beyond our summary, we again direct the reader to the latest collection of excellent reviews specifically dealing with the transcriptional or posttranscriptional regulation of individual miRNAs as well as with qualitative modulations of single RNAi factors.[17–24]

## II. Control of Individual miRNAs and RNAi Components

As implied in the introduction, one can simplify a very elaborate scenario and basically distinguish two principal layers on which regulation of the RNAi machinery can occur, that is, quantitatively or qualitatively (Fig. 1). While the first comprises alterations in the expression of primary miRNAs or RNAi factors, the second refers to modulations of their respective activities. In reality, these two layers are of course interconnected, since miRNA processing represents a hybrid layer between expression and activity that is moreover also tightly coupled to the quantity and quality of the enzymatic RNAi machinery. Hence, any changes in central RNAi components such as Drosha or Dicer will inevitably affect cellular miRNA processing and thus activity, regardless of initial pri-miRNA expression levels. To make matters even more complex, one can further distinguish miRNA dysregulations on either the individual or global level. This is again tightly coupled with the quality and quantity of the rest of the canonical RNAi machinery in the cell, but also involves a variety of additional non-RNAi factors that control specific miRNAs by mediating differential pri-miRNA transcription or processing.

In the following, we briefly list representative examples for each layer of control to provide the reader with a better impression of the amazing complexity underlying RNAi regulation in humans, while we again refer to further reviews for more comprehensive background information.[17–24] In addition, Fig. 1 provides an overview of the main mechanisms discussed in the following, including examples of affected miRNAs and selected references. The only exceptions that we will not describe are quantitative global dysregulations of key RNAi factors because they are the main focus of this review and will accordingly be discussed in much greater detail in Sections III–V.

## A. (Post-)Transcriptional Control of miRNAs

Perhaps the most radical way for a cell to modulate the expression of a given miRNA is alteration in the respective DNA copy numbers. Indeed, as recently reviewed thoroughly and extensively by Esquela-Kerscher and Slack,[28] many miRNA-encoding genes are located at fragile chromosomal sites or regions undergoing changes in copy numbers, such as duplications or deletions, or other types of rearrangements in cancer. Secondly, transcription of pri-miRNAs is often tightly regulated as clearly evidenced by the well-characterized tissue-specific expression of many human miRNAs, such as miR-122 which is exclusively found in the liver.[29]

Generally, because the promoters (where known) of many miRNA genes closely resemble those for genes encoding canonical protein-encoding mRNAs, they are likely controlled by the same set of transcription factors and hence undergoing comparable dysregulations in disease. A prominent example is the upregulation of miR ∼ 17–92 transcription mediated by the c-Myc proto-oncogene via E-boxes present in the miR ∼ 17–92 promoter, potentially explaining why both are frequently codysregulated in tumors.[30] Moreover, c-Myc also downregulates the transcription of other miRNAs such as let-7, as, for instance, observed in Burkitt lymphoma cells.[31] Interestingly, let-7 itself targets the *c-myc* gene, and c-Myc additionally upregulates Lin28, which in turn further suppresses let-7, altogether providing a beautiful example for the highly intricate feedback loops that can control miRNA transcription (see Section II.B). Another important example in this category is the tumor suppressor p53, which activates transcription of the miR-34 family, which then, in turn, promotes cell cycle arrest and apoptosis, hence potentiating the effects of p53 on the cell.[32–36]

A third fundamental mechanism acting on the level of miRNA genes—next to copy number alterations and differential transcription—is epigenetic control, again akin to the regulation of conventional mRNA-encoding genes. Accordingly, numerous miRNA loci were found to be hypermethylated and thus silenced in multiple cancers, while others were hypomethylated and thereby activated, such as let-7a in lung adenocarcinoma.[37–42] Alternatively, miRNA expression can also be regulated, positively or negatively, via histone acetylation, as, for instance, reported for miR-1 or miR-27a in various cancers.[43–47]

## B. Differential Processing of miRNAs

Following transcription, further processing of the pri-miRNA to the pre-miRNA and then to the mature miRNA offers a multitude of opportunities to the cell for additional regulation and miRNA-specific fine-tuning. One mechanism has in fact just been mentioned, which is inhibition of let-7 miRNA biogenesis via Lin28, an RNA-binding protein that promotes pluripotency and inversely correlates with mature let-7 expression during development. The underlying process seems to be highly complex and involves Lin28 binding to conserved bases in the terminal loop of pre-let-7, as well as recruitment of a noncanonical poly(A) polymerase (TUT4, terminal uridylyl transferase), which then adds a uridine tail to the 3′ end of pre-let-7 and thus renders it unavailable for Dicer cleavage. In addition, Lin28 was also reported to directly repress Drosha and Dicer activities by unclear means, again highlighting the typically high complexity and intricacy of miRNA regulation pathways in humans (and other species).[48–53]

A second example for miRNA-specific control is provided by the heterogeneous nuclear ribonucleoprotein A1 (hnRNPA1), which appears to exclusively stimulate maturation of miR-18a, although this miRNA is actually embedded together with five other miRNAs in the miR ~ 17–92 cluster.[54–56] The mechanism involves hnRNPA1 binding to the conserved loop of miR-18a, resulting in a structural change of hairpin conformation that creates a more favorable Drosha/DGCR8 binding and cleavage site. Such a chaperone-like function for certain factors involved in miRNA biogenesis might actually be relatively widespread considering that about 14% of human miRNAs contain highly conserved loop sequences that may serve as landing platforms for regulatory proteins.[56] Indeed, another example in this category is KSRP (KH-type splicing regulator protein) that recognizes

G-rich regions present within the terminal loop of a subset of pri-miRNAs, such as let-7a and miR-206, and synergistically promotes their Drosha/DGCR8- and Dicer-mediated processing.[23,57]

While further analogous factors and mechanisms have lately been reported elsewhere, such as repression of let-7 processing by NF90 and NF45 proteins,[58] another noteworthy example that should be pointed out here is the recent discovery of Dicer-independent, Ago-2-dependent biogenesis of miR-451.[59–62] This pathway, in which pre-miR-451 bypasses Dicer cleavage and is instead directly loaded into Ago-2 for further processing, is the latest entry into an ever-expanding repertoire of noncanonical miRNA biogenesis processes that contradict the classical dogma of RNAi. This curious list also includes small nucleolar RNAs (snoRNAs) and tRNA-derived small RNAs (tsRNAs) that bypass Drosha, mirtrons that likewise do not require Drosha, as well as ac-pre-miRNAs that are preprocessed by Ago-2 (which cleaves the passenger strand) prior to Dicer-mediated trimming.[63–70] While significant questions about the mechanism and function remain in all these recently discovered exceptional cases, it is probably safe to assume that, like conventional miRNAs, these noncanonical miRNAs and other small RNAs also play particular and presumably vital roles for the cell.

Yet another intriguing possibility for posttranscriptional regulation of specific miRNAs is provided in the form of RNA editing mediated by ADAR (adenosine deaminases that act on RNA) enzymes, which typically catalyze the conversion of adenine to inosine (A to I). The resulting inosine shares base-pairing properties with guanosine and hence preferentially pairs with cytosine instead of thymine. These changes in the primary miRNA sequence can alter the secondary structure and thus prevent efficient processing of the miRNA by Drosha or Dicer, as for instance reported for miR-142 or miR-151.[71–73] In some cases (e.g., miR-142), the edited and unprocessed miRNA is then rapidly degraded by the inosine-specific ribonuclease Tudor-SN (Tudor staphylococcal nuclease).[71] Moreover, A to I editing can also affect target recognition and shift specificities, namely in cases where it occurs in the miRNA seed sequence (nucleotides two to seven which predominantly mediate binding to the target mRNA). An example is miR-376, which is edited by both ADAR1 and ADAR2, resulting in several modified versions, one of which is capable of binding and suppressing a certain mRNA (encoding phosphoribosyl pyrophosphate synthetase 1—PRPS1) that is not recognized by the unedited wild-type miR-376.[73] However, it is still controversial whether A to I editing is a widespread phenomenon across miRNAs or rather an individual curiosity. As already indicated, the ultimate resolution of this question may be particularly complicated by the fact that editing can result in aberrant miRNA forms that are readily degraded and thus escape cloning and detection.

Besides, analogous to A to I editing, alterations in miRNA processing and/or target specificity can also originate from single nucleotide polymorphisms (SNPs) and point mutations within the pri-miRNA. Again, there are already several known examples where these phenomena have been linked to cancers, such as a polymorphism in miR-146 that is associated with increased risk of papillary thyroid cancer.[74] Akin to editing, it is currently unclear how frequent and relevant these alterations really are, but the fact that some of them can already be conclusively linked to human disease certainly warrants their further investigation in the future.

## C. Regulation on the Mature miRNA or mRNA Target Level

Even when expressed and processed, miRNAs can be further regulated by various means, affecting their inherent stability and ensuing half-life, or their interaction with the mRNA target. Regarding the first, surprisingly little is currently known about the half-life and degradation of most human miRNAs, yet there has been recent evidence that the four Argonaute proteins regulate miRNA abundance posttranscriptionally probably by binding and thus stabilizing mature miRNAs. Because Ago proteins are limiting in human cells,[75,76] this could create a feedback mechanism by which the availability of Ago proteins titrates the expression of mature miRNAs and hence ensures the formation of functional, stoichiometric miRNA–RISC complexes. This also highlights the importance for the cell to regulate Ago proteins themselves, as briefly outlined in Section II.D.

Besides Ago proteins, other factors may more specifically influence the stability of individual miRNAs by various means. One example is the cytoplasmic poly(A) polymerase GLD-2 which 3′ monoadenylates and thus stabilizes miR-122 after Dicer cleavage and unwinding in human liver cells.[77] Notably, because several other miRNAs were also found to contain single adenosines on their 3′ end, and, because addition of either adenosine or uridine is presumed to alter the stability of both protein-coding as well as ncRNAs, this mechanism may actually not be specific but instead extend well beyond miR-122. Another example along this line is the DNA/RNA-binding protein translin (TSN) that also increases *in vivo* miR-122 stability in male late-stage germ cells (which curiously enough express miR-122, though it is widely believed to be a liver-specific miRNA).[78] Last but not least, it should be mentioned that miR-122 expression levels have also recently been found to decrease in cells treated with interferon $\beta$ (IFN$\beta$), as one of about 30 miRNAs that responded (positively or negatively) to IFN.[79] While it remained unclear whether this effect occurred transcriptionally or also involved posttranscriptional alterations, for example, in miR-122 stability, this report is remarkable as it provides a crucial example for modulation of cellular miRNAs as components of the innate immune system in mammals.

In addition, mature miRNA functionality can also be regulated on the level of the mRNA target. One possibility includes the above-mentioned SNPs, which cannot only occur in the miRNA itself, but also in the 3′UTR of an mRNA where they can likewise disrupt the binding of a certain miRNA to its designated target. One clinically relevant example is an SNP in the KRAS 3′UTR which perturbs binding of let-7 miRNA and hence increases the risk of non-small-cell lung cancer (NSCLC).[80] Moreover, not only can target sites in mRNA be altered, but they can also be eliminated altogether owing to the alternative use of cleavage and polyadenylation sites which likewise abolish the miRNA functionality. This is indeed observed in proliferating cells which are frequently characterized by shortened 3′UTRs as well as by globally down-regulated overall miRNA expression.[81,82] One can easily imagine that these two effects—fewer miRNA-binding sites and fewer mature miRNAs—may synergistically contribute to this particular pathogenesis of cancer.

Furthermore, interesting to note in this context is a recent study by Poliseno *et al.*, who discovered that pseudogenes (genomic loci that resemble real genes) can act as decoys that quantitatively sequester cellular miRNAs and hence derepress their actual target, that is, the cognate gene.[83] This was mainly demonstrated for the PTEN tumor suppressor gene, heterozygosity of which leads to multiple human cancers, and which is heavily regulated by different miRNAs. Poliseno and colleagues now found that the PTENP1 pseudogene is fully processed into an mRNA that bears only 18 mismatches to the cognate PTEN mRNA, including one that prevents translation, and a highly homologous (albeit about 1 kb shorter) 3′UTR with perfectly conserved binding sites for a number of PTEN-targeting miRNAs (including members of the miR ∼ 17–92 cluster). Indeed, as shown in a series of elegant experiments, overexpression of this pseudogene UTR was capable of derepressing miRNA-mediated inhibition of PTEN itself, suggesting that the pseudogene acted as an miRNA decoy. Moreover, screening of available cancer databases (including colon and prostate) revealed a population of samples with detectable copy number losses occurring specifically at the PTENP1 locus, supporting the notion that it is under selective pressure to be reduced in human cancers. Finally, the group demonstrated that this coregulation of gene and pseudogene extends well into other known cancer-related pairs, implying that this novel mechanism may be more widespread than initially expected in this study. In fact, as speculated by the authors, it may also comprise additional long ncRNA transcripts in cells, such as large intergenic noncoding RNAs (lincRNAs) or even the entire transcript of coding genes.[83]

Finally, yet another means by which cells regulate miRNAs on the target mRNA level is competition for the actual binding site. One conclusive example first reported by Witold Filipowicz's group is the cationic amino acid transporter 1 (CAT-1) mRNA whose binding by miR-122 results in translocation to cellular P-bodies and gene silencing.[84] Strikingly, CAT-1 is derepressed in liver cells undergoing stress via binding of HuR, an AU-rich-element binding protein, to the 3′UTR of the CAT-1 mRNA. This induces release of the suppressed CAT-1 mRNA from the P-bodies and recruitment to polysomes, showing that miRNA-mediated mRNA inhibition is not final but can actually be counteracted and reversed under certain circumstances.[84] Another similar example originally discovered in zebrafish is the RNA-binding Dead end 1 (Dnd1) protein which can derepress mRNAs containing U-rich miRNA-binding sites.[85,86] Besides, the same group (Reuven Agami) most recently reported yet another intriguing example for the role of RNA-binding proteins as modulators of miRNA activity on the mRNA level. In this latest study,[87] they found that the p27 tumor suppressor is highly expressed in quiescent cells because its inhibitors miR-221 and miR-222 are underactive toward the p27 3′UTR. Yet, growth factor stimulation resulted in upregulation and phosphorylation of Pumilio-1 (PUM1), which then bound to the p27 3′UTR and induced local changes in RNA structure that favored association with miR-221 and miR-222, altogether causing p27 suppression and rapid entry to the cell cycle. Based on this and prior studies, one can readily predict that many further intriguing examples for RNA-binding proteins that act as molecular or structural switches to modulate miRNA-mediated gene silencing on the target mRNA level await their discovery.

## D. Activity Alterations of RNAi Factors

Regulation of the RNAi pathway by control of individual RNAi factors is certainly the most complex layer amongst all that are discussed here, unfortunately making it impossible to comprehensively even mention all the different mechanisms and players that have been described to date. As this particular topic has been in the focus of a number of recent brilliant reviews,[17–24] however, we gladly direct the reader to this literature for a complete and thorough discussion and will only highlight a few selected examples in the following.

These include, for instance, several factors that modulate Drosha activity, such as the DEAD-box RNA helicases p67 (DDX5) and p72 (DDX17).[88–90] Originally identified as components of the larger of two Drosha complexes and then shown to be also associated with DGCR8, they are currently believed to act redundantly as factors that promote Drosha processing of a subset of miRNAs, such as miR-21. The mechanisms remain unclear but may involve structural alterations upon direct binding akin to some other proteins discussed above, or recruitment of further proteins which are the actual effectors.

One such class of factors might be the p68-interacting SMAD proteins (*s*mall and *m*others *a*gainst *d*ecapentaplegic homolog), which are signal transducers that function as transcriptional activators in response to TGF-β signaling. As shown by Davis and colleagues, treatment of vascular smooth muscle cells (VSMCs) with TGF-β and its family member[18,91] BMP4 (bone morphogenetic protein 4) results in induction of miR-21 which in turn suppresses PDCD4 (programmed cell death protein 4) and stimulates muscle cell differentiation. Notably, this specific upregulation of miR-21 involves SMAD-dependent enhancement of Drosha activity, perhaps via promoted association of Drosha with the pri-miR-21 hairpin, but the exact details of this mechanism also require clarification.

Besides control via non-RNAi factors, a set of intriguing reports also identified an intricate auto-regulatory loop by which Drosha activity is modulated by DGCR8, and vice versa. Briefly, Narry Kim's group found that DGCR8 stabilizes Drosha through a direct interaction, while Drosha in turn cleaves two hairpins in the DGCR8 mRNA that resemble miRNAs, thus causing DGCR8 mRNA degradation.[92] As a consequence, the two RNAi factors regulate each other in a tightly coupled double-negative feedback loop that likely contributes to the homeostatic control of miRNA biogenesis, next to the other factors mentioned above and below. Besides, this model raises interesting questions as to whether Drosha regulates other hairpin-containing miRNAs in a similar manner, or whether they, vice versa, compete with authentic pri-miRNA for Drosha and thus attenuate global miRNA biogenesis, perhaps reminiscent of the function of adenoviral VA RNAs (see Section IV).

As in the case of Drosha and DGCR8, one of the mechanisms controlling Dicer is also through interaction with another RNAi factor, namely, TRBP. Interestingly, Paroo *et al.* found that TRBP is regulated by the mitogen-activated protein kinase (MAPK) pathway, since it interacts with phosphorylated Erk kinase (one of the MAPKs that is critically involved in cellular proliferation and survival) and itself becomes phosphorylated at four serine residues.[93] In turn, phosphorylated TRBP was elevated and then led to an increase in Dicer levels as well, likely via protein stabilization. Not unexpectedly, this concerted elevation in Dicer and TRBP ultimately caused an upregulation of numerous miRNAs, in particular those fostering cell proliferation and growth, thereby mechanistically linking an essential signaling pathway and the miRNA machinery.

Phosphorylation may also be a major mechanism regulating yet another set of key RNAi factors, that is, the four human Argonaute proteins. Thus far, two different groups have independently reported phosphorylation of Ago-2, at distinct sites and with potentially unique outcomes. In the first study, Zeng *et al.* reported Ago-2 phosphorylation at serine 387 in response to activation of the MAPK p38 pathway under cellular stress conditions.[94,95] While this phosphorylation did not alter Ago-2 levels, it facilitated Ago-2 localization to cellular P-bodies, cytoplasmic areas associated with mRNA decay and turnover, as well as translational regulation. However, the physiological significance of this observation or its potential dysregulation in disease remained unclear, and it should also be pointed out that P-bodies or Ago-2 localization to these areas are likely not mandatory for RNAi. Together, this makes it hard to assess the actual relevance of serin-387 phosphorylation for Ago-2 function at this point.

More recently, Gunter Meister's group uncovered a second phosphorylation site in Ago-2 and postulated exciting underlying functions and mechanisms.[96] This site, tyrosine 529 (Y529), is located in the small RNA 5′end-binding pocket located in the Ago MID domain and is conserved in all four human Ago proteins as well as in numerous other species, immediately implying its potential relevance. Interestingly, when the group then mutated Y529 to a glutamate (Y529E) and hence introduced a negatively charged environment similar to a phosphorylated Y529, they found that the resulting mutant localized mainly to the diffuse cytoplasm and no longer to P-bodies, reminiscent of the findings by Zeng and colleagues. Furthermore, the Y529E mutant had also lost its slicing activity (i.e., it could no longer cleave a target mRNA) and it could also no longer be loaded with a small RNA, tempting the authors to suggest that the glutamate mimicking phosphorylation had created a repulsive force against the negatively charged 5′ phosphate of the small RNA. Albeit the amount of wild-type Ago-2 that was actually found to be phosphorylated on Y529 was then relatively small, Ruedel *et al.* concluded that phosphorylation within the 5′end-binding pocket of the Ago MID domain might act as a molecular switch that promotes or inhibits small RNA loading into Ago proteins.[96] Which protein kinase mediates this phosphorylation and how it reaches the Y529 residue within the binding pocket, or whether the phosphate is actually directly transferred from the loaded small RNA all, remain intriguing questions for future work.

Next to phosphorylation via (unknown) kinases, we shall only briefly mention that Ago proteins were found to potentially interact with a myriad of further proteins, at least some of which might also modulate their activity or stability.[97] These include heat shock protein 90 (HSP90, potentially chaperones Ago proteins for small RNA loading),[98] 14-3-3 proteins (interact with Ago-2 with unclear function),[99] TRIM32 (interacts with Ago-1 and potentiates the activity of selected miRNAs),[100,101] or Lin41 (binds and ubiquitylates Ago-2, thus regulating its turnover).[102,103] Another recently published interaction and modification that should finally be mentioned here is hydroxylation at proline 700 by the type I collagen prolyl-4-hydroxylase C-P4H(I). As discovered by Qi and coworkers, this results in stabilization of Ago-2 but has no detectable effect on the catalytic slicing activity of Ago-2.[104] Interestingly, C-P4H(I) was found to also hydroxylate the other three Ago proteins, with varying effects on their P-body localization and stability, providing one of the very few reports to date hinting at differential regulation of the four human Ago proteins. Again, however, a number of vital questions remain to be answered in future studies, such as that about the actual physiological (or pathological) relevance of Ago-2 hydroxylation, or the general question that was asked in Section I and that is still valid despite the sum of studies on RNAi regulation already published to date, namely, what in the cell controls the regulator, or who watches the watchmen?

# III. Global Dysregulation of Key RNAi Factors in Human Disease

In the previous two sections, we have briefly introduced cellular miRNA mechanisms as highly intricate and tightly controlled processes whose dysregulation frequently causes diseases in humans, and highlighted a series of regulatory events that modulate the biogenesis and activity of individual miRNAs or fine-tune the function of selected RNAi factors. However, the major focus of this review and Sections III–V is on rapidly emerging findings that human disorders, in particular cancers and neurological diseases (Section III) or viral infections (Section IV), are associated with global and quantitative dysregulations of key RNAi factors, as well as RNAi-based therapeutic strategies (Section V). Therefore, we first comprehensively review the respective literature from the past 5 years for each of the major RNAi factors, namely, Drosha, Exportin-5, Dicer, TRBP, Ago, and other proteins. Note that all main findings are also summarized in Fig. 2.

## A. Drosha

One of the very first evidences for Drosha dysregulation in human cancers came from Sugito *et al.*, who studied expression levels of Dicer, DGCR8, and Drosha in esophageal squamous cell carcinoma (ESCC), one of the most lethal malignancies in Japan (5-year survival rate of less than 30% after curative surgery) and a frequently recurring and metastasizing cancer.[105] In this work, the authors compared the expression of the three RNAi factors in 73 ESCC tissues with corresponding normal esophageal epithelium by real-time PCR (RT-PCR), and they moreover particularly examined Drosha expression in 27 cell lines. Interestingly, the patients could be clearly and consistently divided into two groups with low and high Drosha expression, and these clades then showed significantly distinct relationships with the clinicopathologic characteristics of ESCC. Notably, the cases with low Drosha expression exhibited greater

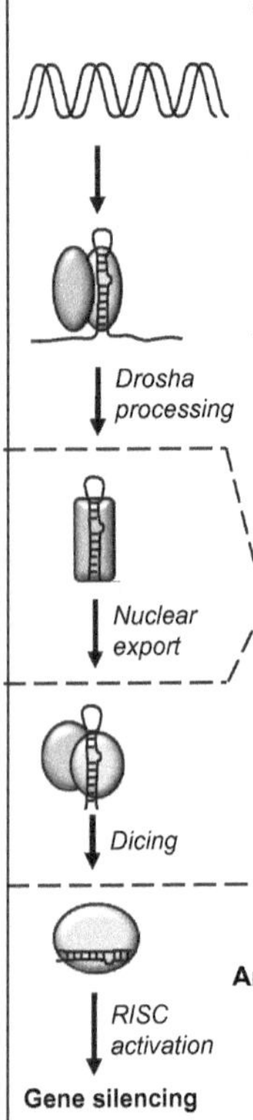


| Factor | Up-regulation | Down-regulation | Unchanged |
|---|---|---|---|
| Drosha | Esophageal cancer (58)<br>Ovarian cancer (60)<br>Cervical cancer (59)<br>Salivary gland tumor (61)<br>Skin cancer (62)<br>Gastric cancer, esp. early stages (65)<br>Schizophrenia (67, 68)<br>Huntington's disease, early timepoints (69) | Breast cancer (64)<br>Ovarian cancer (63)<br>Huntington's disease, late timepoints(69) | Breast cancer (90)<br>Ovarian cancer (66)<br>Prostate cancer (71)<br>Age-related macular degeneration (95) |
| DGCR8 | Salivary gland tumor (61)<br>Prostate cancer (70)<br>Schizophrenia (67, 68) | Lung cancer (72) | Esophageal cancer (58)<br>Cervical cancer (59)<br>Prostate cancer (71)<br>Breast cancer (90)<br>Huntington's disease (69)<br>Age-related macular degeneration (95) |
| Exportin-5 | Prostate cancer (71, 73) | Lung cancer (72) | Huntington's disease (69) |
| Dicer | Prostate cancer (70, 71)<br>Gastric cancer (65)<br>Oral cancer (79)<br>Salivary gland tumor (61, 78)<br>Skin cancer, SCC (62)<br>Ovarian cancer (60)<br>Myeloid leukemia (84)<br>Schizophrenia (67, 68) | Lung cancer (72, 74, 77)<br>Breast cancer (64, 90, 91, 92)<br>Liver cancer (86)<br>Salivary gland tumor (78)<br>Skin cancer, BCC (62)<br>Ovarian cancer (63, 87, 88)<br>Age-related macular degeneration (95) | Esophageal cancer (58)<br>Huntington's disease (69) |
| Argonautes/ RISC | Testicular cancer, seminomas (100, 101)<br>Prostate cancer (71)<br>Breast cancer (90, 91, 98)<br>Colon cancer (99)<br>Gastric cancer (102)<br>Pancreatic cancer (103)<br>Soft-tissue sarcoma (104) | Lung cancer (72) | Testicular cancer, non-seminomas (100, 101)<br>Age-related macular degeneration (95) |

FIG. 2. Summary of reported dysregulations in the human RNAi pathway as discussed in more detail in the text. Numbers in brackets are references.

absence of lymph-node metastases. Furthermore, the probability of survival was significantly lower for patients with high Drosha levels, and Drosha abundance was also identified as a predictive factor for poor prognosis.

Together, these findings strongly suggested that expression levels of Drosha have a substantial effect on postoperative survival of ESCC patients. Further support for this concept came from the analyses of esophageal cancer cell lines, many of which were found to express more Drosha than a normal human esophageal mucosa cell line (Het1A). Indeed, Drosha knockdown in these cancer cell lines using specific siRNAs resulted in up to 80% reductions in the number of cells after 72 h. Last but not least, immune-histochemical analyses of ESCC samples also revealed aberrantly increased Drosha expression and thus further strengthened the correlation with tumorigenesis.

Particularly strong staining was observed in the tumor periphery, tempting the authors to speculate that Drosha upregulation might be involved in tumor invasion to surrounding tissues. Of note, no clear relationships between expression and proliferation, survival, or prognosis were found for the other two RNAi factors studied in this work, namely, Dicer and DGCR8.

Very similar findings were subsequently also reported by Muralidhar *et al.* in models and samples of cervical squamous cell carcinomas (SCC), which represent the majority of all new cervical carcinoma cases.[106] Development of these cancers requires a combination of infection with high-risk papillomaviruses (HPV16 or 18) and acquisition of genomic abnormalities. Interestingly, among the latter, in particular copy-number gain and amplification of chromosome 5p occur in over 50% of advanced SCC and is also frequently seen in other carcinomas, including those of the head and neck or lung. In this work, the authors now showed that it also occurs rapidly and specifically (no further genomic differences were noted during different passages) in a cervical carcinogenesis cell culture model (W12 cell line) and that it was associated with the acquisition of a growth advantage and enhanced ability to invade collagen, resembling microinvasive SCC. Strikingly, global mRNA expression analyses revealed that the 5p gene exhibiting the most significant increase in expression levels was that encoding Drosha, located at chromosome 5p13.3 and showing an almost twofold upregulation compared to earlier cell passages with normal 5p numbers. Consistent with this, Drosha gain was then also seen in almost 60% of clinical specimens of pretreatment SCC and 80% of SCC cell lines, while Drosha loss was never observed. Notably, Drosha copy-number gain was absent from SIL (squamous intraepithelial lesion, a condition that precedes cancer), implying that Drosha gain is a late step in cervical carcinogenesis. Also of note, copy-number gains and overexpression were infrequent for other key RNAi factors examined in this study, including DGCR8, Exportin-5, and Dicer.

Interestingly, this group then also investigated expression of miRNAs in correlation with Drosha levels in 10 SCC cell lines and 8 clinical specimens, and found 16 miRNAs to be differentially expressed in the cell lines as compared to normal keratinocytes or SCC lines without Drosha overexpression. Moreover, 4 of these 16 miRNAs (miR-31, -21, -193b, and -29a) were then also upregulated in SCC clinical specimens with high Drosha expression. Even though the authors noted that the same miRNAs have previously been implicated in other cancers, such as colorectal cancers or leukemias, they unfortunately did not further study the potential role of these four candidates for SCC. Moreover, there was no experimental explanation as to why other miRNAs actually decreased in the presence of Drosha overexpression, although there might be a myriad of reasons, including differential sensitivity to, or dependence on, Drosha for processing (e.g., mirtrons,[66–68]) or feedback mechanisms between multiple miRNAs. Regardless, together with the study of Sugito *et al.*,[105] these early pieces of work already clearly hinted at a distinct correlation between aberrant expression of RNAi factors (Drosha in these cases) and progressive disease.

With particular respect to Drosha, a series of more recent papers further substantiated the relationship between upregulation and tumorigenesis in various cancer models and clinical specimens. One of them was a study by Flavin *et al.*, who characterized alterations in the expression of the genes encoding Drosha as well as Dicer and eIF6 (a factor recruited by RISC with unclear function in RNAi) in ovarian serous adenocarcinomas (OSC), the most common histotype amongst epithelial ovarian cancers.[107] Using RT-PCR to examine 50 formalin-fixed, paraffin-embedded OSC samples, the group identified a 2.7-fold upregulation of Drosha relative to normal whole ovary, and similar numbers were also reported for eIF6 and Dicer (2.9- and 1.6-fold, respectively). However, while the eIF6 and Dicer levels could be clearly correlated with clinicopathological features (see Section III.D), these studies were not extended to Drosha, leaving the statistical relevance of its consistent upregulation for OSC unclear in this study. Alas, there were no data on miRNA expression profiles, precluding mechanistic insights into possible functional effects of Drosha (and Dicer plus eIF6) upregulation on cellular miRNA or gene expression.

Such additional studies were instead conducted by Zhang *et al.*, who likewise noted consistent increases (at least twofold in 7 out of 10 cases studied) in the expression of multiple components of the miRNA processing machinery, namely Drosha as well as Dicer, DGCR8, and p68, in pleomorphic adenomas of the salivary gland (benign tumors with relatively low recurrence rates).[108] Moreover, in 16 matched pairs of adenomas and normal tissues, the group also identified 17 miRNAs that were specifically elevated in the tumors (including miR-21, a prototype oncomir), whereas only 5 miRNAs were repressed, implying a good correlation between a global disturbance (upregulation) of the miRNA biogenesis machinery and levels of actual miRNAs. Not unexpectedly, the authors finally also noted that the miRNAs which they had found to be dysregulated might control factors in the WNT, MAPK, and JAK-STAT (Janus kinase-signal transducer and activator of transcription) pathways, which have all been extensively implicated in many diverse tumor phenotypes before.

Another group of cancers in which Drosha was found to be upregulated is epithelial skin cancers, that is, basal cell carcinomas (BCCs) and SCCs.[109] Curiously, though, the changes (measured by RT-PCR) were highly significant only when compared to unrelated healthy subjects, but it was no longer so when compared to intraindividual controls (i.e., sites of healthy skin from the same patient with the tumor). This may suggest that patients with BCC/SCC show generally altered Drosha expression levels in the tumor as well as in adjacent healthy tissue (from which the control samples were taken), even though sound experimental proof for this hypothesis was lacking in this paper. Likewise, it remained unclear whether the Drosha upregulation had any effect on miRNA expression in the tumor (or healthy) tissues.

Notably, the observations of elevated Drosha levels in SCC of the skin are highly reminiscent of the above-mentioned similar findings by Muralidhar *et al.* in cervical SCC.[106] Moreover, both studies consistently noted a striking absence of Drosha upregulation in premalignant specimens (actinic keratoses and SIL). These congruences thus further support the general idea that Drosha gain is a rather late step in human carcinogenesis.

Although all papers described thus far have reported an upregulation of Drosha in human malignancies, it is curious to note that a similarly large number of additional studies have also found the opposite. In one of the first examples along this line, Merritt *et al.* measured Drosha (and Dicer, see Section III.D) mRNA levels in a large cohort of specimens of invasive epithelial ovarian cancer from 111 patients.[110] In contrast to the above studies, they actually found a decrease in Drosha mRNA in 51% of all tumors (measured by RT-PCR and confirmed by immunohistochemistry, as well as also noted in ovarian cancer cell lines), which correlated with suboptimal cytoreductive surgery (a poor prognostic factor) and reduced median overall survival, all observations which clearly contradict those in other studies and models (e.g., Sugito *et al.*[105]). Yet, while high expression of Drosha mRNA was associated with increased survival of patients with ovarian cancer, this correlation did not hold for other tumors that the group then also investigated, in particular lung and breast cancers. It was then speculated that these divergent expression patterns of Drosha (and other factors, see below) among different tumors and studies, as well as their differential association with clinical and pathological variables, may be explained by the pleiotropic interactions of the various components of the RNAi machinery. Hence, alterations in levels of some factors (e.g., Drosha) may be compensated for by mutations or changes of other cofactors (such as DGCR8 in case of Drosha, see Section III.B).

Confirmation for both a downregulation of Drosha in specific cancers as well as for the lack of association between Drosha levels and outcome at least in patients with breast cancer came from a very recent study by Dedes and colleagues.[111] This group investigated Drosha (and Dicer, both by RT-PCR) expression in 200 primary breast cancer samples and found Drosha downregulation in 18% of all cases. Interestingly, the decrease in Drosha was significantly associated with high histological grade and high proliferation rates, and the same subset of tumors also showed a marked (15%) concurrent downregulation of Drosha and Dicer. Also, cases with Drosha downregulation exhibited a trend for shorter disease-free survival, yet Drosha levels did not reach statistical significance as predictors for metastasization or survival, and there were also no other robust associations with the outcome. Important to note is that all patients in this study were uniformly treated with adjuvant anthracycline-based chemotherapy and/or endocrine therapy (accounting for up to 70% of all patients with early breast cancer), which may hamper a direct comparison with the discrepant results on Drosha/Dicer as prognostic factors in the above-mentioned study by Merritt and colleagues (in which patients received distinct systemic therapy regimens).[110] Moreover, complicating the resolution of the discrepancies between the two studies is that different sample types and technologies were used to isolate and quantify Drosha/Dicer levels. Still, the results from this latest study by Dedes *et al.* clearly show that Drosha is downregulated in certain breast cancers and thus strongly suggest that distinct components of the miRNA machinery are dysfunctional in tumor subgroups.[111]

In support of this, a similarly complex scenario was also reported in another recent study by Tchernitsa *et al.*, who sytematically characterized Drosha expression in 332 gastric cancers (the second leading cause of cancer-related deaths worldwide).[112] Compared to nonneoplastic mucosa, Drosha was generally upregulated in tumor tissue, but, curiously, the exact levels varied with the tumor type (higher in intestinal vs. diffuse tumors) as well as with the local tumor growth. In fact, it was lower in more advanced stages, suggesting that gastric cancer cells regulate Drosha expression during cancer development and progression, thus supporting the findings and conclusions from the Merritt and Dedes groups in their specific models. Further, there was also a nonsignificant trend toward reduced Drosha (and Dicer) expression in node-positive (i.e., metastasizing) tumors. Importantly, in the case of gastric cancers, low Drosha expression may also again serve as a prognosticator since, in a follow-up study on 190 patients, those with Drosha-negative cancer cells had significantly shorter survival times. Finally, the group also measured miRNA expression in intestinal tumors of six patients and found an upregulation of 18 miRNAs (including a large number well known to be increased in gastric cancers), while only two were downregulated, in line with the increased Drosha levels in the intestinal tumor subtype.

Last but not least, adding to the overall complexity and the controversy around Drosha's role as key player in tumorigenesis and/or as prognostic marker are a number or reports according to which Drosha levels are never altered at all in certain cancers. One of the first studies along this line was from Lu *et al.*, who profiled miRNAs across a large panel of over 334 samples representing diverse human tissues and tumor types.[8] While the group reported a general downregulation of miRNAs in tumors compared with normal tissue, consistent with numerous other data, concomitant mRNA profiling did not reveal substantial decreases in the expression of the factors of the miRNA processing machinery, including Drosha.

Similar findings were also reported by Zhang *et al.*, who characterized miRNA as well as Drosha (and Dicer) expression in human epithelial ovarian cancer (18 cell lines and four primary samples).[113] Indeed, almost 90% of all detectable miRNAs were downregulated in malignant versus healthy lines or tissue, and even 100% were reduced in primary human ovarian cancer specimens. Yet, there was no evidence for concurrent significant downregulation (RNA and protein levels) of Drosha (or Dicer) in any of the tumor samples, and also not in a meta-analysis of two relevant public cDNA microarrays. Instead, DNA copy-number alterations and epigenetic silencing were proposed as likely important mechanisms underlying the widespread reduction in miRNA expression. Moreover, and further strikingly differing with the above-described data from Merritt *et al.*,[110] the group also found no correlation between Drosha expression and patient survival.

Finally and not surprisingly, first evidence is currently accumulating that Drosha dysregulation is also critically involved in other human diseases beyond cancer, in particular neuropathologies. In one example, Santarelli *et al.* studied the role of miRNAs in schizophrenia and therefore profiled them in postmortem brain tissues of the dorsolateral prefrontal cortex Brodmann area 46 from a cohort of 37 matched pairs of patients with schizophrenia and nonpsychiatric control subjects.[114] Differential expression analyses revealed a total of 28 miRNAs to be dysregulated in the patients, with 25 up- and only three miRNAs downregulated. Concurrent with the predominant miRNA increases, Drosha as well as Dicer and DGCR8 were also robustly elevated, and there was also a significant correlation between Drosha and Dicer upregulation. Together, these findings strongly support a role of miRNAs in the neuropathology of schizophrenia and imply global dysregulation of the miRNA processing machinery as a key mechanism, akin to the above-reviewed reports in cancer models.

This concept was indeed further validated by a similarly designed study from Beveridge *et al.*, who also studied schizophrenia-associated alterations in miRNA and RNAi factor expression in human postmortem brain samples (Brodmann areas 9 and 22).[115] As in the Santarelli study,[114] they found substantial upregulation of 26 or 59 miRNAs in the two investigated brain areas, respectively. In addition, Beveridge *et al.* also noted robust upregulation of Drosha (and DGCR8) in the same specimens.[115] Unfortunately though, neither study attempted to unravel the potentially common molecular mechanisms underlying the consistent increases in Drosha in different regions of the brain in patients with schizophrenia.

Another example in this category of neurological disorders for which alterations of miRNAs and their associated machinery were studied is Huntington's disease (HD), a genetic neurodegenerative disease caused by abnormal CAG expansions in the *huntingtin* gene. Accordingly, Lee and colleagues measured the expression of miRNAs and all essential RNAi factors in two transgenic mice and one chemically induced rat HD models.[116] To briefly recapitulate the complex miRNA expression patterns that were observed, it should suffice to note that, especially in the mouse models, the vast majority of detected miRNAs was downregulated at the onset of HD phenotypes (e.g., locomotory dysfunctions in rotarod performance tests), whereas it was the opposite at earlier time points. These data correlated with a dynamic dysregulation particularly of Drosha expression which was higher than in wild-type mice at early time points, and lower later at phenotype onset in older mice. Notably, at least in one of the mouse models, Drosha was in fact the only key RNAi factor that showed dysregulation, while DGCR8, Dicer, Exportin-5, and Ago-2 were unchanged. Also noteworthy is that, although they are completely different diseases, the fluctuations in Drosha expression reported here are reminiscent of the changes found in gastric cancer with tumor type and growth,[112] suggesting that dynamic alterations in the basic RNAi machinery might represent a common mechanism in distinct human pathologies.

## B. DGCR8

Considering that DGCR8 is a cofactor for Drosha (together they form the microprocessor), it is not surprising that, in most cases to date that reported DGCR8 dysregulation in human diseases (or models thereof), this was noted concurrently with alterations in Drosha. This makes biological sense since Drosha and DGCR8 have evolved to tightly coregulate each other via an intricate double-negative feedback circuit in which DGCR8 stabilizes Drosha through a direct interaction, whereas, in turn, Drosha cleaves two pri-miRNA-like structures in the DGCR8 mRNA and thus interferes with DGCR8 expression (see also Section II.D).[92] For more information on this intriguing and in this form perhaps unique auto-regulatory loop by Drosha and its partner DGCR8, the reader is referred to the original literature[92] as well as the recent excellent and comprehensive reviews.[17–24]

With respect to human cancers, DGCR8 has frequently been found dysregulated in the same manner as Drosha, as mentioned. The relevant studies have thus already been described in the previous section and will not be discussed at length again. One example worth noting again, however, is the work by Zhang *et al.*, who studied salivary gland pleomorphic adenomas and found frequent upregulations of both Drosha and DGCR8.[108] Nonetheless, Drosha and DGCR8 alterations and outcomes do not always fully coincide, as exemplified by the also already-mentioned study by Sugito *et al.* on ESCC.[105] While high levels of Drosha expression were associated with low probability of survival, high rates of metastases, and generally poor prognosis for these patients, such clear correlations were absent for DGCR8 although it showed similar nonsignificant trends. Likewise, while Muralidhar *et al.* observed Drosha copy-number gains resulting it its overexpression in cervical SCC, this phenomenon was infrequent for DGCR8 in the same samples.[106]

Occasionally, some reports also focused exclusively on DGCR8 (and other factors) while lacking data on Drosha, leaving open whether the two factors were also concurrently altered in these cases. One example is a study by Ambs *et al.*, who profiled miRNAs and quantified some of the key RNAi components in 60 primary prostate tumors and 16 nontumor prostate tissues.[117] The two factors that were notably upregulated in the tumor samples were Dicer and DGCR8 while Drosha was not further mentioned, leaving a possible coregulation or a potential functional consequence of the DGCR8 increase unclear. Notably, when Chiosea *et al.* had earlier analyzed the expression of Drosha and DGCR8 in 64 patients with prostate adenocarcinoma (PCa; metastatic or organ-confined), they had found no evidence for upregulation of these two factors as compared to normal prostate, while all other RNAi components analyzed (Dicer, Exportin-5, and Ago-2) were elevated.[118] Besides, the same group subsequently analyzed essential RNAi factors in various tumor stages during the formation of non-small-cell lung carcinoma and found a substantial downregulation of DGCR8 in stage I lung adenocarcinoma versus normal lung, but not in earlier bronchioalveolar carcinoma (BAC).[119] Alas, there were again no correlating data on Drosha expression or on a possible prognostic role of the alterations in DGCR8.

As with Drosha, DGCR8 dysregulations were moreover also examined in the context of diseases other than cancer, such as in the above-mentioned study by Santarelli *et al.* on the role of miRNAs in schizophrenia.[114] As noted, Drosha and DGCR8 were concomitantly upregulated in postmortem brain tissues of the dorsolateral prefrontal cortex Brodmann area 46, concurrent with and perhaps explaining the increases in miRNA expression. This was also observed in the other mentioned study from the same group (Beveridge *et al.*), even though here the upregulation was even more pronounced for DGCR8, at least in one of the two brain regions investigated (superior temporal gyrus).[115] In this chapter, the group also made some interesting speculations on the potential mechanism underlying the DGCR8 increase, and hypothesized that it could derive from low copy repeat-induced microduplications and subsequent elevated gene dosages. A second possibility raised is that polymorphisms could have destroyed the pri-miRNA-like structures within the *DGCR8* gene that are recognized and cleaved by Drosha as part of the Drosha/DGCR8 auto-regulatory feedback loop (see above). Accordingly, such presumptive mutations (for which no experimental evidence was provided) could hinder feedback inhibition and result in DGCR8 elevation. Finally, Drosha and DGCR8 coregulation was also noted in the different Huntigton's disease models studied by Lee *et al.*, with the sole exception of one particular mouse strain (R6/2) that showed lower Drosha levels than wild-type mice, but unchanged DGCR8 expression, for reasons unknown.[116]

## C. Exportin-5

Unlike Drosha and DGCR8, data on the dysregulation of Exportin-5 in diseases other than viral infections (see Section IV) are very sparse, mainly because few groups to date have actually included Exportin-5 in their quantitative studies. One of them was the already-mentioned group of Chiosea and colleagues which analyzed gene expression in 64 patients with organ-confined PCa as well as four patients with metastatic PCa.[118] Congruent with earlier work by Varambally *et al.*,[120] they noted significantly increased Exportin-5 expression in metastatic PCa versus normal prostate, and a 1.6-fold elevation in metastatic versus organ-defined disease, implying a dual role of Exportin-5 in tumor growth and metastasization. Unfortunately, the group did not investigate alterations of miRNAs in this initial study and thus failed to provide a direct mechanistic link between upregulation of Exportin-5 plus other RNAi components with tumorigenesis.

In a similar subsequent study, the same team then investigated the expression of various RNAi factors, including Exportin-5, in different stages of lung cancer.[119] A robust threefold downregulation of Exportin-5 was consistently noted both in nonmucinous BAC (noninvasive neoplastic lesions) as well as in the ensuing invasive adenocarcinoma. Yet, there were again no miRNA data in this paper that could have validated possible phenotypic consequences of Exportin-5 reduction, such as impaired nuclear export or stability of miRNA precursors. Likewise, the potential prognostic value of Exportin-5 alterations during lung cancer progression also remained unclear.

Besides cancer, Exportin-5 may also be dysregulated in other diseases, as exemplified in the already-mentioned report by Lee and colleagues where Exportin-5 mRNA levels were altered (increased) in at least one of the two murine Huntingon's disease models.[116]

## D. Dicer

In marked contrast to Exportin-5, the literature on Dicer dysregulation especially in human cancers is already vast and expands rapidly, making it impossible to comprehensively review all relevant studies within this chapter. Instead, we will summarize hallmark findings and emerging models from the different studies and otherwise refer the reader to the original papers (note that all references are also provided in Fig. 2) for additional details.

One of the very first indications that Dicer can be upregulated in some human cancers came from Rajiv Dhir's group in their 2006 paper, which has already been mentioned above.[118] Briefly, while studying the expression of RNAi components in 68 patients with PCa, the team found a 2.5-fold transcriptional upregulation of Dicer in metastatic versus organ-confined PCa. Moreover, compared to normal prostate tissue or benign prostatic hypertrophy specimens, immunohistochemical analyses of PCa revealed a series of marked changes, including expression in both basal and luminal cells of the gland. Notably, Dicer expression increased with the clinical stage of the PCa and pathological TNM stages. (Note: TNM describes the extent of cancer in a patient's body, with T indicating the tumor size and whether it has invaded nearby tissue, N the regional lymph nodes that are involved, and M distant metastases.) Accordingly, highest Dicer levels were seen in PCa with a Gleason score > 7. (*Note*: the Gleason grading system serves to evaluate the prognosis of men with prostate cancer based on microscopic analyses; Gleason scores range from 2 to 10, with 10 implying the worst prognosis.) Together, these data suggest that Dicer elevation parallels the progression from intraepithelial neoplasia to PCa and in particular to the aggressive metastatic forms. The underlying mechanism remained unclear, but it was speculated that Dicer upregulation could be induced by genomic instability (amplification) of its location on the subtelomeric region of chromosome 14 (14q32.13). The group furthermore hypothesized that Dicer elevation might result in a global increase in miRNAs in PCa, but supportive data were not provided in this chapter.

Full corroboration of this work and conclusion came from a more recent paper by Ambs *et al.*, who analyzed Dicer expression in 60 primary prostate tumors and 16 nontumor prostate tissues, and likewise found an upregulation (1.6-fold) of Dicer in the tumor specimens.[117] Also, additionally corroborating the data from Chiosea *et al.*[118] was their finding that Dicer was more highly expressed in prostate tumors with a high Gleason score,[7–9] although the differences to those with lower scores were marginal (1.2-fold).

Besides prostate, the second tumor that a number of early studies have focused on is cancers of the lung, one of the leading causes of cancer-related deaths worldwide. It is also a canonical example for a tumor for which Dicer's role for progression and prognosis is highly and controversially debated, akin to the situation in breast and ovarian cancers (see further below). One of the first to study Dicer expression in lung cancer were Karube *et al.*, who examined 67 NSCLC cases that had undergone potentially curative surgical resection.[121] Their key finding was that cases with low Dicer expression showed significantly greater prevalence of poorly differentiated tumors compared to those with high Dicer levels, implying a relationship of Dicer with clinicopathological characteristics of lung cancer. Indeed, the group moreover noted a reduced probability of survival for the patient cohort with low levels of Dicer expression (roughly 18-fold hazard ratio for earlier death for low vs. high Dicer expression), leading them to conclude that the latter represents a significant predictive factor for poor prognosis, independent of disease stage. It should be noted, however, that at least some of these initial data remained rather speculative and require validation with an independent larger cohort. Furthermore, the mechanism underlying Dicer reduction was unclear, although the group could eliminate promoter hypermethylation as a possible cause. Instead, loss of heterozygosity on the long arm of chromosome 14 (where Dicer resides) was hypothesized, in line with the frequent finding of this genetic aberration in lung and various other human cancers. Regardless of mechanism, Dicer downregulation in lung cancers corroborated earlier findings from the same group and others that at least some miRNAs, such as let-7a-1 and let-7f-1, are frequently reduced in lung tumors, and that deliberate overexpression of let-7 can significantly inhibit growth of lung cancer cells in culture.

Partial confirmation for these results and hypotheses was indeed provided in a study by Chiosea *et al.*, who reported a more detailed differential examination of Dicer expression in the various histologic subtypes and stages of lung cancer. Congruent with the pioneering work by Karube *et al.*,[121] this group likewise found a Dicer reduction particularly in invasive stage I and II lung adenocarcinoma.[119] Moreover, they identified loss of heterozygosity of the Dicer-encoding region on the long arm of chromosome 14 as likely cause in up to 80% of stage II adenocarcinomas. Interestingly, beyond these similarities, Chiosea *et al.* also noted a transient Dicer upregulation in noninvasive lung cancer precursors (atypical adenomatous hyperplasia and BAC), implying that Dicer regulation is dynamic and that the stoichiometry of the miRNA machinery varies along the progression sequence of lung cancers. This study's finding that high Dicer expression is limited to prognostically favorable noninvasive precursors is indeed again in line with that of Karube *et al.*,[121] who concluded that lower Dicer levels are associated with shorter postoperative survival in lung carcinoma patients. Finally, noteworthy about this work are the potential implications for RNAi gene therapies, especially in view of data from our own group[75,122,123] and others that the miRNA machinery is prone to saturation by exogenous RNAi triggers. Hence, the sum of these studies suggests that future RNAi therapies will critically benefit from a thorough quantitative analysis of all RNAi components in a given tumor type and stage in order to determine the susceptibility to a certain RNAi trigger (e.g., siRNA vs. shRNA) and minimize potential therapy-induced adverse effects (see also our outlook in Section VI).

More recently, Hill *et al.* studied the role of Dicer for a specific further type of lung cancer, namely, pleuropulmonary blastoma (PPB).[124] Particularly notable about this rare pediatric tumor is that it is often part of an inherited cancer syndrome, with about 20% of all affected children having a family history of neoplasia. Intriguingly, Hill *et al.* were now able to map the PPB locus to chromosome 14q with a family-based linkage study. As mentioned above in the context of the other lung cancer studies, 14q encompasses the Dicer-encoding region, and, indeed, Hill and colleagues then found heterozygous germline mutations in the Dicer gene in 11 independent affected families. These truncations resulted in the loss of the two catalytically important

C-terminal RNase III domains, likely causing loss of function of the truncated Dicer variant. Surprisingly, immunohistochemical analyses then showed that expression of the wild-type allele was retained in the mesenchymal tumor cells but lost in benign-appearing tumor-associated lung epithelium. This tempted the authors to speculate that PPB may arise through a novel mechanism of non-cell-autonomous cancer initiation, possibly including altered miRNA-based regulation of diffusible growth factors that promote mesenchymal cell proliferation. While presently lacking direct evidence, this model is certainly intriguing and justifies further investigations.

Next to prostate and lung cancers, head and neck cancers form a third class of tumors for which evidence for an association with Dicer dysregulation has been reported. About 6% of head and neck cancers are salivary gland malignancies, which are in turn best represented by mucoepidermoid carcinoma (MEC) as the most common primary type. The prognostic value of Dicer expression for these specific tumors was assessed by Chiosea *et al.*, who measured and compared Dicer levels in 78 MECs by immunohistochemistry.[125] Notably, 60% of the studied specimens showed abnormalities in Dicer expression, with a nearly equal distribution of cases exhibiting an up- or downregulation, respectively. Overall, these dysregulations were found significantly more frequently in high-grade stage III/IV tumors of patients over 55 years of age, and they correlated with poorer disease-specific survival. Alas, like in several other studies describing abnormalities in expression of key RNAi factors, the underlying mechanisms as well as the implications for miRNA expression remained unclear, making these very important topics for future work.

Of note, consistent Dicer upregulation was also reported in the already-mentioned study by Zhang *et al.*[108] in pleomorphic salivary gland adenomas, along with tumor-specific increases in 17 miRNAs, implying a correlation between global disturbance of the miRNA biogenesis machinery and miRNA levels and thus hinting at direct mechanistic links.

Upregulation of Dicer in head and neck tumors was further confirmed in a more recent study by Jakymiw *et al.*, who analyzed Dicer expression in oral squamous cancer cell lines and tissues and found up to 24-fold increases as compared to normal primary gingival epithelial cells or tissue, respectively.[126] Importantly, the group also shed light on the cellular mechanisms underlying this dysregulation by demonstrating that upregulation did not occur on the level of Dicer transcripts, but was rather possibly linked to reductions in a certain miRNA (let-7b) known to suppress Dicer.[127–129] Experimental proof for this inverse correlation of Dicer and let-7b was obtained by transfecting the miRNA into oral cancer cells, resulting in the predicted increased posttranscriptional downregulation of Dicer and the ensuing significant inhibition of cell proliferation. It should be noted, however, that a second miRNA also known to control Dicer (let-7a) was unchanged in the cells, and that in two of the four analyzed cell lines, Dicer mRNA was also increased. Together, this raises the possibility that additional regulatory mechanisms determine Dicer expression in this and other tumors, and that our mechanistic picture of Dicer regulation in cancer is still far from complete. One possible further circuit discussed by the authors involves other let-7 family members such as let-7d which also regulates Dicer and has likewise been reported by others to be reduced in head and neck squamous cancer cell lines.[128,130] Generally, the model emerging from this and other studies that cancer-specific Dicer dysregulation might mainly occur posttransciptionally is highly relevant, as it may in fact explain some of the discrepancies in the literature where groups found divergent Dicer alterations even within the same tumor type. As raised in the work by Jakymiw *et al.*,[126] the reason for this might be that at least some of these teams primarily focused on measuring Dicer mRNA rather than protein, implying that the results may not have been fully indicative of the actual Dicer levels in the cell. Last but not least, the observation of substantial Dicer upregulation suggests that strategies to silence Dicer, for example, using specific si or shRNAs, might provide beneficial synergism in conjunction with conventional anticancer therapies.

Four additional examples for cancers exhibiting Dicer dysregulation that should briefly (due to the sparse specific literature to date) be discussed are leukemias as well as skin, liver, or gastric cancers. Regarding the first, Martin *et al.* assessed Dicer expression (albeit again only on the mRNA level, see previous section) in 71 cases of acute myeloid leukemia and found an upregulation (as compared to normal bone marrows) in 86% of all specimens.[131] Yet, there was no significant correlation between Dicer and the clinical outcome, including overall survival, questioning the usefulness of Dicer as prognostic cancer marker at least for this specific tumor type. Concerning skin cancers, one of the first evidences for an involvement of Dicer came from Zhang *et al.*, who noted DNA copy-number gains within a collection of 45 primary cultured melanoma cell lines of up to 20% (vs. about 9% losses).[132] Yet, the group did not dissect the relevance of these findings for the concurrently observed massive dysregulation of miRNA expression in melanoma (and other cancers), leaving unclear whether and to what extent Dicer contributed to the alterations in miRNA levels, or whether these were also mainly copy-number-based. In a different study, Sand *et al.* also noted upregulation of Dicer in skin cancer, but only in SCC and only as compared to intraindividual controls.[109] In case of BCCs, the opposite was found, with Dicer levels being significantly lower compared to both intraindividual as well as healthy controls. No dysregulation of Dicer in either direction was noted in premalignant actinic keratoses, implying that the alterations were indeed linked to skin cell transformation. This possible relationship as well as mechanisms underlying the differential changes in Dicer expression in the different skin cancer types are clearly interesting topics to be addressed in more detail in future studies.

With respect to the liver, Wu *et al.* compared Dicer levels in tumor tissues of 36 patients with hepatocellular carcinoma (HCC) to intraindividual controls from adjacent histologically noncancerous tissue.[133] Their key finding was that Dicer mRNA and protein were significantly lower in malignant tissue versus nonneoplastic counterparts in nearly 95% of all specimens. Unlike several other cancers (see above), there was no association with any clinicopathological characteristics of HCC, including age, sex, or tumor number, size or metastasization. Notably, though this study concluded that Dicer downregulation is typical for HCC, it remained possible that Dicer may also be otherwise dynamically dysregulated at earlier stages of hepatocarcinogenesis, akin to findings, for example, lung cancer (see above). Finally, Dicer expression in gastric cancer was studied in the already-discussed report by Tchernitsa and colleagues who characterized 332 gastric tumors.[112] Identical to their findings for Drosha (Section III.A), Dicer was upregulated in tumor tissue as compared to nonneoplastic mucosa and particularly high in intestinal (vs. diffuse) tumors. However, unlike for Drosha, there was insufficient evidence from a study of 190 subjects that Dicer levels might also serve as prognostic markers for patient survival.

Last but certainly not least, the two most notable examples for tumors associated with Dicer alterations—based on the vast and highly controversial literature, as well as the devastating death toll which they continue to exert on the female population—are breast and ovarian cancers. One of the first studies to assess Dicer expression in both cancers came from Zhang *et al.*, who analyzed 109 ovarian cancer (93 primary tumors and 16 cell lines) and 73 breast cancer (55 primary tumors and 18 cell lines) specimens.[132] As already noted above for their concurrent melanoma study, they found substantial and frequent genomic Dicer alterations including gains in DNA copy number by about 25% (vs. 8% loss) in the ovarian and 10% (vs. 3% loss) in the breast cancer samples. While the obvious suggestion was made that these dysregulations might affect miRNA expression in the two tumor types, a clear correlation was not proved. It, moreover, remained unclear whether and to what extent the changes in DNA copy numbers were reflected on Dicer protein levels in the different tumors.

In a follow-up study, the same group then again profiled miRNAs as well as quantified Dicer (and Drosha) expression in ovarian cancers (106 primary human specimens), and came to the conclusion that Dicer is actually not further dysregulated in this particular cancer when comparing early to late stages (note that no correlation with normal samples was performed anymore in this second study from this group, unlike in their 2006 report, see above).[113] Likewise, there was no detectable correlation between the expression level of either protein (including Drosha) and patient survival. To still explain the concurrently reported widespread and marked downregulation of miRNAs in the late-stage samples, the group postulated alterations in miRNA gene copy numbers as well as epigenetic changes as two possible mechanisms. What remained unclear in this new study, however, was the relationship to the earlier data from the same group according to which Dicer is upregulated in ovarian cancers. It was thus left open whether this upregulation was actually causing an miRNA elevation in earlier tumor stages which then declined later due to other mechanisms (see above) in the absence of further Dicer alterations, or whether the initially reported increase in Dicer was unrelated to miRNA expression to begin with. Clearly, these are important questions, especially considering the ongoing controversy in this particular area emerging from a wealth of further studies.

In one of these, Flavin *et al.* quantified Dicer protein expression in a tissue microarray of 66 ovarian serous carcinomas, and moreover studied Dicer mRNA levels in 50 fixed and embedded corresponding samples.[107] As compared to normal ovary, Dicer was upregulated 1.6-fold on the mRNA level with a tendency toward significance, a change that was not as substantial as that noted for two other key RNAi factors, Drosha (2.7-fold) and eIF6 (2.9-fold). Notably, on the protein level, the authors then found a greater than fourfold and statistically significant increase in Dicer immunoreactivity over normal ovarian surface epithelium. Also, the group noted clear relationships between Dicer expression and clinicopathological features, in a sense that cases with high Dicer levels showed greater absence of lymph-node metastases and greater prevalence of a proliferation index less than 50%. However, the probability of both disease-free and overall survival was not significantly different between patients with low or high Dicer expression levels, which the authors could not explain in view of the above-mentioned correlation with other features. Alas, this study also provided no further insight into possible mechanisms causing Dicer dysregulation, albeit hinting at genomic instability at chromosome 14q (the site of the Dicer locus, see also above) as one likely option.

Strikingly, a series of additional studies came to the exact opposite conclusion, namely, that Dicer is in fact downregulated in ovarian cancers. One of the first was published by Merritt *et al.*[110] just a few months after the Flavin *et al.*[107] report just discussed. In the Merritt study, the authors measured Dicer mRNA levels in invasive epithelial ovarian cancer from 111 patients and found a decrease in Dicer transcripts in 60% of all tumors. Notably, as mentioned before, Drosha was also reduced in over 50% of all tumors and, in fact, 39% of all specimens exhibited decreased levels of both Dicer and Drosha mRNAs. These results were also confirmed on the protein levels using histochemical analyses, and further validated in a panel of ovarian cancer cell lines. Regarding clinical associations, Dicer expression did not correlate with age, tumor grade, or response to chemotherapy, yet low Dicer levels were significantly associated with advanced tumor stage and reduced median overall survival, making decreased Dicer mRNA levels an indicator of poor prognosis. Interestingly, while this did not hold true for Drosha alone (see also above), the combination of low Dicer and low Drosha levels had an even greater association with decreased survival. Besides, the same correlation between high Dicer expression and increased disease-free survival was also noted in the same study for breast cancer in a cohort of 129 patients. Finally, the authors showed that the reduction of Dicer results in decreased activities of exogenously expressed shRNAs, but it remained unclear whether and to what extent this allows conclusion of the activities of endogenous miRNAs, especially since at least some cellular miRNAs might bypass Dicer altogether.

A second study that basically validated the findings and conclusions of Merritt *et al.*[110] was published by Faggad and colleagues who investigated Dicer protein expression in tissue samples from 140 patients with primary ovarian carcinomas.[134] Again, Dicer was downregulated in a subgroup of these patients, and low Dicer expression was linked to poor overall survival (roughly 52 vs. 84 months in Dicer-positive patients, reaching borderline prognostic significance). While these data are principally in line with Merritt *et al.*'s work,[110] Faggad *et al.* additionally studied miRNA and mRNA expression profiles in Dicer-negative versus -positive tissues. As expected, they noted a reduction in about 96% of all consistently detectable miRNAs in the absence of Dicer as well as 29 upregulated genes (17 of which were targets of the suppressed miRNAs); yet the origin of the also detected 60 downregulated mRNAs remained unclear. Interestingly, the group also obtained evidence for an association of Dicer and estrogen receptor (ER) expression, since both were simultaneously downregulated and the ER levels moreover dropped upon experimental silencing of Dicer. While the underlying mechanism remained unclear, it is intriguing to note that others have also observed correlations between the expressions of ER and key RNAi machinery in other tumors, in particular breast cancer (see below).

Pampalakis *et al.* reported a third study also describing Dicer mRNA downregulation in 16 malignant ovarian specimens as compared to 10 normal tissues or 8 benign tumors, with particularly striking reductions in tumors of higher grade and higher stage.[135] While interesting, validation in a larger cohort of patient material and appropriate controls is certainly indicated considering the small sample numbers in this study. Unfortunately, like many other studies, this report remained mainly descriptive and lacked mechanistic explanations for the observed dysregulation of Dicer (or further key RNAi factors in the other reports). Instead, the authors speculated that the reasons are probably complex and may involve a combination of genomic alterations, epigenetic modifications, and alternative promoter usage. Clearly, unraveling and fully understanding the actual molecular mechanisms hence remain important and urgent goals in this field, as this will not only yield valuable insights into the biology of RNAi regulation in disease, but will ideally also provide novel avenues for therapeutic intervention (see also Section VI).

A final curiosity was added to the ongoing controversy about Dicer's role in ovarian carcinoma in a very recent paper by Vaksman *et al.*, in which they compared Dicer (as well as Ago, see Section III.F) expression in primary tumors versus effusions from patients with disease spread beyond the ovary.[136] Because the cohort of tumors and matched controls was rather small, and because the presented data on Dicer protein expression were not fully consistent between Western blot and immunofluorescence analyses, it should only briefly be mentioned that the authors noted higher Dicer (and Ago-1) levels in the effusions. The potential underlying mechanisms and clinical implications remained unclear but are certainly interesting topics for more comprehensive follow-up studies.

Predominant downregulation of Dicer has also been commonly observed in a final devastating cancer that should be discussed here, that is, breast cancer. Blenkiron *et al.* were amongst the first to comprehensively quantify Dicer levels in five subtypes of breast tumors: luminal A, luminal B, basal-like, HER2+, and normal-like (differing in their mRNA expression profiles as well as their clinicopathological characteristics).[137] Their key finding was a downregulation of Dicer (on the mRNA level) in the more aggressive basal-like, HER2+, and luminal B type tumors. Interestingly, Ago-2, which was quantified in parallel, showed the exact opposite pattern (see also Section III.F), while Drosha, DGCR8, and the three other human Argonautes remained unchanged in the five tumor types. Dicer expression was moreover higher in the majority of tumor samples with a known positive ER status as compared to ER-negative samples, while Ago-2 again showed the reverse pattern. How this dysregulation of key RNAi machinery, in particular the opposite effects on Dicer and Ago-2, could explain, or correlate with, the observed complex and differential miRNA expression in the various tumor types was unfortunately not conclusively addressed in this study (however, see below). Regardless, this work provided the first evidence that Dicer dysregulation is involved in the etiology of human breast cancer, a theory that was convincingly solidified in subsequent independent studies.

In one of them, Cheng *et al.* compared miRNA activities in ER-negative versus ER-positive breast cancer subtypes and along these lines also examined potential differential expression of Dicer (plus Ago-1 and -2).[138] Consistent with Blenkiron *et al.*,[137] they found a significant downregulation of Dicer (and TRBP as well) in ER-negative samples and, vice versa, an upregulation of Ago genes. The mechanism was once again not clarified, but one hypothesis raised was that the MAPK signaling cascade, which is one of the major signal transduction pathways in ER-negative breast cancers, could enhance the signal of IFNI, which in turn represses Dicer. Moreover, there was evidence presented that some of the miRNAs that showed higher activities in ER-negative samples could target Dicer's 3′UTR, thus furthermore contributing to the reduction in Dicer expression. In addition to the data on Dicer expression, a second curious and intriguing pivotal finding of this study was that a substantial portion of all detected miRNAs were markedly more active in ER-negative cells, despite a lack of a clear correlation with their physical expression levels. Together with the above-noted observation of anticorrelated Dicer and Ago expression, this clearly implies that there is no necessary link between mature miRNA expression and miRNA/RISC activity.

This conclusion that the ultimate ability of a given miRNA to repress its cellular target may not always or strictly be reflected by its expression level has far-reaching implications. Mainly, it can help to explain and consolidate some of the seemingly counterintuitive or even contradictory results from further studies in breast cancer or other tumors. One example is the study by Blenkiron *et al.*[137] just discussed, where the authors found that the most differentially expressed miRNAs were in fact downregulated in ER-negative samples; yet these changes did not correlate with alterations in the mRNA levels of their direct targets. The additional data from Cheng *et al.*[138] now imply that miRNAs might exert a stronger inhibitory effect in ER-negative cells due to the upregulation of Ago-2/RISC, and despite their reduced (Dicer-dependent) abundance. More generally, this model strongly suggests that future studies aiming at linking miRNA expression profiles to a certain disease should take into account the potential differential expression of core RNAi machinery as well. Consequently, they should appreciate that, on one hand, this may induce miRNA activities that are discordant with the physical miRNA abundancies, implying that merely profiling miRNA signatures may not be sufficient to conclude on the actual effects on the transcriptome of the cell; on the other, this potential anticorrelated expression of RNAi factors itself should be regarded as a crucial signature for a certain disease (or physiological state), likewise implying that solely studying and quantifying an individual factor (e.g., Dicer or Ago) will yield an incomplete picture, and that it is the entirety of all components (e.g., Dicer and Ago) that must be considered (again, see our outlook in Section VI).

In a third concurrent study, Grelier *et al.* not only confirmed Dicer downregulation in breast cancer, but they were also able to correlate these consistently observed alterations with clinicopathological tumor features.[139] To briefly summarize this very comprehensive and complex work, the authors studied various models of breast cancer progression and metastasization and found that Dicer was significantly decreased in (1) more advanced tumor stages, (2) two bone metastasis subpopulations derived from a human breast cancer cell line, (3) mesenchymal phenotype cells, and (4) a panel of further cell lines established form invasive ductal carcinoma or from metastases. Regarding the clinical and prognostic value, it was then noted that lower Dicer expression was significantly associated with lymph-node metastases and with the more aggressive luminal B, HER2+, and basal-like tumor subtypes, in full agreement with and validating the study by Blenkiron and coworkers.[137] Moreover, Dicer was lower in ER-negative tumors, also confirming the independent data by the Blenkiron and Cheng groups (see above).[137,138] Last but not least, Dicer mRNA (but not protein) levels were predictive for patients' general survival and also served as a prognostic marker for the formation of metastases, with low levels more frequently found in patients with metastatic relapse. A question that remained open, however, was why there was only a 72% concordance between Dicer mRNA and protein levels, and why only mRNA, but not protein levels, is prognostically relevant with respect to patient outcome. As speculated by the authors, at least the mRNA–protein discordance may be explained by the various regulatory mechanisms controlling Dicer transcription and translation, including specific miRNAs and expression of different Dicer splice variants. Nonetheless, the definition of the variables explaining the association of Dicer mRNA levels and survival or metastasis, respectively, clearly needs more work.

Strong validation for all these data and conclusions was provided in a comprehensive study by Martello *et al.*,[140] along with additional intriguing insights that mechanistically link the different observations in the Blenkiron, Cheng, and Grelier studies.[137–139] The group first showed that Dicer translation is regulated via miR-103/107, which stood out amongst all miRNAs for the presence of eight evolutionarily conserved binding sites in the Dicer 3′UTR, implying cooperative binding and a biological role. It should be noted here that miR-103 was in fact also studied in the Blenkiron as well as the Cheng studies,[137,138] and found in the latter to be more active in the more aggressive ER-negative breast cancer types that were concurrently characterized by low Dicer levels. In fact, Martello *et al.* then also showed that deliberate upregulation of miR-107 leads to a global reduction in cellular miRNAs, and this was phenocopied by direct Dicer knockdown, as expected.[140] Moreover, they next found an inverse correlation between miR-103/107 levels and Dicer protein expression in a set of breast cancer as well as colon cancer cells lines. For reasons unclear, this particular finding is in fact contrary to the observations by Cheng *et al.*,[138] who noted a drop in miR-103 levels in the more aggressive ER-negative tumors. Still, further corroborating the earlier data, Martello *et al.* also discovered a good direct correlation of miR-103/107 levels in a collection of patients with breast cancer with their probability to develop metastases; that is, the high miR-103/107 group had low Dicer levels yet high risk of metastasization and overall poor outcome.[140] The same was confirmed when interrogating public gene expression datasets from over 1000 patients with breast cancer. Importantly, and contradicting the Grelier *et al.* study,[139] this significant association between Dicer protein levels and clinical relapse was not found on the Dicer mRNA level, which makes sense considering the proposed mechanism of specific posttranscriptional Dicer modulation via miRNAs. On a cellular basis, Martello and colleagues then demonstrated that Dicer downregulation by miR-103/107 unleashes aggressive cell behavior, including enhanced motility and invasion, fostered metastasis *in vivo* (mice), as well as typical epithelial-to-mesenchymal transitions.

The intriguing model proposed in this study, and perhaps applicable to tumors other than breast cancers as well, is that tumor cells manage to strike a balance between globally downregulated miRNA expression yet maintenance of sufficient RNAi capacities to sustain basic cellular functions, such as stemness and mitosis. As suggested by Martello *et al.*,[140] breast cancer cells might solve this dilemma by using miR-103/107 to suppress Dicer only to a certain threshold level already sufficient to foster transformation while not perturbing cellular functions required for tumor growth. Key to this balance might be the inherent robustness of this miniature network: miR103/107 control Dicer, yet at the same time their biogenesis depends on Dicer action, thus preventing complete deletion of Dicer in tumors and subsequent adverse elimination of all miRNAs. While exciting and plausible in theory, and certainly supported by the data in this study, one should also not forget the findings from the Blenkiron and Cheng groups[137,138] on the anticorrelation of Dicer and Ago-2 levels, which implied that even physically less abundant miRNAs can be more active in Dicer low but Ago-2-high cancer cells, and vice versa. Hence, a stirring goal for future work should be the further clarification of the quantitative relationships of miRNAs and key RNAi machinery, especially Dicer and Ago-2, in breast cancers and other tumors as well, to fully unravel the mechanistic links and prognostic values of each component.

The most recent data on the role of Dicer expression levels in breast cancer come from a study by Dedes *et al.*, which has already been described above in the context of Drosha.[111] In their analyses of 200 primary breast cancer samples, the authors found Dicer downregulation in 46% of all cases. This was significantly associated with lack of expression of ER and other markers of luminal differentiation and with high-grade basal-like phenotypes exhibiting high proliferation rates, corroborating the findings from the previous studies (see above). However, as also already mentioned above for Drosha, Dicer downregulation did not reach statistical significance as predictor for metastasization or survival, and this particular study also yielded no other robust associations with the outcome. The reasons for this discrepancy with a number of previous studies (in Merritt *et al.* as well as Grelier *et al.*[110,139]) may be complex and manifold, and possibly include the fact that all patients in the Dedes study were uniformly treated with adjuvant anthracycline-based chemotherapy and/or endocrine therapy, whereas patients in the other studies either received distinct systemic therapy regimens, or this information was not provided at all. Further discrepancies comprise different means of normalization in the various studies plus inconsistencies as well as technical issues regarding the quantification and use of Dicer mRNA or protein levels. Nonetheless, one can certainly conclude from all these different reports to date that distinct components of the RNAi machinery are dysregulated and potentially dysfunctional in different molecular subgroups of breast cancer, and that there is accumulating evidence that, in at least some cases, Dicer levels may serve as prognosticators for patient outcome.

A final notable example for Dicer dysregulation in human cancer is provided by a recent study by Potenza and colleagues in which the authors discovered a novel splice variant in neuroblastoma cells lacking exon 27.[141] This exon-skipping removes part of the coding region and changes the reading frame, creating a premature stop codon and hence yielding a truncated "t-Dicer" which is 93 amino acids shorter than wild-type Dicer and differs in the last 41 residues. Though not experimentally proved in this work, the authors then speculated, on the basis of other mutational studies with Dicer and reports of further natural truncation mutants, that t-Dicer should exhibit an unaltered substrate affinity and only a moderately decreased rate of RNA cleavage. Interestingly, a screen of further cancer cell lines as well as of numerous human tissues revealed the t-Dicer appears to be exclusively expressed in neuroblastoma cell lines. Also, evaluation of 55 primary neuroblastomic tumors of different clinical stages and with different histology revealed expression of t-Dicer in over 50% of all samples, without an obvious correlation with the stage of the tumor. However, it was found significantly more frequently (66% vs. 35%) in a certain aggressive tumor subtype (neuroblastoma Schwannian stroma poor) than in those with a favorable prognosis (ganglioneuroblastoma intermixed Schwannian stroma rich), implying a correlation with the differentiation state of the neuroblast cells. Whether this restricted expression of the truncated Dicer mutant is really causatively involved in tumorigenesis, or is just a curious by-product, still remained open in this interesting study and hence awaits further clarification, as does the question about the potential product resulting from t-Dicer activity. One possibility that was raised by the authors is that t-Dicer action yields a partially and aberrantly processed pre-miRNA, which might then interfere with wild-type Dicer functionality, perhaps akin to some of the saturation events observed in the context of viral infections (see Section IV).

As is the case for all other key RNAi factors, Dicer expression is also dysregulated not merely in human cancers but also in a variety of other diseases, though the respective literature is still in its infancy as compared to the cancer data. One notable example is schizophrenia for which an association with Dicer upregulation has been reported by independent groups. Two relevant studies by

Santarelli *et al.* as well as Beveridge *et al.* have in fact already been mentioned above in the context of Drosha dysregulation.[114,115] To briefly reiterate, the Santarelli work revealed a significant increase in Dicer mRNA expression in the dorsolateral prefrontal cortex Brodmann area 46 from a cohort of schizophrenia patients that correlated well with the upregulation of certain miRNAs.[114] While the authors provided no data on a possible relationship with clinicopathological variables, they speculated that Dicer elevation might have crucial implications for structural and functional plasticity of the synapse and might thus contribute to the pathophysiology of schizophrenia. The same conclusion was reached by Beveridge *et al.*, who also found Dicer upregulation in Brodmann areas 9 and 22 in human postmortem brain samples, along with elevations of numerous miRNAs as well as other key RNAi proteins (Drosha and DGCR8) in the same specimens.[115] Finally, another example in the category of neurological disorders for which Dicer alterations were studied is HD, as also already discussed earlier in the context of Drosha. However, in the relevant study by Lee *et al.*, the results on Dicer were less clear as Dicer was only downregulated (at late time points) in one of the two transgenic mouse strains investigated and unaltered in the other, thus preventing a convincing conclusion on the possible involvement of Dicer fluctuations in HD at this point.[116]

Last but not least, another most recent and highly intriguing example for Dicer's role outside cancer was provided by Kaneko *et al.*, who reported on the role and results of Dicer downregulation in age-related macular degeneration (AMD), a leading cause of blindness.[142] Here, the authors studied a certain untreatable advanced form of AMD, namely, geography atrophy (GA) which is characterized by extensive atrophy (cell degeneration) of the retinal pigment epithelium (RPE) and whose pathogenesis is still enigmatic. Interestingly, Kaneko *et al.* have now found that Dicer, but neither Drosha, nor DGCR8, nor Ago-2 is reduced (mRNA and protein) in the macular RPE in human donor eyes with GA compared to control eyes. The Dicer depletion was specific for GA and not a generic damage response found in any dying retina, as shown via analyses of eyes affected by several other retinal diseases. Moreover, experimental ablation of Dicer by various means induced RPE cell degeneration, resulting in dysmorphologies resembling those of human GA eyes. Unexpectedly, depletion (exogenously or in knockout mice) of any other major component of the RNAi machinery (Drosha, DGCR8, TRBP, Ago-1 through -4) could not recapitulate RPE degeneration, strongly suggesting that the Dicer effect involves a very specific mechanism and is not due to global miRNA dysregulation. Indeed, in a series of comprehensive and thoroughly controlled experiments, the authors finally identified a marked increase in the abundance of certain *Alu* transcript sequences that causally induced RPE cell death (via an unclear mechanism that might involve caspase-3 and apoptosis) and ultimately GA and which was specifically due to deficient Dicer processing.

This provocative study obviously raises a series of important questions and theories for follow-up work, including whether similar miRNA-independent effects have actually been missed in the past in other reports describing dysregulation of the RNAi machinery in human diseases (see above). In addition, many mechanistic details on the role of Dicer and *Alu* repeats in GA (and other diseases) remained open, such as the molecular reasons for Dicer downregulation, the potential existence of a feedback loop between Dicer and *Alu*, and the origin of the increased *Alu* sequences. Clearly, the new discovery of Dicer's role as an miRNA-independent cytoprotective surveillance factor implies that our picture of the function and importance of key RNAi factors in human disease is currently far from complete and that we may await many further exciting surprises.

## E. Argonaute/RISC

As mentioned in Section I, humans express a total of eight Argonaute family members, comprising four PIWI and four genuine Argonaute proteins. For both subtypes, a series of reports have already provided evidence for dysregulation in human diseases (mostly cancers), yet the literature is not as comprehensive as for Dicer. This may initially be surprising, considering that, in particular, Ago proteins are major central components of the RISC complex and as such theoretically highly prone to quantitative dysregulation. It is perhaps even more astounding in view of several reports that all eight Argonaute proteins are expressed differentially in various tissues and developmental stages,[143–148] again implying that transcriptional control is tight and crucial, and as such may be dysregulated in diseases. One possible explanation might be, however, that Ago proteins appear to be also extensively modified and controlled on a posttranslational level, and heavily interact with various cellular proteins that might additionally modulate their function. Several of these mechanisms and outcomes have already been mentioned earlier (Section II. D), and the reader is again referred to a number of recent excellent reviews for further details.[1–3,9,143,144,149,150]

A very early indication that Ago protein expression might indeed be dysregulated in human cancers dates back to the original isolation and initial characterization of Ago-1 in 1999. In this report, Koesters and colleagues used differential cDNA cloning to identify candidate genes that might be involved in the development of Wilms tumor, a childhood renal tumor caused by molecular defects in the differentiation of cells in the embryonic kidney.[151] During this study, Koesters *et al.* noted that the *Q99* gene (later characterized as EIF2C1 or Ago-1) was upregulated in tumors harboring a mutation of the Wilms tumor suppressor gene *WT1* as compared to tumors expressing wild-type WT1. Yet, a series of important questions remained open in this original report in fact to date, including the specificity of the original cDNA signals that were quantified in this paper (and may have actually originated from cross-hybridization with other human Ago genes) as well as the actual role of potentially upregulated Ago-1 in Wilms tumorigenesis.

Upregulation of Argonaute family members has subsequently been reported in a series of further studies and for different tumors as well, including breast, ovarian, prostate, and colon cancers. Regarding breast and ovarian cancers, Zhang *et al.* were amongst the first to note frequent copy-number alterations in the genes of all four Ago proteins (Ago-1 to -4).[132] Specifically, in a cohort of 109 ovarian specimens, they found gains in Ago-2 DNA copy number by nearly 52% (vs. 6% losses), and 12% versus 5% gains for Ago-1, -3, and − 4. This was different in a set of 73 breast cancers, which showed an equal distribution of Ago-2 gains and losses (both 18%), whereas the other three Ago members were more frequently lost (18% vs. 4%). Interestingly, the pattern was again different in a third tumor also studied in this paper, namely melanoma, where all four Ago members exclusively exhibited losses in DNA copy numbers by 33%. Yet, unlike for breast and ovarian cancers, the fact that only primary cultured cell lines were investigated left the relevance for primary tumor tissue unclear. Likewise the contribution of the noted quantitative and discordant (between the three tumors) dysregulations of the four Ago proteins for the miRNA expression profiles of breast, ovarian, and skin (melanoma) tumors remained open. As discussed by the authors, it may at least represent a likely additional factor controlling miRNA expression, next to epigenetic, transcriptional, posttranscriptional, and translational mechanisms.

Subsequent to this 2006 paper, a number of other groups validated the upregulation of Ago-2 in breast cancer, including some studies already discussed above. In one of them, Blenkiron *et al.* comprehensively quantified levels of various RNAi factors in five subtypes of breast tumors: luminal A, luminal B, basal-like, HER2+, and normal-like.[137] As noted earlier in Section III.D, a curious central finding of this study was the inverse correlation of Ago-2 and Dicer expression patterns, with Ago-2 being high (and Dicer low) in the more aggressive basal-like, HER2+, and luminal B type tumors. Similarly, Ago-2 was upregulated (and Dicer down) in ER-negative tumor samples. In contrast to Ago-2, none of the other three Ago proteins showed any significant changes in expression in the various cancer subtypes. While the functional consequences or prognostic values of these Ago dysregulations remained open in this study, it is still tempting to speculate that alterations of the core RNAi machinery (including Dicer) are involved in the etiology of human breast cancer. Particularly intriguing in this context is the fact that only Ago-2 appears to be altered in its expression in breast cancer, which may hint at a role outside the miRNA pathway (considering that all four Ago proteins act in this pathway and are somewhat redundant) and/or point to the importance of Ago-2 in miRNA biogenesis. Examples for the latter have already been mentioned in Section II.B and include a specific role of Ago-2 in the generation of ac-pre-miRNAs or in miR-451 biogenesis.[64]

On the other hand, in their comparison of ER-positive and negative breast cancer subtypes, Cheng *et al.* not only validated the upregulation of Ago-2 in ER-negative samples but moreover found the same for Ago-1.[138] Importantly, identical to the Blenkiron study,[137] Cheng and colleagues also found an anticorrelation for Ago and Dicer proteins, with the latter being significantly downregulated in ER-negative tumors. As discussed earlier (Section III.D), this suggests that cancer pathology may generally be influenced by the global dysregulation of multiple RNAi factors and raises caution regarding attempts to exclusively link etiologies to changes in miRNA expression profiles or alterations of individual RNAi components.

In fact, additional support for this proposal comes from a study by Chiosea *et al.*, who provided one of the few reported examples to date for Ago downregulation in human cancers, in this case, advanced lung adenocarcinomas.[119] Reminiscent of the data by Cheng *et al.* in breast cancer,[138] Chiosea and colleagues also noted concurrent dysregulation (5.1- and 3.4-fold decreases) of Ago-1 and Ago-2, and an anticorrelation with Dicer which was substantially (10-fold) increased in the same samples. Besides, this finding is also highly relevant from a therapeutic standpoint considering that conventional RNAi triggers (siRNAs or shRNAs) depend on Ago-2 for their activity, suggesting that future clinical RNAi-based intervention strategies will have to be tailored to individual cancers and adapted in those exhibiting Ago-2 reductions (see also Sections V and VI).

Further mechanistic insight into the above-mentioned striking upregulation of Ago-2 specifically in ER-negative breast cancers was then provided by Adams *et al.*, who proposed and confirmed a role for epidermal growth factor receptor (EGFR) signaling.[152] First, they validated significant upregulation of Ago-2 in ER-negative breast cancer cell lines as well as in five tumors (vs. three in ER-positive). As also noted by Zhang *et al.*,[132] but in contrast to Blenkiron *et al.*,[137] none of the three other Ago proteins was dysregulated, hinting at a specific role for Ago-2. Adams *et al.* could next eliminate gene amplification as the underlying reason, but instead hypothesized constitutive activation of a relevant cell-signaling cascade.[152] To briefly recapitulate their comprehensive data, they indeed identified EGFR/MAPK signaling as the major cascade and showed that Ago-2 elevation occurred via an as-yet unknown pathway triggering posttranslational Ago-2 protein stabilization. In addition, they obtained experimental evidence that stable Ago-2 overexpression in less transformed ER-positive breast cancer cells stimulated cell cycle progression and enhanced proliferation, reduced cell–cell adhesion, and enhanced migratory/invasive capabilities, which are all indicators of tumorigenic transformation. The direct functional relationship between Ago-2 upregulation and tumorigenesis remained unresolved however, despite limited additional evidence for increased activity of a specific miRNA in the Ago-2-transfected cells. It likewise remains an interesting goal for future work to identify further players in the EGFR/MAPK pathway that directly interact with the Ago-2 gene, promoter, or protein to cause the observed upregulation. Two intriguing possibilities are that either a specific E3 ubiquitin ligase acts as a tumor suppressor and targets Ago-2 for proteasomal degradation in an EGF-suppressible manner, or that EGF alters the intracellular localization of Ago-2 and thereby increases its half-life.

Clearly, all these questions and models raised by the different studies require further investigation and validation, and this should be given high priority considering the devastating death toll that breast cancer continues to take on women. In particular ER-negative tumors, which represent about one-third of invasive breast carcinomas, provide little treatment options due to their resistance to conventional antihormone adjuvant therapies and their increased proliferation propensity irrespective of estrogen signaling, altogether invoking poor clinical prognoses. Unraveling the potential role of dysregulated RNAi components in the development of this particular cancer is thus not only interesting from a biological standpoint but also provides reasonable hope to find new clinically relevant targets and thus finally effective therapeutic strategies.

As in the case of breast cancer, others have also validated the finding by Zhang *et al.*[132] that Ago-2 is upregulated in ovarian cancer. In one study, Vaksman and coworkers compared the miRNA-regulating machinery in primary ovarian tumors versus effusions from patients with disease spread beyond the ovary.[136] All three factors studied—Dicer, Ago-1, and -2—were upregulated (on the mRNA levels, and Ago-2 additionally on the protein level) in the effusions suggesting a role in tumor progression. However, as already noted above, larger numbers of patients and matched controls than in this study will be needed to eventually unravel the molecular mechanisms for the global upregulation of the RNAi machinery and to begin to distill the potential clinical roles of the individual components.

Prostate cancer represents yet another example of a human tumor disease tentatively associated with Ago-2 elevation. One of the first reported evidences was found in a 2006 study by Chiosea *et al.*, who compared the expression of the RNAi machinery in a total of 68 patients, comprising 64 cases with organ-confined and four with metastatic disease.[118] As in the case of Dicer (see also Section III.D), mRNA levels of Ago-1 and − 2 were also upregulated in the metastatic specimens, namely by 1.8- and 3-fold, respectively. However, because this study focused mostly on Dicer, the potential clinicopathological relevance of Ago-1/2 dysregulation was not further explored. A similar conclusion can be drawn from a study by Ambs *et al.*, who analyzed Ago-2 expression in 60 primary prostate tumors and 16 nontumor prostate tissues and likewise found a 1.3-fold upregulation in the tumor specimens, especially in those with a high Gleason score of 7–9.[117] Again, the clinical relevance for patients with prostate cancer was not resolved or discussed in this work, leaving the unraveling of these parameters important goals for future studies.

Finally, a recent study by Li *et al.* addressed the role of Argonaute proteins in human colon cancer, one of the most frequent and most lethal malignancies worldwide.[153] Notable about this work is that Li and colleagues not only investigated the participation of the four canonical Ago proteins, but concurrently also that of the four PIWI members. Therefore, they constructed a tissue microarray including 75 specimens from colon cancer and 75 from adjacent noncancer tissue and then assayed Ago/PIWI protein expression by immunohistochemistry using in-house antibodies. Overall, it was noted that all eight proteins were more abundant in the tumorous tissues and, moreover, that the positive expression of Ago2- to -4 as well as of PIWIL4 was associated with the presence of distant metastases. Furthermore, expression of Ago-1 and PIWIL2 significantly correlated with the occurrence of colon cancer tissue in colonic lesions, while the other six Argonaute proteins showed no association, suggesting a specific potential of Ago-1 and PIWIL2 as novel colon cancer markers with early diagnostic significance. While generally considered outstanding as the only study to date exploring the role of all eight Argonaute proteins in human cancer, a number of questions remain unanswered and require validation with a larger and independent patient cohort. These include the prognostic value of the different Ago/PIWI proteins as well as, in particular, the biological mechanism explaining the role of selected Ago/PIWI members in the development and progression of human colon cancer.

Dysregulation of PIWI members of the Argonaute family has in fact also been observed in a number of other human cancers before, including a 2002 study by Qiao and colleagues who molecularly characterized PIWIL1 (also called Hiwi) and its expression in testicular germ cell tumors (TGCTs) of adolescents and adults.[154] These tumors, which are the most frequently occurring malignancy in Caucasian males between 15–45 years of age, originate from an embryonic germ cell. Approximately 50% of TGCTs are pure seminomas exhibiting typical germline phenotype, whereas another 40% are nonseminomas that mimic somatic embryonic development. In their study, Qiao *et al.* found a significant enhancement of PIWIL1 mRNA expression in 12 out of 19 sampled testicular seminomas as compared to normal adult testicular parenchyme, but not in nonseminomas, spermatocytic seminomas, or non-germ-cell (somatic) tumors of the adult testis. Notably, the exact level of PIWIL1 expression was neither correlated to the number of tumor cells present in a given sample nor to a particular mitotic or apoptotic state of seminomas. One possible explanation for these findings worth studying further is that other PIWI members may be dysregulated as well in the malignant progression from embryonic germ cell to seminoma.

In fact, Lee *et al.* found that PIWIL2 (also called Hili) mRNA was also enhanced in 9 out of 10 studied testicular seminomas but not in nonseminomatous tumors or in the surrounding normal testicular tissue.[155] The fact that these results closely mimic the expression pattern of PIWIL1 detected in the Qiao study[154] suggests a possible synergistic effect of *piwi* genes on the initiation and progression of TGCTs. Noteworthy, enhanced expression of PIWIL2 was then also found in a large panel of further human cancers, including prostate, breast, gastrointestinal, ovarian, and endometrial tumors, as well as in different mouse tumors comprising rhabdomyosarcoma and medulloblastoma. This implies that PIWIL2 might belong to the category of cancer-testis antigens (CTAs) and hence be a target for active specific immunotherapy. Mechanistically, PIWIL2 elevation induced the expression of antiapoptotic Bcl-$X_L$, which in turn correlated with an increase in Stat3 (signal transducer and activator of transcription 3), possibly explaining the observed inhibition of apoptosis and enhanced proliferation in PIWIL2-expressing cells. How these alterations were specifically triggered and regulated by PIWIL2 remained open, but it could at least be concluded that the aberrant expression of PIWIL2 makes this Argonaute member an attractive candidate as a prognostic and diagnostic cancer marker.

Other than malignancies originating from germ cells, PIWI proteins can also be dysregulated in, and contribute to, epithelial solid tumors, as reported by Liu and colleagues for gastric cancer as an example.[156] Initial characterization at the mRNA and protein levels revealed that PIWIL1 was highly expressed in 7 out of 10 human gastric cancer tissue samples (vs. only 2 out of 10 surrounding noncancerous tissues). In a more detailed comparison of tissues representing the various stages of tumor progression, the authors then found a pattern clearly correlating with precancerous development, with expression ratios ranging from 10% to 36% and finally 76% in normal gastric tissues, atrophic gastritis, and intestinal metaplasia to gastric cancers, respectively. PIWIL1 expression was also associated with the proliferating activity of the tumor cells, suggesting that it might be a potential target for cancer gene/RNAi therapies. Besides, it will of course be interesting to analyze the possible relationship between PIWIL1 expression and clinicopathological parameters, including the stage and grade of gastric cancers as well as prognosis.

A second example for a human somatic cancer associated with PIWIL1 upregulation, as reported by Grochola and colleagues, is ductal adenocarcinomas of the pancreas (PDAC).[157] Typical for this dismal type of cancer are a late clinical presentation and a poor overall prognosis, as well as a number of biological features, such as high invasive and metastatic potential plus resistance to chemotherapy and radiation, which suggest involvement of stem-cell-like cells and hence a role of PIWI proteins. Indeed, Grochola *et al.* found PIWIL1 mRNA upregulation in roughly 71% of microdissected PDAC tissues and elevation in PIWIL1 protein levels in nearly 30% of all samples. There was no general impact of PIWIL1 mRNA or protein elevation on survival, yet, as compared to a subgroup of men exhibiting intermediate PIWIL1 expression, male patients with lower or higher mRNA levels had a significantly increased risk of tumor-related death. Interestingly, this disparity was not observed either when the same differential analyses were performed in female subgroups, or on the PIWIL1 protein levels, hinting at a more complex correlation between PIWIL1 expression and tumorigenesis in the pancreas, including a surprising and currently ill-understood gender dependency that is certainly interesting to study further.

As a final example, Taubert *et al.* examined PIWIL1 expression in a cohort of 65 primary tumor samples from patients with soft-tissue sarcomas (STS), which is another cancer suspected to involve stem cells and their dysregulation.[158] Because PIWIL1 could in fact be detected in all 65 samples, they were then further subgrouped into 32 low, 20 medium, and 13 high expressing cohorts. Very similar to the Grochola study,[157] Taubert *et al.* then noted a correlation between dysregulated (i.e., lower or higher than the intermediate group) PIWIL1 expression and increased risk of tumor-related death, though the significance was reached only for the higher group. In turn, this implied that a medium PIWIL1 expression level might represent a normal physiological status in a mesenchymal cell, a theory that clearly requires more validation using actual normal tissues for comparison. While this study also did not address potential molecular mechanisms, the sum of reports on PIWI dysregulation to date strongly suggests that, of the four human PIWI members, at least PIWIL1 and PIWIL2 are associated with development and/or progression of various malignancies, implying that PIWI alterations in cancer will be an interesting and prospering future research area.

## F. Other Factors

Akin to Exportin-5 (see Section III.B), literature on dysregulation of the remaining RNAi factors (i.e., interaction partners of Drosha, Dicer, or RISC) in human disease is still relatively sparse and, where examined, they were in fact frequently unaltered (contrary to other RNAi components studied in parallel). Examples include the lack of dysregulation (always compared to respective normal tissue) of DDX5 and DDX17 in the cerebral cortex of patients with schizophrenia,[115] DCP1 in a mouse model of HD,[116] or of PACT, TRBP, and Gemins 3/4 in prostate cancer specimens.[118]

On the other hand, the latter study in fact reported an upregulation of TNRC6B (one of the three paralogs of GW182 in humans, see Section I) in prostate cancer versus normal tissue that was further increased twofold in more aggressive metastatic specimens.[118] Curiously, the same group later found the opposite: that is, a marked TNRC6B downregulation, both in nonmucinous

BAC as well as in stage I lung adenocarcinoma.[119] Generally, alterations in TNRC6B are highly intriguing, as the TNRC6A–C family has recently been identified as a functional key component of miRNA-induced RISC complexes[9] (see also Section I), implying that disease-associated TNRC6 dysregulations might exert substantial global effects on miRNA targets. Moreover, there is also recent evidence from plants that the interaction of GW182 with Argonaute proteins is a bottleneck that can be deliberately and quantitatively perturbed by pathogenic viruses,[159–161] indicating multiple levels of GW182/TNRC6 control in physiology and disease. It should thus be an exciting goal for future studies to further address the role and potential dysregulation of the three individual human TNRC6 members in additional cancer models and specimens, as well as in other disorders.

Another RNAi factor that has been sporadically included in quantitative analyses of the RNAi machinery is the already-mentioned TRBP. While unaltered in some cases, it was found to be substantially (9.4-fold) upregulated in advanced stage I lung adenocarcinoma in a study by Chiosea and colleagues.[119] This increase actually paralleled an almost identical 10-fold elevation of Dicer in the same samples, which may make biological sense considering that TRBP may contribute to the stabilization of Dicer protein. Notably, the most pronounced dysregulation was observed in this study for HSP90, which dropped by over 14-fold in the stage I lung adenocarcinoma, for unknown reasons and with unclear consequences for patient outcome or prognosis. Correlated alterations (in this case downregulation) of Dicer and TRBP were then also reported by Cheng *et al.* in ER-negative breast cancer specimens,[138] substantiating the possibility that the Dicer–TRBP complex is co-(dys)regulated at least in some human cancers.

Finally, yet another component of the RNAi machinery that was intermittently found to be altered in cancers is Tudor-SN, a transcriptional coactivator whose specific function in RNAi remains largely elusive. In one example, Tsuchiya *et al.*[162] investigated the involvement of Tudor-SN in colon carcinogenesis and found twofold and greater upregulation (compared to noncancerous counterparts) of Tudor-SN mRNA in 28 out of 32 human colon cancer tissues, even in early stage lesions and without a clear correlation with the different clinicopathological stages of the cancer. As then shown experimentally, one of the potential underlying mechanisms is enhanced cell proliferation and loss of contact inhibition upon Tudor-SN overexpression, explaining the possible role of this RNAi factor in colon carcinogenesis, but leaving the association with miRNAs or other RNAi components still unclear.

A second example for Tudor-SN involvement in human cancers was provided by Kuruma *et al.* in their investigation of specimens from 62 prostate cancer patients.[163] Notably, upregulated Tudor-SN expression was detected in 97% of all cases, while it was much weaker or absent in high-grade prostatic intraepithelial neoplasia or benign tissues. The intensity of Tudor-SN immunostaining moreover correlated with Gleason score and high prostate specific antigen (PSA), but not with the pathological T stage. In a long-term follow-up study, increased Tudor-SN expression was also not an independent predictor for biochemical failure after radical prostatectomy. While all these initial conclusions probably need validation in larger sample populations, the intriguing first evidence for the diagnostic potential of Tudor-SN in prostate cancer specimens provided by this study certainly warrants further analyses.

## IV. Quantitative Interference of Viruses with Host RNAi Factors

As detailed in the accompanying chapter by Roberts, Lewis, and Jopling in this issue, as well as in a number of comprehensive and excellent reviews recently published elsewhere,[164–168] human pathogenic viruses heavily interact with their host's RNAi machinery and exploit and reprogram the latter in multiple ways. This includes deliberate dysregulation of cellular miRNAs (which themselves control host dependency or restriction factors), binding of host-encoded miRNAs to viral RNAs/transcripts in order to foster virus replication or latency, as well as expression of virally encoded miRNAs which in turn alter host gene expression in favor of the virus. Moreover, the emerging picture is that a large number of viruses have evolved suppressors of RNAi silencing (SRS) that can modulate the activity of central RNAi components, or bind and thus sequester cellular miRNAs in order to support viral replication or to block host mechanisms directed against the virus. It should be noted, however, that the existence of SRS in human viruses continues to be debated, as does the general role of RNAi as an antiviral defense mechanism in mammals. Again, the reader is referred to a series of recent outstanding review articles for more details on this intriguing ongoing controversy.[164–168]

In the context of the present chapter, we will briefly review the small subset of the latest literature describing evidence for a mainly quantitative involvement or dysregulation of core RNAi factors in human viral infection. In this regard, a prototypic virus is HIV-1, which is well known for its extensive mutual interactions—both qualitatively and quantitatively—with the human host miRNA machinery. Relevant evidence includes early findings by Bennasser *et al.*[169,170] that HIV-1 Tat protein can act as an inhibitor of human Dicer activity and thus block Dicer-dependent RNAi triggers, that is, shRNAs and miRNAs. Note, however, that this was again more recently strongly debated by Lin and Cullen who reported no effect of Tat overexpression or HIV-1 infection on host miRNA biogenesis or shRNA activity.[171] Curiously, these studies not only contradict each other, but also another report by Triboulet *et al.* who described substantial dysregulations (up and down) of various cellular miRNAs in HIV-1-infected cells.[172] Regardless of these unresolved discrepancies, a notable finding in the first Bennasser *et al.* study with regard to the present review was that neither Tat expression nor HIV-1 infection perturbed cellular Dicer expression, arguing against an active dysregulation on the transcriptional or translational levels.[170]

This was, however, again questioned in a very recent report by Coley and colleagues who analyzed the effects of HIV-1 infection in monocytes and monocyte-derived macrophages (MDMs).[173] Compared to the HeLa cells used in the Bennasser study,[170] monocytes and MDMs are more relevant in the HIV-1 context since both can harbor productive as well as latent infections. Accordingly, they are part of the viral reservoir that is essential for chronic infection and is responsible for a major deal of HIV-1 resurgence after cessation of highly active antiretroviral therapy. Interestingly, the authors in the Coley study now found that monocytes are deficient for key miRNA processing proteins, evidenced by barely detectable (by Western blotting) levels of Drosha, Ago-1, and Ago-2, plus a complete absence of Dicer.[173] Treatment of the monocytes with PMA or M-CSF, which triggers differentiation into MDMs, then resulted in marked upregulation of all four RNAi factors as well as in concurrent increases in miRNA expression, as could be expected. Strikingly, when this experiment was repeated in the presence of HIV-1 (an electroporated infectious clone), Drosha was still induced, but Dicer remained conspicuously absent in the resulting MDM. Concomitantly, a majority of miRNAs were no longer upregulated either, together suggesting that HIV-1 suppresses host RNAi during its normal course of infection of monocytes and MDMs by downregulating Dicer expression and thus blocking an essential step in miRNA biogenesis. Subsequent mutational analyses implied that it is in particular the HIV-1 Vpr protein that mediates the inhibition of Dicer expression, perhaps in conjunction with the Nef protein which gave a less robust but similar effect. It, unfortunately, remained unresolved how these HIV-1 proteins actually mediate their inhibitory effect, but the authors could show that a specific miRNA, miR-106a, is expressed in undifferentiated monocytes and suppresses Dicer via binding to its 3′UTR. Whether HIV-1 Vpr/Nef dysregulates this particular miRNA during the course of infection, and by which mechanism, thus remain very interesting questions for future studies.

Last but not least, the authors also investigated the source of the miRNAs which they had detected in the undifferentiated monocytes and which were initially hard to explain due to the absence of Dicer in these cells. Interestingly, they then found that PIWIL4, one of the four human PIWI proteins, is present in monocytes and obtained evidence that it might substitute for Dicer in processing miRNAs. Since this was only shown for a single miRNA (miR-572), this conclusion certainly requires broader validation. Finally, interesting to note was that unlike Dicer, PIWIL4 protein levels were neither affected by differentiation into MDM nor by HIV-1 infection, showing that HIV-1 differentially dysregulates various components of the RNAi machinery.

An alternative second mechanism by which HIV-1 might quantitatively interfere with host miRNA pathways was proposed by Bennasser and colleagues.[174] Briefly, they showed that a surfeit of TAR (transactivation responsive sequence, a stem-bulge-loop RNA structure present twice in all HIV-1 transcripts and required for recognition by HIV-1 Tat protein for viral transcription to occur potently) can sequester TRBP, thereby shunting it away from the Dicer/RISC complex and hence blocking the endogenous RNAi pathway. As predicted, they then also found that titration of TRBP via overexpressed TAR resulted in a reduction of three specific miRNAs, seemingly supporting the global inhibition of the miRNA/RNAi pathway. Important to point out is, however, that all these conclusions were drawn from rather artificial experiments involving deliberate overexpression of isolated HIV-1 RNAs and proteins, and that the biological relevance of these findings in the actual infectious context remains heavily debated.

A related phenomenon involving quantitative sequestration or saturation of the RNAi machinery by a virally encoded RNA decoy was also repeatedly reported in the context of another virus, namely, human adenovirus (a family of double-stranded DNA viruses causing upper respiratory infections in children and adults). Because the literature on this specific virus–host interaction is already rather complex, and because the underlying mechanism does not truly involve dysregulations of expression of the RNAi machinery (the overarching topic of this review), we will only briefly review some of the hallmark studies and refer the reader to the original literature and references therein for further details.

A first seminal paper highlighting the possibility that human adenovirus may purposely overwhelm and thus inhibit cellular RNAi machinery came from Bryan Cullen's group in 2004.[175] In this work, Lu and Cullen focused on the so-called VA1 RNA, a roughly 160-nucleotide long highly structured RNA that is produced at extremely high levels of about $10^8$ molecules per cell during adenoviral replication and is used by the virus to block protein kinase R (PKR) activation and to thus enhance viral mRNA translation. In a series of elegant experiments, Lu and Cullen demonstrated that VA1 overexpression in cells blocked the activity of coexpressed miRNAs as well as shRNAs, but had a lesser effect on a synthetic pre-miRNA stem loop, and none on an siRNA. Altogether, this implied that VA1 RNA inhibits two independent steps in the RNAi pathway: that is, Exportin-5-mediated nuclear export of pre-miRNAs and shRNAs, as well as their further processing by Dicer in the cytoplasm. Indeed, experimental overexpression of Exportin-5 was able to largely rescue the VA1-mediated inhibitory effect, supporting the idea that Exportin-5 is rate-limiting (see also Section V below) and that adenoviral VA1 RNA outcompetes endogenous miRNAs for this limiting Exportin-5 pool. Finally, Lu and Cullen showed that VA1 RNA also binds Dicer quantitatively and specifically, and speculated that they thereby additionally compete with cellular pre-miRNAs.

In particular, the Dicer-related findings were then validated and extended in a follow-up study by Andersson and colleagues that also provided more insight into the mechanisms governing VA RNA-mediated quantitative RNAi suppression.[176] Having recapitulated the original finding from the Cullen lab[175] that adenovirus infection annulls shRNA activity in cells, they went on to demonstrate that in fact both adenoviral VA RNAs, VA1 and VA2, can individually block shRNA-mediated RNAi. Mechanistically, this was found to involve recognition of the two VA RNAs as competitive Dicer substrates, which were actually cleaved into functional siRNAs and incorporated into RISC, and hence, by sheer amount in adenovirus-infected cells, likely act to out-titrate the cell's capability to process its own miRNAs. It remained open in this study, however, whether the putative siRNAs resulting from VA1/2 processing bind actual cellular or viral targets in adenovirus-infected cells, and whether their abundant expression truly represents an effort by the virus to quantitatively block cellular RNAi or rather serves other more conventional functions, such as to inhibit the IFN-induced antiviral defense system.

The first question regarding the potential biological role of the two VA RNAs was in fact addressed and partially answered in a series of more recent follow-up studies, including additional work from the same group as well as various reports from Puri Fortes' lab.[177–179] Since this specific topic is beyond the scope of the present chapter, it suffices to summarize that there is indeed independent and comprehensive evidence for a functional RNAi-like role for VA1/2 RNAs in modulating cellular gene expression (including at least 30 candidates related to cell growth, apoptosis, transcription, and other essential processes) in adenovirus-infected cells,[177] and that this activity seems to be required for virus viability. Important to note in the context of this review is that none of these studies has provided evidence to date for a quantitative saturation and thus global dysregulation of RISC by the VA RNAs as well, suggesting that their inhibitory activity on the RNAi machinery really focuses on titrating Exportin-5 and blocking Dicer.

One final study that should be highlighted in this regard is a very recent report, again by Bennasser *et al.*, that shed surprising new light on the mechanisms underlying the VA RNA effects and on the (dys)regulation of the RNAi machinery in general.[180] The initial hallmark finding in this study was that deliberate knockdown of Exportin-5 resulted in an unexpected coreduction of Dicer expression and vice versa; that is, Exportin-5 overexpression also increased Dicer levels in parallel. Curiously, this was not due to dysregulation of Dicer mRNA abundancy; instead, the authors found that Exportin-5 acts as a karyopherin responsible for Dicer mRNA nuclear export. Hence, Exportin-5

countermand the presumably antiviral effect of RNAi, in particular by blocking the activity and/or expression of key RNAi factors. From a translational standpoint, a second interesting idea emerging from this work is that deliberately dysregulating these RNAi factors by exogenous means may in turn improve the production of gene transfer vectors derived from natural viruses. The feasibility of this novel approach is indeed supported by a growing body of evidence, starting with various studies with wild-type viruses or variants thereof. These include early findings by Wang *et al.* that silencing Dicer in Huh-7 cells using a specific shRNA augmented luciferase activity from an appropriately tagged and cotransfected subgenomic HCV replicon as well as increased formation of replicon-positive cell clones.[190]

Similarly, Triboulet *et al.* reported that siRNA-mediated knockdown of Drosha and Dicer in peripheral blood mononuclear cells (PBMCs) from HIV-1-infected donors resulted in faster viral replication kinetics as compared to virus from cells transfected with control siRNAs.[172] The same was observed in Drosha/Dicer-depleted established T-cell lines and also for individual knockdowns of the two RNAi factors, showing that both contribute to the HIV-1 inhibition. Importantly, a mechanistic link was also made in this study based on one particular miRNA cluster (miR ~ 17–92) that was specifically downregulated in HIV-1-infected cells and that in turn represses PCAF, which is an important cofactor of the viral HIV-1 Tat protein. Hence, it was speculated that Drosha/Dicer ablation may have reduced the miR ~ 17–92 cluster and thus upregulated and activated the PCAF/Tat complex, resulting in enhanced viral replication. While reasonable, the questions why 11 miRNA were actually upregulated in infected cells and what happened to these miRNAs and their targets in Drosha/Dicer-depleted cells remained unanswered.

The central finding that Dicer attenuates HIV-1 was nonetheless validated in a report by Qian *et al.* who also observed an up to fourfold increase in HIV-1 replication and virion production following siRNA-mediated Dicer knockdown.[192] Mechanistically, this involved upregulation of *de novo* HIV-1 Gag protein synthesis independently of general effects on cellular protein synthesis and instead probably attributable to host miRNAs. Notably, this study also validated several other findings noted above, including the capability of adenoviral VA RNAs to serve as a decoy substrate for Exportin-5, Dicer, and RISC, as well as the putative role of HIV-1 Tat as an SRS.[169,170,176,178]

Further confirmation for a suppression of HIV-1 by the host RNAi/miRNA machinery came from Chable-Bessia *et al.*, who knocked down a whole series of RNAi factors—RCK/p54, GW182, LSm-1, XRN1, Drosha, and DGCR8—and in each case found marked upregulation of HIV-1 virus production.[193] Moreover, they found colocalization of HIV-1 mRNAs and Ago-2 plus RCK/p54 within cellular P-bodies, and were able to prove physical association of these components via immunoprecipitation experiments. Because RCK/p54 and LSm-1 knockdowns resulted in the disruption of P-bodies, the authors then concluded that accumulation of HIV-1 mRNA in P-bodies limits viral replication, possibly explaining the sum of their findings. Finally, they translated their findings into PBMCs from latently HIV-1 infected human subjects under antiretroviral treatment and showed that siRNA-mediated Drosha, DGCR8, or RCK/p54 knockdown likewise reactivated virus in these cells, which has obvious clinical implications with respect to urgently needed strategies to eradicate persisting HIV-1.

Last but not least, as already mentioned, the Moelling group showed that Dicer depletion also enhances IAV infection,[186] and extended these findings in a subsequent study to vesicular stomatitis virus as well as poliovirus.[194] The specific cells used in all these studies were Vero cells that lack IFN-α/β genes, which could readily explain their hypersusceptibility to infection with different viruses after additional Dicer knockdown. Combined with the facts that Vero cells are well characterized, infected with a large variety of clinically relevant viruses, easy to handle, and already extensively used in biotechnological approaches, this suggests that Dicer-ablated Vero cells might represent a highly useful cell culture system for growing wild-type or recombinant viruses.

Generally, the lessons learned from this study and all the others described earlier indeed lay a strong foundation for attempts to rationally improve virus production systems by purposely dysregulating central RNAi factors—either by directly knocking them down (or out), or by overexpressing viral SRS that interfere with their function. This concept was in fact convincingly validated in a recent study from the Berkhout lab, where they stably expressed various SRS (ebola virus VP35, vaccinia virus E3L, IAV NS1, or Carnation Italian ringspot virus P19) and, in all cases, observed a significant increase in wild-type HIV-1 production, as well as up to sixfold improvements in yields of recombinant lentiviral (i.e., HIV-1-derived) gene therapy vectors.[195] Very similar marked increases were then also noted for adenoviral vectors as a representative of DNA viruses, and even up to 100-fold improvements were finally reported for the Sindbis virus (which is developed as alphaviral vectors in human gene therapy).

From a translational view, these are exciting findings, as they imply that this strategy may help to reduce costs and time to produce viral vectors or vaccines in large scale for clinical trials, including all relevant classes, from retro-, lenti-, adeno-, and alphaviral vectors perhaps even to some others, such as adeno-associated viruses (AAV). In this context, it would then be highly interesting to test whether the efficiency of the overall approach can be further improved by combining the SRS strategy together with the previously discussed constitutive knockdown of individual RNAi factors, which may result in additional potentiation of viral replication and of vector/vaccine yields. Last but not least, as always, all this will certainly benefit from more work to fully unravel the biological mechanisms underlying all the observed effects, which will then particularly help to tailor specific strategies to individual viruses. In fact, and perhaps unsurprisingly, de Vries *et al.*[195] have already reported the first evidence that particular combinations of SRS and virus are more efficient than others, such as the juxtaposition of ebola virus VP35 and lenti- or adeno-viral vectors, providing a strong rationale for further studies in this promising new direction.

## V. Adverse Effects in Gene/RNAi Therapy Studies

Ironically, while we just described how deliberately interfering with the key host RNAi machinery can be exploited to potently improve virus/vector production, there is also a wealth of evidence that RNAi-expressing viral vectors may vice versa globally and adversely perturb miRNA biogenesis and function in their recipient cells and hosts. This particular phenomenon, by which highly abundant vector-expressed RNAi triggers (typically shRNAs) saturate and thus block the endogenous miRNA machinery, was first described in 2006 by Grimm and colleagues in Mark Kay's lab in livers of adult mice,[123] and was subsequently validated by several groups and in other tissues. Because we have recently already extensively reviewed the original study as well as all relevant ensuing literature until 2009, we refer the reader to our other review article for background information and more details.[196] Here, we will briefly summarize the key conclusions from these initial studies and then go on to discuss the latest reports that further substantiate the risks of global RNAi dysregulation by high-level shRNA expression, and that also highlight important and promising novel approaches to overcome these concerns and hence improve the clinical safety of RNAi.

The early 2006 study by Grimm *et al.*[123] came as a surprise to the entire field, as it appeared amidst a flurry of reports suggesting that RNAi holds tremendous promise as a novel clinical modality to treat a variety of human diseases, from cancers to viral infections. In fact, the authors of this study also targeted a viral pathogen, namely, the hepatitis B virus (HBV), with specific shRNAs expressed from a potent (particularly in the liver) AAV vector.[123] Indeed, when administered at a moderate dose, one of their vectors, expressing a 19-mer against the HBV sAg gene from a U6 promoter, yielded efficient and sustained (over 5 months) HBV suppression without noticeable adverse effects. However, a large set of over 30 additional shRNA vectors caused liver toxicities, and over 20 vectors even killed the mice within 2 weeks after vector injection. What eventually became apparent upon thorough analyses of the affected mice was that most of the toxic vectors had expressed extremely high amounts of shRNAs, while some endogenous miRNAs were concurrently reduced, including the liver-specific and normally highly abundant miR-122. Moreover, the authors then obtained evidence that this accumulation of exogenous shRNAs had adversely clotted the RNAi pathway, in particular the Exportin-5 karyopherin which exports shRNAs as well as cellular miRNAs out of the nucleus. This was backed up by supporting findings from Bryan Cullen's group who had made similar observations with transfected shRNAs in tissue culture,[181,182] though these experiments were only transient and provided no evidence for the actual toxicity caused by this molecular competition. Also interesting to note in this context are, again, the very latest data by Bennasser and colleagues according to which cellular Exportin-5 and Dicer levels are coupled,[180] suggesting that a secondary adverse effect from Exportin-5 saturation by excess shRNAs might also be Dicer inhibition, which will clearly be interesting to study further, especially in the physiologically relevant animal context.

The two seminal general conclusions from this study—that overabundance of extrinsic RNAi triggers can cause cytotoxicity and even organ failure and lethality, and that at least part of the explanation is quantitative saturation of endogenous RNAi components and subsequent global dysregulation of miRNA biogenesis and function—were then later validated by a number of groups and in different systems. Most notable are those from the groups of John Rossi and of Beverly Davidson who have both made vital and unique contributions to further unraveling the underlying mechanisms as well as to improving the safety of clinical RNAi. Again, because these studies have also been extensively described in our recent review, it should suffice to summarize here that Castanotto *et al.* from Rossi's lab were able to show that saturation can also occur on the level of RISC, in addition to Exportin-5, and that accordingly, excess siRNAs (which do not depend on Exportin-5) can also overload a critical part of the cellular machinery.[197] Moreover, they found that, when an siRNA was embedded in, and expressed from, an miRNA context, it was less prone to saturating the RNAi pathway as compared to an shRNA, which they related to the slower and hence presumably favorable kinetics of RISC loading with an miRNA template.

These potentially beneficial properties of miRNA expression vectors, as compared to shRNA templates, were in fact corroborated by Boudreau, McBride, and colleagues in Davidson's lab, who repeatedly reported decreased *in vivo* toxicity from these vectors in mouse brain as compared to shRNA vectors expressing the same siRNA sequence.[198–200] However, as also discussed in more detail in our other review chapter,[196] the generation of potent miRNA vectors is not straightforward at this point and the resulting constructs can vary in their siRNA expression, and hence efficacy, to a greater extent than what is typically observed with conventional shRNA vectors.

Worth noting in this context is moreover the fact that a presumable particular benefit of miRNA-based RNAi vectors, namely the option to employ RNA polymerase II (Pol II) promoters including the many tissue-specific and regulatable variants, in fact also exists for typical shRNA cassettes. This was, amongst

Exportin-5 expression. Hence, mice in which mature miR-9 was increased (for unknown reasons) after shRNA expression showed reduced levels of Ago-2, and vice versa. Whether this implies a currently unknown auto-regulatory feedback mechanism between shRNAs, miRNAs, and Ago-2, or whether it reflects a nonspecific effect due to neuronal cell loss and astroglial reaction, and whether and how these changes are related to or even causally involved in toxicity, are all certainly highly intriguing questions for future work.

Regardless of these unresolved issues, this latest study as well as the previous work clearly highlights the risks of high-level shRNA expression, and concurrently underscores the necessity to carefully pick the experimental model system to evaluate this risk as well as shRNA/vector efficacies. This includes the general and frequent observation that lack of cell death or other adverse effects in cultured cells do not translate into safety of the same RNAi trigger once administered at high doses and over long periods in animals, stressing the need for *in vivo* studies to thoroughly address these issues. It also includes the particular finding in the Ehlert study[202] that even animal models can vastly differ in their experimental outcome when used to assess shRNA toxicities, for reasons that are not fully understood at this point but that may involve inherent differences in their RNAi machinery, making them more or less susceptible to adverse saturation effects. Luckily, as noted in the work by Ehlert *et al.*[202] as well as in our own 2006 study,[123] these toxic effects do not require the presence of an actual target for the RNAi trigger, permitting their further study and ultimate resolution in wild-type animals that are usually easier to obtain than transgenic or otherwise genetically modified experimental littermates.

Being aware of the grave risks associated with high-level therapeutic RNAi expression, a number of groups, including our own, have recently set out to alleviate these concerns by inventing and evaluating novel strategies to maintain shRNA efficacy while eliminating saturation, or at least minimizing it to tolerable degrees. One strategy already mentioned above is embedding into an miRNA context, which may slow loading into RISC and hence reduce competition with endogenous miRNAs, though this mechanism remains hypothetical at this point. Another potent approach is to utilize weaker and more specific promoters to drive the therapeutic shRNA sequences, a strategy for which successful proof of concept has, for instance, been provided in the previously mentioned study by Giering and coworkers.[201] Another example along this line came from Chen *et al.*, who employed the H1 promoter to drive anti-HBV shRNAs in HBV-transgenic mice.[204] As compared to U6, H1 is substantially weaker, readily explaining why Chen and colleagues could administer high vector doses of $1 \times 10^{12}$ AAV particles to the mice and still not observe toxicity over a follow-up period of 22 months.

In a more comprehensive *in vivo* analysis, our own group has most recently tested two alternative RNA polymerase III (Pol III) promoters, H1 and 7SK, for the expression of various shRNAs that we had previously assumed to cause minor toxicity in mice.[75] One of them was a 19-mer directed against human alpha-1-antitrypsin (hAAT) that had given approximately 70% stable knockdown when expressed from the potent U6 promoter in hAAT-transgenic adult mice.[123] In our new direct side-by-side comparison, the two weaker Pol III promoters exhibited a slower onset of hAAT knockdown, but were ultimately more efficient and suppressed more than 90% of hAAT protein levels for over 3 months, while the U6 group declined to about 65%. This was paralleled by striking differences on the DNA and shRNA levels, where the U6 vector showed substantial and rapid losses over time, whereas the H1 and 7SK variants persisted. A straightforward explanation was that low-level shRNA expression from the latter caused marginal, if any, cytotoxicity, while the stronger U6 promoter produced an excess of shRNAs that became toxic and locally killed individual hepatocytes, resulting in liver regeneration and loss of the nonintegrating AAV/shRNA vectors, along with loss of RNAi. For further validation in our clinically relevant HBV mouse model of chronic HBV infection, we likewise expressed our best anti-HBV 19-mer under the H1 or 7SK promoters and again observed a delayed, but eventually more potent (> 95%, vs. 80% with U6) and more stable (over 1 year) HBV suppression in the adult animals. Taken together, these two sets of experiments by Grimm *et al.*[75] thus strongly indicate that use of marginally active Pol III promoters is a very simple yet potent and safe strategy to achieve long-term *in vivo* RNAi, which might be particularly interesting and promising for persisting targets such as a number of human viruses, including HBV but also HCV or HIV-1.

On the other hand, one can readily imagine therapeutic or biological applications where rapid and utmost potent silencing of a disease- or phenotype-associated gene has higher priority than limiting toxicity. Obvious examples include aggressive cancers that may particularly require immediate vector/RNAi efficacy, or high-throughput *in vitro* RNAi screens that typically likewise aim at fast and strong responses to the exogenous RNAi trigger (si or shRNAs). In these cases, rather than downregulating the RNAi inducer to a minimal effective level, it may be much more important and helpful to vice versa upregulate the potentially rate-limiting factors in the recipient cell or host. Based on this rationale, we extended the 2006 study by Grimm *et al.*,[123] in which we had already identified Exportin-5 as a saturable factor and screened the complete set of key RNAi machinery for additional components that are present at limiting amounts in cells and mice.[75] A central finding was that, next to Exportin-5, Ago-2 is restricting the activity of exogenous RNAi triggers, as evidenced by substantial (up to 10-fold) increases in siRNA- or shRNA-mediated knockdowns when Ago-2 was concurrently overexpressed in cultured cells as well as (with shRNAs) in livers of mice. This result is highly important, as it enlightens a series of previous and new curious observations in our latest study as well as in earlier reports. For one, it explains why initial attempts to overcome *in vivo* toxicity by merely upregulating

Exportin-5 in mice had actually increased toxicities and lethalities, because it had most likely further elevated cytoplasmic shRNA levels which then adversely saturated Ago-2. Besides, it also helps to explain the earlier findings by Castanotto *et al.*[197] as to why siRNAs can also compete with sh or miRNAs, since all three share Ago-2 (or more generally, RISC) for activity. Indeed, we then experimentally proved that deliberate upregulation of Ago-2 in cultured cells not only increases the efficiency of individual si or shRNAs, but moreover alleviated competition between different si/shRNAs.[75]

From a translational standpoint, the most relevant conclusion from this study may however be that concurrent upregulation of the two presumably major rate-limiting factors, Exportin-5 and Ago-2, substantially improved *in vivo* RNAi, by potentiating and extending knockdown efficiencies while concomitantly reducing hepatotoxicities. As mentioned above, this may translate into highly potent and useful clinical strategies for treatment of diseases such as cancer where achieving maximum knockdown in minimal time and for rather short periods (e.g., to destroy metastases) may be critical. On the other hand, it must be pointed out that the long-term consequences of such increases of crucial RNAi machinery are unknown at this point, and that in particular the previously discussed findings of an association of permanent global dysregulation of RNAi factors with cancers raise concerns about possible adverse effects. While these have not been observed during the length of our latest *in vivo* experiments (4 months), it will thus be critical to thoroughly assess these possibilities in future studies, which should then also include close examinations of potential fluctuations in endogenous miRNA expression and activities.

Until then, another conclusion that can be drawn from our most recent study[75] and that has further substantial relevance for RNAi applications is that Ago-2 is also rate-limiting in many routinely used cell lines (in fact, all that we have tested to date; unpublished observations). This result, which is also congruent with data from an independent study by Diederichs *et al.*,[76] has two consequences: One is that it appears reasonable to deliberately upregulate Ago-2 expression in cell lines in order to broadly enhance the efficiency of typical RNAi triggers as they are used in individual or high-throughput knockdown or screening experiments, that is, siRNAs and shRNAs. As a matter of fact, our lab has recently engineered a series of standard human cell lines to stably overexpress a codon-optimized version of Ago-2, and, indeed, these cells show significantly higher knockdown efficiencies for an overwhelming majority of siRNAs and shRNAs as well as for various targets, including different viruses and p53 (Boerner and Grimm, unpublished observations).

A second consequence is that, owing to the rate-limiting nature of several central RNAi components, their saturation by highly abundant exogenous RNAi triggers may perturb authentic miRNA function and gene regulation, resulting in aberrant results in *in vitro* RNAi knockdown experiments, including large-scale screens. In fact, Khan *et al.* published a comprehensive meta-analysis in which they assessed genome-wide transcript responses from 151 previously reported siRNA transfection experiments and concluded that competition and saturation effects between extrinsic and intrinsic RNAi triggers can have substantial impacts on the outcomes of such studies.[205] Most notable is that a large number of genes were actually upregulated in these prior experiments, which can, in hindsight, be explained by inadvertent inhibition of cellular miRNAs that in turn controlled these genes. Important to point out again is that these adverse dose-dependent effects were likewise observed with expressed hairpin RNAs as well as with fully processed siRNAs, further underscoring the role of Ago-2/RISC as a major rate-limiting component in cultured cells. Importantly, one of the genes that is most frequently upregulated was TNRC6C, which encodes a central component of the miRNA machinery, implying that this particular dysregulation may subsequently cause widespread secondary perturbations of cellular gene expression due to global alterations in miRNA efficacies.

Also noteworthy was that several other upregulated genes coded for proteins were involved in cell cycle control, raising further warning flags also for *in vivo* scenarios with respect to potential induction of tumorigenesis in animals or patients as a result of global saturation and dysregulation of the RNAi machinery. In fact, such mechanisms may also have contributed to the outcome of another recent study from our lab in which we found that even marginal expression of vector-encoded shRNAs can accelerate tumorigenesis in transgenic mice that are prone to developing liver cancer.[122] While we reasoned that this phenomenon was mainly due to minimal shRNA-induced liver damage and subsequent overrepopulation with hepatocytes expressing a potent oncogene (*c-myc* in this particular mouse model), it is well possible, and an interesting hypothesis for future analyses, that the above-noted additional effects—perturbation of TNRC6 and global dysregulation of miRNA activities, resulting in upregulation of cell cycle programs—may have further contributed to our phenotypes. Altogether, these findings and hypotheses thus once again strengthen the necessity to gather a much more quantitative "systems biology" understanding of human RNAi processes to increase our knowledge of these fundamental biological processes as well as to help avoid adverse effects in *in vitro* and *in vivo* applications. As comprehensively discussed by Khan *et al.*,[205] these looming studies should then include attempts to quantify the concentrations in the cell of all components of the RNAi pathways, of the intrinsic and extrinsic miRNAs and RNAi triggers, of the specific target mRNAs, and of the numbers thereof, as well as the kinetics of trigger–target binding and turnover. Clearly, additional efforts to synergistically complement these experimental studies with mathematical predictions and computational models will help in succeeding with this important endeavor (see also Section VI).

As if to furthermore underscore the necessity of these joint interdisciplinary efforts, Pan *et al.* most recently published their findings that two commonly used lentiviral shRNA libraries caused perturbations of cellular miRNA expression which resulted in many false-positive hits in screens for HCV host factors.[206] Intriguingly, this coincided with almost 70% reductions of endogenous miR-122 levels in the liver cell line used in these studies (Huh-7), as well as with further substantial dysregulation of five other hepatic miRNAs. This unintentional effect was directly dependent on the dose of the perturbing shRNA, thus strongly supporting the global saturation model originally introduced by our group and others and underscoring the above-mentioned needed efforts to better understand the quantitative relationships and physiological coregulation of the human RNAi machinery.

Toward this goal, a particularly unique and powerful approach finally worth mentioning—that should complement classical cellular, molecular, and biochemical analyses and may also have vital practical ramifications—is screening for small molecules or compounds that modulate the RNAi pathway. While the past 5 years indeed have provided ample and growing evidence for the feasibility and usefulness of this strategy, the relevant findings will not be discussed in detail here as these modulators typically affect the activity or interaction of single RNAi factors, including specific miRNAs such as miR-21 and miR-122,[207] rather than altering their absolute levels or exerting other quantitative effects which are at the focus of the present chapter. Moreover, a recent review to which we referred the reader already provides an excellent and complete overview over the sum of published strategies prior to 2009.[208]

Nonetheless, three very recent studies should briefly be pointed out here because of the reported global effects on RNAi and their particular relevance for the potential development of future small-molecule-based RNAi/cancer therapies. In one of them, Watashi *et al.* screened a library of 530 compounds containing many previously characterized drugs for candidates that would block RNAi, as measured by the rescue of a luciferase target in the presence of an antiluciferase shRNA in cultured cells.[209] This yielded two compounds—PLL (polylysine) and TPF (trypaflavine)—that suppressed luciferase knockdown in a dose-dependent fashion while showing minimal effects in control reactions employing nonspecific shRNAs. Owing to the analogy of PLL with the also highly basic HIV-1 Tat polypeptide which is presumed to block Dicer (see above), the authors then speculated and indeed subsequently proved that PLL acts at the Dicer level by inhibiting its capability to bind and then process hairpin RNAs. In contrast, the other compound, TPF, perturbed association of siRNAs (or miRNAs) with Ago-2 and likely with the other three human Ago proteins as well, without dysregulating the actual expression levels of the four Ago variants. Moreover, TPF also disturbed the interactions of Ago-2 and TRBP, thus potentially affecting the loading of si/miRNAs into RISC. As could be expected, in particular PLL (which acts on the Dicer level, see above) but also TPF was then found to alter the cellular miRNA expression profile and the activities of individually tested miRNAs.

Importantly, Watashi *et al.* then also assessed their two compounds in two different miRNA-dependent tumor models and found promising efficacies. In the first model, they initially established cell lines stably overexpressing two miRNAs (miR-93 or miR-130b) previously found to be involved in leukemogenesis, and then showed that implantation of these cells into nude mice produced gross *in vivo* tumors. As hoped for, pretreatment of these cells with PLL and TPF not only reduced their colony forming ability in soft agar but also significantly diminished their propensity to form tumors upon implantation. In the second model, they consistently showed that PLL/TPF treatment of HTLV-1-positive adult T-leukemic cell lines (which overexpress miR-93 and miR-130b) substantially reduced cell viability, as compared to normal PBMC which remained unaffected. While certainly highly promising with respect to potential further use of these and other compounds (see below) as novel chemical inhibitors of RNAi processes that are involved in tumorigenesis, it must not be overlooked that the mechanisms of action are far from understood at this point, preventing (amongst other hurdles) immediate clinical use of these molecules. Also important to consider is that such global deliberate alterations of the RNAi machinery will not only affect single tumor-related miRNAs but will of course always have broad and commensurate effects on many further miRNAs and their targets, raising the same concerns about nonspecific and potentially adverse outcomes as discussed above in the context of stable Exportin-5 or Ago-2 upregulation strategies.

As a final example, two studies are briefly described that reported the discovery and application of yet another small compound and one that, unlike PLL or TPF, actually enhances RNAi rather than blocking it. In the initial study, Shan *et al.* used a cell-based shRNA-dependent assay similar to that of Watashi *et al.* to screen a collection of 2000 compounds and natural products approved by the U.S. Food and Drug Administration.[210] One molecule they found, which enhanced RNAi in a dose-dependent, specific, and relatively nontoxic manner, was enoxacin (Penetrex), a member of a family of synthetic antibacterial compounds based on a fluoroquinolone skeleton that are in clinical use to treat infections from gonorrhea to urinary tract infections. However, enoxacin was in fact the only family member amongst all tested in this study that showed the enhancing RNAi effect or any phenotype at all. Notably, enoxacin did not alter the expression levels of core RNAi factors, including Dicer, TRBP, and Ago-2, but also had only marginal effects on global gene expression which is counterintuitive for a drug broadly affecting RNAi and hence presumably prone to perturbing miRNA-dependent mRNA profiles. Perhaps this conundrum could be explained by the additional observation that enoxacin treatment only altered the steady-state levels of 15 out of 157 investigated miRNAs, and only about twofold, implying that these rather subtle changes were insufficient to exert substantial effects on the cellular transcriptome. Mechanistically, the mode of enoxacin action was postulated to involve multiple layers, including Dicer-mediated small RNA processing, miRNA interaction with TRBP, and finally RISC loading, hinting at a very complex and not fully understood scenario. Finally, the authors proved that enoxacin is also active *in vivo* by demonstrating enhanced (from 20% to 60%) RNAi in Gfp-transgenic mice infected with an anti-Gfp shRNA-expressing lentiviral vector and concurrently treated with enoxacin.

Picking up on these concluding findings in the Shan study,[210] Melo *et al.* very recently suggested the possible therapeutic application of enoxacin as a novel strategy to globally restore the dysregulated miRNAome of cancer cells and provided several proofs of concept in animal cancer models.[211] First, however, they examined the putative anticancer properties of enoxacin in the context of 12 cancer cell lines from seven malignancies and indeed found a very potent and specific growth-inhibitory effect that was due to $G_2$/M arrest and subsequent apoptosis. Curiously, they then noted that 122 out of 731 analyzed miRNAs were dysregulated in enoxacin-treated cells, which is a substantially higher proportion than what was observed in the original Shan study (see above, 15 out of 157). Of these 122 altered miRNAs, a clear majority of those that were upregulated (100 out of 122) had potential tumor suppressor features according to previous studies, and downregulation of their respective target oncoproteins was also noted, together providing a likely mechanistic explanation. Besides, the group then also validated the original finding by Shan *et al.* that at least one mechanism of enoxacin action included binding to TRBP and increasing its affinity for miRNAs. Importantly, an *in vivo* cancer-specific antiproliferative effect was then also documented in nude mice xenografted with colorectal cancer cells. In these mice, daily enoxacin injections over 4 weeks resulted in potent tumor growth inhibition, characterized by necrosis in tumor tissues and the concurrent absence of any toxicities in normal tissues. Moreover, in two further models, enoxacin was capable of blocking liver and lung macro- and micrometastases.

Most importantly, these results were then finally also repeated in mice orthotopically implanted with human colorectal tumors, in which daily enoxacin treatment over a period of 15 days resulted in significant and specific reductions in tumor growth and weight. This correlated with a predicted overall upregulation amongst the 731 analyzed miRNAs that shifted the miRNAome to a profile that more closely resembled normal colon tissue, suggesting that treatment had indeed restored a more physiological miRNA transcriptome in these human tumors. Because many human cancers are characterized by global dysregulations in their core RNAi machinery (see above) and ensuing alterations in miRNA expression, frequently evidenced by widespread downregulations, the use of enoxacin or future drugs with similar enhancing properties that can restore miRNAomes is certainly tempting and promising. Nevertheless, one must, of course, again not ignore the potential pitfalls already discussed above in the context of the RNAi-suppressing drugs PLL and TPF, namely, that any external modulation of central RNAi factors is highly prone to causing a multitude of pleiotropic and hard-to-predict or difficult-to-control effects in the cell which may go well beyond the intended phenotypes. Thus, it will surely take another few years and a battery of additional thorough experiments in different animal models before we should attempt the clinical evaluation of these novel drugs, despite the fact that their potential is undisputedly high and currently unique.

## VI. Outlook: Relevance for Molecular Biology and Clinical Translation of RNAi

As if the past decade had not already been exciting enough with the discovery of RNAi in humans and the invaluable implications for biological as well as applied research, the past 5 years have provided yet another wave of enthusiasm across the community, as enormously important insights have been gained into the phenomena linking RNAi dysregulation to major human pathologies. As a result, we can now much better appreciate that, like in many other biological or medical area, we need to widen our focus and should approach regulatory RNAi mechanisms and their alterations in disease or therapy from a much broader systems biology perspective. In other words, the field will likely advance most rapidly if we continue to understand miRNAs and their entire biogenesis and processing machinery as a whole, rather than a collection of separate factors that can be individually dysregulated and corrected therapeutically. Instead, what we are facing are highly intricate and interwoven cellular networks in which perturbation of a single component will affect many other players simultaneously, and in which each factor is vice versa regulated by a multitude of partners inside and even outside the RNAi network. Important to point out, this realization does of course not question the validity of earlier and future approaches to fundamentally study or therapeutically regulate individual miRNAs or RNAi factors, but we perhaps need to be more aware of the overall picture and of the global effects in the cell.

With this in mind, we believe that this increasingly complex and complicated concept of RNA–protein networks with multiple feedback loops, intermolecular interactions, and control mechanisms poses two major challenges for us as biologists or therapists: from a biological standpoint, we reckon that an utmost critical next (or already initiated) step will be to further open up our experimental minds and progressively foster the inclusion of quantitative and theoretical aspects into our benchwork. The old saying that the whole is greater than the sum of its parts may perfectly describe the massive enigma that we are still facing and that we all wish to untangle and understand. In our opinion, we as biologists will only reach this important and intriguing goal if we combine our efforts with other disciplines and feed our experimental data into mathematical models and predictions which we will then in turn validate and improve. Only then will we hopefully finally be able to recognize and comprehend how, for instance, dysregulation of a single RNAi factor such as Dicer can have such immense and diverse consequences for miRNA and gene homeostasis, or why initial aberrant expression or activity of just one miRNA may already suffice to change, kill, or transform an entire cell with all its other networks.

From a medical standpoint, the situation and challenges are very similar, as we also need to better understand and appreciate the RNAi network in all its complexity, rather than view it as a collection of individual parts that one can extrinsically modulate at will. Again, we certainly do not question the power and promise of therapeutic strategies that aim at correcting single dysregulated miRNAs, especially since there is already a plethora of evidence for the potential of this approach to battle certain cancers or viral infections. Still, our wish is to raise caution and create the awareness that tinkering with a single component of the cellular RNAi network will inevitably affect a vast number of additional factors, even if it may take a while before these effects become apparent. This concern is perhaps most impressively evidenced by our own prior findings with shRNA-expressing vectors in livers of adult mice, where we only came to appreciate their massive toxicity following weeks of high-level expression and associated perturbance of the endogenous RNAi machinery, whereas no such effects had ever become apparent in previous transient experiments in cultured cells (a general discordance that is unfortunately frequently observed in biology).[75,122,123,201]

Almost without doubt, such potential grave adverse effects would likely be even more pronounced if one would try and correct global dysregulations of key RNAi factors, such as disease-associated decreases of Drosha or Dicer, systematically by simply overexpressing the defect proteins in human cells or tissues. Such a strategy may be useful in very specialized scenarios, such as a transient deliberate upregulation of Ago-2 to temporarily boost shRNA efficacies as suggested by our own work, but in most other cases the outcomes would be completely unpredictable. In this respect, it should be obvious that the current generation of small compounds, despite their undisputed usefulness for fundamental research, is far from specific and characterized enough to be considered as potential clinical therapeutics that could safely restore certain RNAi dysregulations. The same is true for the present classes of gene therapy vectors expressing RNAi triggers or factors which, despite their potency, specificity, and presumable safety in tissue culture or small animal models, clearly require much more thorough testing especially under long-term conditions and in larger animal models before one can reasonably think about their widespread clinical translation.[212,213]

Last but not least, also very important will be to continue research into alternative RNAi-related technologies for gene silencing that may ultimately prove to be even safer and/or more efficient than our current standard tools and vectors, or that may provide additional and unique benefits which allow their deliberate tailoring to certain target cells not amenable to conventional RNAi strategies. One highly promising approach might, for instance, be represented by tethering strategies that were originally developed as an RNAi surrogate by the Filipowicz lab and that permit the direct coupling of silencing machinery (e.g., Ago or TNRC6 proteins) to a target mRNA in the absence of an actual small RNA trigger.[16,214,215] With this particular property, therapeutic tethering vectors should obviously be particularly interesting in the context of tumor cells lacking essential processing factors, such as Dicer.

Luckily, all this essential work has most likely already been launched in many labs worldwide and will surely progress at breathtaking speed like all fundamental and applied RNAi research over the past decade, convincing us that really in the not-too-distant future it is finally we—as basic researchers and clinical doctors—who can safely and reasonably take on the role as external watchers of the inherent (miRNA) watchmen.

## Acknowledgments

The authors greatly appreciate the funding of research and personnel in the Grimm lab by the Heidelberg University Cluster of Excellence CellNetworks as well as by the Chica and Heinz Schaller (C.H.S.) foundation. Additional support by the Department of Infectious Diseases (Virology), Heidelberg University Hospital, as well as the FRONTIERS program at Heidelberg University is likewise thankfully acknowledged. S. M. moreover appreciates a Ph.D. fellowship from the Heidelberg University Cluster of Excellence CellNetworks. The authors sincerely apologize to all individual scientists and groups whose exciting research and findings could not be cited in this review due to space constraints, but they once again refer the reader to the excellent further review articles referenced throughout this review and additionally found in the literature which highlight their important work.

## References

1. Eulalio A, Huntzinger E, Izaurralde E. Getting to the root of miRNA-mediated gene silencing. *Cell* 2008;**132**:9–14.
2. Huntzinger E, Izaurralde E. Gene silencing by microRNAs: contributions of translational repression and mRNA decay. *Nat Rev Genet* 2011;**12**:99–110.

47. Scott GK, Mattie MD, Berger CE, Benz SC, Benz CC. Rapid alteration of microRNA levels by histone deacetylase inhibition. *Cancer Res* 2006;**66**:1277–81.
48. Heo I, Joo C, Cho J, Ha M, Han J, Kim VN. Lin28 mediates the terminal uridylation of let-7 precursor microRNA. *Mol Cell* 2008;**32**:276–84.
49. Heo I, Joo C, Kim YK, Ha M, Yoon MJ, Cho J, et al. TUT4 in concert with Lin28 suppresses microRNA biogenesis through pre-microRNA uridylation. *Cell* 2009;**138**:696–708.
50. Newman MA, Thomson JM, Hammond SM. Lin-28 interaction with the Let-7 precursor loop mediates regulated microRNA processing. *RNA* 2008;**14**:1539–49.
51. Piskounova E, Viswanathan SR, Janas M, LaPierre RJ, Daley GQ, Sliz P, et al. Determinants of microRNA processing inhibition by the developmentally regulated RNA-binding protein Lin28. *J Biol Chem* 2008;**283**:21310–4.
52. Rybak A, Fuchs H, Smirnova L, Brandt C, Pohl EE, Nitsch R, et al. A feedback loop comprising lin-28 and let-7 controls pre-let-7 maturation during neural stem-cell commitment. *Nat Cell Biol* 2008;**10**:987–93.
53. Wulczyn FG, Smirnova L, Rybak A, Brandt C, Kwidzinski E, Ninnemann O, et al. Post-transcriptional regulation of the let-7 microRNA during neural cell specification. *FASEB J* 2007;**21**:415–26.
54. Guil S, Caceres JF. The multifunctional RNA-binding protein hnRNP A1 is required for processing of miR-18a. *Nat Struct Mol Biol* 2007;**14**:591–6.
55. Michlewski G, Caceres JF. Antagonistic role of hnRNP A1 and KSRP in the regulation of let-7a biogenesis. *Nat Struct Mol Biol* 2010;**17**:1011–8.
56. Michlewski G, Guil S, Semple CA, Caceres JF. Posttranscriptional regulation of miRNAs harboring conserved terminal loops. *Mol Cell* 2008;**32**:383–93.
57. Trabucchi M, Briata P, Garcia-Mayoral M, Haase AD, Filipowicz W, Ramos A, et al. The RNA-binding protein KSRP promotes the biogenesis of a subset of microRNAs. *Nature* 2009;**459**:1010–4.
58. Sakamoto S, Aoki K, Higuchi T, Todaka H, Morisawa K, Tamaki N, et al. The NF90–NF45 complex functions as a negative regulator in the microRNA processing pathway. *Mol Cell Biol* 2009;**29**:3754–69.
59. Cheloufi S, Dos Santos CO, Chong MM, Hannon GJ. A dicer-independent miRNA biogenesis pathway that requires Ago catalysis. *Nature* 2010;**465**:584–9.
60. Cifuentes D, Xue H, Taylor DW, Patnode H, Mishima Y, Cheloufi S, et al. A novel miRNA processing pathway independent of Dicer requires Argonaute2 catalytic activity. *Science* 2010;**328**:1694–8.
61. Yang JS, Lai EC. Dicer-independent, Ago2-mediated microRNA biogenesis in vertebrates. *Cell Cycle* 2010;**9**:4455–60.
62. Yang JS, Maurin T, Robine N, Rasmussen KD, Jeffrey KL, Chandwani R, et al. Conserved vertebrate mir-451 provides a platform for Dicer-independent, Ago2-mediated microRNA biogenesis. *Proc Natl Acad Sci USA* 2010;**107**:15163–8.
63. Ghildiyal M, Zamore PD. Small silencing RNAs: an expanding universe. *Nat Rev Genet* 2009;**10**:94–108.
64. Diederichs S, Haber DA. Dual role for argonautes in microRNA processing and posttranscriptional regulation of microRNA expression. *Cell* 2007;**131**:1097–108.
65. Miyoshi K, Miyoshi T, Siomi H. Many ways to generate microRNA-like small RNAs: non-canonical pathways for microRNA production. *Mol Genet Genomics* 2010;**284**:95–103.
66. Ruby JG, Jan CH, Bartel DP. Intronic microRNA precursors that bypass Drosha processing. *Nature* 2007;**448**:83–6.
67. Okamura K, Hagen JW, Duan H, Tyler DM, Lai EC. The mirtron pathway generates microRNA-class regulatory RNAs in Drosophila. *Cell* 2007;**130**:89–100.

68. Berezikov E, Chung WJ, Willis J, Cuppen E, Lai EC. Mammalian mirtron genes. *Mol Cell* 2007;**28**:328–36.
69. Haussecker D, Huang Y, Lau A, Parameswaran P, Fire AZ, Kay MA. Human tRNA-derived small RNAs in the global regulation of RNA silencing. *RNA* 2010;**16**:673–95.
70. Ender C, Krek A, Friedlander MR, Beitzinger M, Weinmann L, Chen W, et al. A human snoRNA with microRNA-like functions. *Mol Cell* 2008;**32**:519–28.
71. Yang W, Chendrimada TP, Wang Q, Higuchi M, Seeburg PH, Shiekhattar R, et al. Modulation of microRNA processing and expression through RNA editing by ADAR deaminases. *Nat Struct Mol Biol* 2006;**13**:13–21.
72. Kawahara Y, Zinshteyn B, Chendrimada TP, Shiekhattar R, Nishikura K. RNA editing of the microRNA-151 precursor blocks cleavage by the Dicer–TRBP complex. *EMBO Rep* 2007;**8**:763–9.
73. Kawahara Y, Zinshteyn B, Sethupathy P, Iizasa H, Hatzigeorgiou AG, Nishikura K. Redirection of silencing targets by adenosine-to-inosine editing of miRNAs. *Science* 2007;**315**:1137–40.
74. Jazdzewski K, Murray EL, Franssila K, Jarzab B, Schoenberg DR, de la Chapelle A. Common SNP in pre-miR-146a decreases mature miR expression and predisposes to papillary thyroid carcinoma. *Proc Natl Acad Sci USA* 2008;**105**:7269–74.
75. Grimm D, Wang L, Lee JS, Schurmann N, Gu S, Boerner K, et al. Argonaute proteins are key determinants of RNAi efficacy, toxicity, and persistence in the adult mouse liver. *J Clin Invest* 2010;**120**:3106–19.
76. Diederichs S, Jung S, Rothenberg SM, Smolen GA, Mlody BG, Haber DA. Coexpression of Argonaute-2 enhances RNA interference toward perfect match binding sites. *Proc Natl Acad Sci USA* 2008;**105**:9284–9.
77. Katoh T, Sakaguchi Y, Miyauchi K, Suzuki T, Kashiwabara S, Baba T. Selective stabilization of mammalian microRNAs by 3′ adenylation mediated by the cytoplasmic poly(A) polymerase GLD-2. *Genes Dev* 2009;**23**:433–8.
78. Yu Z, Hecht NB. The DNA/RNA-binding protein, translin, binds microRNA122a and increases its in vivo stability. *J Androl* 2008;**29**:572–9.
79. Pedersen IM, Cheng G, Wieland S, Volinia S, Croce CM, Chisari FV, et al. Interferon modulation of cellular microRNAs as an antiviral mechanism. *Nature* 2007;**449**:919–22.
80. Chin LJ, Ratner E, Leng S, Zhai R, Nallur S, Babar I, et al. A SNP in a let-7 microRNA complementary site in the KRAS 3′ untranslated region increases non-small cell lung cancer risk. *Cancer Res* 2008;**68**:8535–40.
81. Sandberg R, Neilson JR, Sarma A, Sharp PA, Burge CB. Proliferating cells express mRNAs with shortened 3′ untranslated regions and fewer microRNA target sites. *Science* 2008;**320**:1643–7.
82. Ji Z, Lee JY, Pan Z, Jiang B, Tian B. Progressive lengthening of 3′ untranslated regions of mRNAs by alternative polyadenylation during mouse embryonic development. *Proc Natl Acad Sci USA* 2009;**106**:7028–33.
83. Poliseno L, Salmena L, Zhang J, Carver B, Haveman WJ, Pandolfi PP. A coding-independent function of gene and pseudogene mRNAs regulates tumour biology. *Nature* 2010;**465**:1033–8.
84. Bhattacharyya SN, Habermacher R, Martine U, Closs EI, Filipowicz W. Stress-induced reversal of microRNA repression and mRNA P-body localization in human cells. *Cold Spring Harb Symp Quant Biol* 2006;**71**:513–21.
85. Kedde M, Agami R. Interplay between microRNAs and RNA-binding proteins determines developmental processes. *Cell Cycle* 2008;**7**:899–903.
86. Kedde M, Strasser MJ, Boldajipour B, Oude Vrielink JA, Slanchev K, le Sage C, et al. RNA-binding protein Dnd1 inhibits microRNA access to target mRNA. *Cell* 2007;**131**:1273–86.

87. Kedde M, van Kouwenhove M, Zwart W, Oude Vrielink JA, Elkon R, Agami R. A Pumilio-induced RNA structure switch in p27-3′ UTR controls miR-221 and miR-222 accessibility. *Nat Cell Biol* 2010;**12**:1014–20.
88. Fukuda T, Yamagata K, Fujiyama S, Matsumoto T, Koshida I, Yoshimura K, et al. DEAD-box RNA helicase subunits of the Drosha complex are required for processing of rRNA and a subset of microRNAs. *Nat Cell Biol* 2007;**9**:604–11.
89. Fuller-Pace FV. DExD/H box RNA helicases: multifunctional proteins with important roles in transcriptional regulation. *Nucleic Acids Res* 2006;**34**:4206–15.
90. Fuller-Pace FV, Ali S. The DEAD box RNA helicases p68 (Ddx5) and p72 (Ddx17): novel transcriptional co-regulators. *Biochem Soc Trans* 2008;**36**:609–12.
91. Davis BN, Hilyard AC, Lagna G, Hata A. SMAD proteins control DROSHA-mediated microRNA maturation. *Nature* 2008;**454**:56–61.
92. Han J, Pedersen JS, Kwon SC, Belair CD, Kim YK, Yeom KH, et al. Posttranscriptional crossregulation between Drosha and DGCR8. *Cell* 2009;**136**:75–84.
93. Paroo Z, Ye X, Chen S, Liu Q. Phosphorylation of the human microRNA-generating complex mediates MAPK/Erk signaling. *Cell* 2009;**139**:112–22.
94. Rudel S, Meister G. Phosphorylation of Argonaute proteins: regulating gene regulators. *Biochem J* 2008;**413**:e7–9.
95. Zeng Y, Sankala H, Zhang X, Graves PR. Phosphorylation of Argonaute 2 at serine-387 facilitates its localization to processing bodies. *Biochem J* 2008;**413**:429–36.
96. Rudel S, Wang Y, Lenobel R, Korner R, Hsiao HH, Urlaub H, et al. Phosphorylation of human Argonaute proteins affects small RNA binding. *Nucleic Acids Res* 2010;**39**:2330–43.
97. Hock J, Weinmann L, Ender C, Rudel S, Kremmer E, Raabe M, et al. Proteomic and functional analysis of Argonaute-containing mRNA-protein complexes in human cells. *EMBO Rep* 2007;**8**:1052–60.
98. Johnston M, Geoffroy MC, Sobala A, Hay R, Hutvagner G. HSP90 protein stabilizes unloaded argonaute complexes and microscopic P-bodies in human cells. *Mol Biol Cell* 2010;**21**:1462–9.
99. Stoica C, Carmichael JB, Parker H, Pare J, Hobman TC. Interactions between the RNA interference effector protein Ago1 and 14-3-3 proteins: consequences for cell cycle progression. *J Biol Chem* 2006;**281**:37646–51.
100. Loedige I, Filipowicz W. TRIM-NHL proteins take on miRNA regulation. *Cell* 2009;**136**:818–20.
101. Schwamborn JC, Berezikov E, Knoblich JA. The TRIM-NHL protein TRIM32 activates microRNAs and prevents self-renewal in mouse neural progenitors. *Cell* 2009;**136**:913–25.
102. Rybak A, Fuchs H, Hadian K, Smirnova L, Wulczyn EA, Michel G, et al. The let-7 target gene mouse lin-41 is a stem cell specific E3 ubiquitin ligase for the miRNA pathway protein Ago2. *Nat Cell Biol* 2009;**11**:1411–20.
103. Dueck A, Meister G. TRIMming microRNA function in mouse stem cells. *Nat Cell Biol* 2009;**11**:1392–3.
104. Qi HH, Ongusaha PP, Myllyharju J, Cheng D, Pakkanen O, Shi Y, et al. Prolyl 4-hydroxylation regulates Argonaute 2 stability. *Nature* 2008;**455**:421–4.
105. Sugito N, Ishiguro H, Kuwabara Y, Kimura M, Mitsui A, Kurehara H, et al. RNASEN regulates cell proliferation and affects survival in esophageal cancer patients. *Clin Cancer Res* 2006;**12**:7322–8.
106. Muralidhar B, Goldstein LD, Ng G, Winder DM, Palmer RD, Gooding EL, et al. Global microRNA profiles in cervical squamous cell carcinoma depend on Drosha expression levels. *J Pathol* 2007;**212**:368–77.
107. Flavin RJ, Smyth PC, Finn SP, Laios A, O'Toole SA, Barrett C, et al. Altered eIF6 and Dicer expression is associated with clinicopathological features in ovarian serous carcinoma patients. *Mod Pathol* 2008;**21**:676–84.

108. Zhang X, Cairns M, Rose B, O'Brien C, Shannon K, Clark J, et al. Alterations in miRNA processing and expression in pleomorphic adenomas of the salivary gland. *Int J Cancer* 2009;**124**:2855–63.
109. Sand M, Gambichler T, Skrygan M, Sand D, Scola N, Altmeyer P, et al. Expression levels of the microRNA processing enzymes Drosha and dicer in epithelial skin cancer. *Cancer Invest* 2010;**28**:649–53.
110. Merritt WM, Lin YG, Han LY, Kamat AA, Spannuth WA, Schmandt R, et al. Dicer, Drosha, and outcomes in patients with ovarian cancer. *N Engl J Med* 2008;**359**:2641–50.
111. Dedes KJ, Natrajan R, Lambros MB, Geyer FC, Lopez-Garcia MA, Savage K, et al. Down-regulation of the miRNA master regulators Drosha and Dicer is associated with specific subgroups of breast cancer. *Eur J Cancer* 2011;**47**:138–50.
112. Tchernitsa O, Kasajima A, Schafer R, Kuban RJ, Ungethum U, Gyorffy B, et al. Systematic evaluation of the miRNA-ome and its downstream effects on mRNA expression identifies gastric cancer progression. *J Pathol* 2010;**222**:310–9.
113. Zhang L, Volinia S, Bonome T, Calin GA, Greshock J, Yang N, et al. Genomic and epigenetic alterations deregulate microRNA expression in human epithelial ovarian cancer. *Proc Natl Acad Sci USA* 2008;**105**:7004–9.
114. Santarelli DM, Beveridge NJ, Tooney PA, Cairns MJ. Upregulation of dicer and microRNA expression in the dorsolateral prefrontal cortex Brodmann area 46 in schizophrenia. *Biol Psychiatry* 2011;**69**:180–7.
115. Beveridge NJ, Gardiner E, Carroll AP, Tooney PA, Cairns MJ. Schizophrenia is associated with an increase in cortical microRNA biogenesis. *Mol Psychiatry* 2010;**15**:1176–89.
116. Lee ST, Chu K, Im WS, Yoon HJ, Im JY, Park JE, et al. Altered microRNA regulation in Huntington's disease models. *Exp Neurol* 2011;**227**:172–9.
117. Ambs S, Prueitt RL, Yi M, Hudson RS, Howe TM, Petrocca F, et al. Genomic profiling of microRNA and messenger RNA reveals deregulated microRNA expression in prostate cancer. *Cancer Res* 2008;**68**:6162–70.
118. Chiosea S, Jelezcova E, Chandran U, Acquafondata M, McHale T, Sobol RW, et al. Up-regulation of dicer, a component of the MicroRNA machinery, in prostate adenocarcinoma. *Am J Pathol* 2006;**169**:1812–20.
119. Chiosea S, Jelezcova E, Chandran U, Luo J, Mantha G, Sobol RW, et al. Overexpression of Dicer in precursor lesions of lung adenocarcinoma. *Cancer Res* 2007;**67**:2345–50.
120. Varambally S, Yu J, Laxman B, Rhodes DR, Mehra R, Tomlins SA, et al. Integrative genomic and proteomic analysis of prostate cancer reveals signatures of metastatic progression. *Cancer Cell* 2005;**8**:393–406.
121. Karube Y, Tanaka H, Osada H, Tomida S, Tatematsu Y, Yanagisawa K, et al. Reduced expression of Dicer associated with poor prognosis in lung cancer patients. *Cancer Sci* 2005;**96**:111–5.
122. Beer S, Bellovin DI, Lee JS, Komatsubara K, Wang LS, Koh H, et al. Low-level shRNA cytotoxicity can contribute to MYC-induced hepatocellular carcinoma in adult mice. *Mol Ther* 2010;**18**:161–70.
123. Grimm D, Streetz KL, Jopling CL, Storm TA, Pandey K, Davis CR, et al. Fatality in mice due to oversaturation of cellular microRNA/short hairpin RNA pathways. *Nature* 2006;**441**:537–41.
124. Hill DA, Ivanovich J, Priest JR, Gurnett CA, Dehner LP, Desruisseau D, et al. DICER1 mutations in familial pleuropulmonary blastoma. *Science* 2009;**325**:965.
125. Chiosea SI, Barnes EL, Lai SY, Egloff AM, Sargent RL, Hunt JL, et al. Mucoepidermoid carcinoma of upper aerodigestive tract: clinicopathologic study of 78 cases with immunohistochemical analysis of Dicer expression. *Virchows Arch* 2008;**452**:629–35.

167. Grassmann R, Jeang KT. The roles of microRNAs in mammalian virus infection. *Biochim Biophys Acta* 2008;**1779**:706–11.
168. Umbach JL, Cullen BR. The role of RNAi and microRNAs in animal virus replication and antiviral immunity. *Genes Dev* 2009;**23**:1151–64.
169. Bennasser Y, Jeang KT. HIV-1 Tat interaction with Dicer: requirement for RNA. *Retrovirology* 2006;**3**:95.
170. Bennasser Y, Le SY, Benkirane M, Jeang KT. Evidence that HIV-1 encodes siRNA and a suppressor of RNA silencing. *Immunity* 2005;**22**:607–19.
171. Lin J, Cullen BR. Analysis of the interaction of primate retroviruses with the human RNA interference machinery. *J Virol* 2007;**81**:12218–26.
172. Triboulet R, Mari B, Lin YL, Chable-Bessia C, Bennasser Y, Lebrigand K, et al. Suppression of microRNA-silencing pathway by HIV-1 during virus replication. *Science* 2007;**315**:1579–82.
173. Coley W, Van Duyne R, Carpio L, Guendel I, Kehn-Hall K, Chevalier S, et al. Absence of DICER in monocytes and its regulation by HIV-1. *J Biol Chem* 2010;**285**:31930–43.
174. Bennasser Y, Yeung ML, Jeang KT. HIV-1 TAR RNA subverts RNA interference in transfected cells through sequestration of TAR RNA-binding protein, TRBP. *J Biol Chem* 2006;**281**:27674–8.
175. Lu S, Cullen BR. Adenovirus VA1 noncoding RNA can inhibit small interfering RNA and microRNA biogenesis. *J Virol* 2004;**78**:12868–76.
176. Andersson MG, Haasnoot PC, Xu N, Berenjian S, Berkhout B, Akusjarvi G. Suppression of RNA interference by adenovirus virus-associated RNA. *J Virol* 2005;**79**:9556–65.
177. Aparicio O, Carnero E, Abad X, Razquin N, Guruceaga E, Segura V, et al. Adenovirus VA RNA-derived miRNAs target cellular genes involved in cell growth, gene expression and DNA repair. *Nucleic Acids Res* 2010;**38**:750–63.
178. Aparicio O, Razquin N, Zaratiegui M, Narvaiza I, Fortes P. Adenovirus virus-associated RNA is processed to functional interfering RNAs involved in virus production. *J Virol* 2006;**80**:1376–84.
179. Xu N, Segerman B, Zhou X, Akusjarvi G. Adenovirus virus-associated RNAII-derived small RNAs are efficiently incorporated into the rna-induced silencing complex and associate with polyribosomes. *J Virol* 2007;**81**:10540–9.
180. Bennasser Y, Chable-Bessia C, Triboulet R, Gibbings D, Gwizdek C, Dargemont C, et al. Competition for XPO5 binding between Dicer mRNA, pre-miRNA and viral RNA regulates human Dicer levels. *Nat Struct Mol Biol* 2010;**18**:323–7.
181. Yi R, Doehle BP, Qin Y, Macara IG, Cullen BR. Overexpression of exportin 5 enhances RNA interference mediated by short hairpin RNAs and microRNAs. *RNA* 2005;**11**:220–6.
182. Yi R, Qin Y, Macara IG, Cullen BR. Exportin-5 mediates the nuclear export of pre-microRNAs and short hairpin RNAs. *Genes Dev* 2003;**17**:3011–6.
183. Aliyari R, Wu Q, Li HW, Wang XH, Li F, Green LD, et al. Mechanism of induction and suppression of antiviral immunity directed by virus-derived small RNAs in Drosophila. *Cell Host Microbe* 2008;**4**:387–97.
184. Haasnoot J, de Vries W, Geutjes EJ, Prins M, de Haan P, Berkhout B. The Ebola virus VP35 protein is a suppressor of RNA silencing. *PLoS Pathog* 2007;**3**:e86.
185. Li WX, Li H, Lu R, Li F, Dus M, Atkinson P, et al. Interferon antagonist proteins of influenza and vaccinia viruses are suppressors of RNA silencing. *Proc Natl Acad Sci USA* 2004;**101**:1350–5.
186. Matskevich AA, Moelling K. Dicer is involved in protection against influenza A virus infection. *J Gen Virol* 2007;**88**:2627–35.

187. Jopling CL, Norman KL, Sarnow P. Positive and negative modulation of viral and cellular mRNAs by liver-specific microRNA miR-122. *Cold Spring Harb Symp Quant Biol* 2006;**71**:369–76.
188. Jopling CL, Schutz S, Sarnow P. Position-dependent function for a tandem microRNA miR-122-binding site located in the hepatitis C virus RNA genome. *Cell Host Microbe* 2008;**4**:77–85.
189. Jopling CL, Yi M, Lancaster AM, Lemon SM, Sarnow P. Modulation of hepatitis C virus RNA abundance by liver-specific microRNA. *Science* 2005;**309**:1577–81.
190. Wang Y, Kato N, Jazag A, Dharel N, Otsuka M, Taniguchi H, et al. Hepatitis C virus core protein is a potent inhibitor of RNA silencing-based antiviral response. *Gastroenterology* 2006;**130**:883–92.
191. Ji J, Glaser A, Wernli M, Berke JM, Moradpour D, Erb P. Suppression of short interfering RNA-mediated gene silencing by the structural proteins of hepatitis C virus. *J Gen Virol* 2008;**89**:2761–6.
192. Qian S, Zhong X, Yu L, Ding B, de Haan P, Boris-Lawrie K. HIV-1 Tat RNA silencing suppressor activity is conserved across kingdoms and counteracts translational repression of HIV-1. *Proc Natl Acad Sci USA* 2009;**106**:605–10.
193. Chable-Bessia C, Meziane O, Latreille D, Triboulet R, Zamborlini A, Wagschal A, et al. Suppression of HIV-1 replication by microRNA effectors. *Retrovirology* 2009;**6**:26.
194. Matskevich AA, Jung JS, Schumann M, Cascallo M, Moelling K. Vero cells as a model to study the effects of adenoviral gene delivery vectors on the RNAi system in context of viral infection. *J Innate Immun* 2009;**1**:389–94.
195. de Vries W, Haasnoot J, van der Velden J, van Montfort T, Zorgdrager F, Paxton W, et al. Increased virus replication in mammalian cells by blocking intracellular innate defense responses. *Gene Ther* 2008;**15**:545–52.
196. Grimm D. Small silencing RNAs: state-of-the-art. *Adv Drug Deliv Rev* 2009;**61**:672–703.
197. Castanotto D, Sakurai K, Lingeman R, Li H, Shively L, Aagaard L, et al. Combinatorial delivery of small interfering RNAs reduces RNAi efficacy by selective incorporation into RISC. *Nucleic Acids Res* 2007;**35**:5154–64.
198. Boudreau RL, Martins I, Davidson BL. Artificial microRNAs as siRNA shuttles: improved safety as compared to shRNAs *in vitro* and *in vivo*. *Mol Ther* 2009;**17**:169–75.
199. Boudreau RL, Monteys AM, Davidson BL. Minimizing variables among hairpin-based RNAi vectors reveals the potency of shRNAs. *RNA* 2008;**14**:1834–44.
200. McBride JL, Boudreau RL, Harper SQ, Staber PD, Monteys AM, Martins I, et al. Artificial miRNAs mitigate shRNA-mediated toxicity in the brain: implications for the therapeutic development of RNAi. *Proc Natl Acad Sci USA* 2008;**105**:5868–73.
201. Giering JC, Grimm D, Storm TA, Kay MA. Expression of shRNA from a tissue-specific pol II promoter is an effective and safe RNAi therapeutic. *Mol Ther* 2008;**16**:1630–6.
202. Ehlert EM, Eggers R, Niclou SP, Verhaagen J. Cellular toxicity following application of adeno-associated viral vector-mediated RNA interference in the nervous system. *BMC Neurosci* 2010;**11**:20.
203. Martin JN, Wolken N, Brown T, Dauer WT, Ehrlich ME, Gonzalez-Alegre P. Lethal toxicity caused by expression of shRNA in the mouse striatum: implications for therapeutic design. *Gene Ther* 2011;**18**:666–73.
204. Chen CC, Ko TM, Ma HI, Wu HL, Xiao X, Li J, et al. Long-term inhibition of hepatitis B virus in transgenic mice by double-stranded adeno-associated virus 8-delivered short hairpin RNA. *Gene Ther* 2007;**14**:11–9.
205. Khan AA, Betel D, Miller ML, Sander C, Leslie CS, Marks DS. Transfection of small RNAs globally perturbs gene regulation by endogenous microRNAs. *Nat Biotechnol* 2009;**27**:549–55.

206. Pan Q, de Ruiter PE, von Eije KJ, Smits R, Kwekkeboom J, Tilanus HW, et al. Disturbance of the microRNA pathway by commonly used lentiviral shRNA libraries limits the application for screening host factors involved in hepatitis C virus infection. *FEBS Lett* 2011;**585**:1025–30.
207. Young DD, Connelly CM, Grohmann C, Deiters A. Small molecule modifiers of microRNA miR-122 function for the treatment of hepatitis C virus infection and hepatocellular carcinoma. *J Am Chem Soc* 2010;**132**:7976–81.
208. Deiters A. Small molecule modifiers of the microRNA and RNA interference pathway. *AAPS J* 2009;**12**:51–60.
209. Watashi K, Yeung ML, Starost MF, Hosmane RS, Jeang KT. Identification of small molecules that suppress microRNA function and reverse tumorigenesis. *J Biol Chem* 2010;**285**:24707–16.
210. Shan G, Li Y, Zhang J, Li W, Szulwach KE, Duan R, et al. A small molecule enhances RNA interference and promotes microRNA processing. *Nat Biotechnol* 2008;**26**:933–40.
211. Melo S, Villanueva A, Moutinho C, Davalos V, Spizzo R, Ivan C, et al. Small molecule enoxacin is a cancer-specific growth inhibitor that acts by enhancing TAR RNA-binding protein 2-mediated microRNA processing. *Proc Natl Acad Sci USA* 2011;**108**:4394–9.
212. Castanotto D, Rossi JJ. The promises and pitfalls of RNA-interference-based therapeutics. *Nature* 2009;**457**:426–33.
213. Kim DH, Rossi JJ. Strategies for silencing human disease using RNA interference. *Nat Rev Genet* 2007;**8**:173–84.
214. Pillai RS, Artus CG, Filipowicz W. Tethering of human Ago proteins to mRNA mimics the miRNA-mediated repression of protein synthesis. *RNA* 2004;**10**:1518–25.
215. Pillai RS, Bhattacharyya SN, Artus CG, Zoller T, Cougot N, Basyuk E, et al. Inhibition of translational initiation by Let-7 microRNA in human cells. *Science* 2005;**309**:1573–6.

# Index

## A



## B



## C



## D



## E



## G

## N

## P



## R



## S

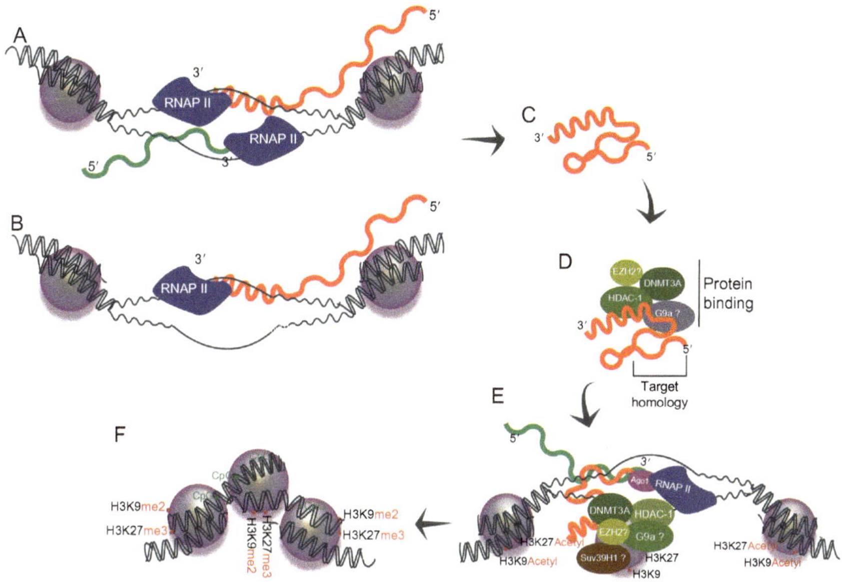


STUART KNOWLING AND KEVIN V. MORRIS, FIG. 1. Endogenous ncRNA-directed epigenetic regulation of gene transcription. (A) Long antisense ncRNAs can be expressed in *cis* or (B) *trans*, where they (C) fold into complex structures that can (D) contain both target homology to other loci or the bidirectionally transcribed loci in the genome as well as possess an ability to bind epigenetic remodeling proteins. The ncRNA-directed epigenetic complexes could then (E) target the silencing of the corresponding homology-containing loci, which can result in (F) heterochromatinization of the targeted site.